American Science in the Age of Jefferson

American Science
in the
Age of Jefferson

J O H N   C.   G R E E N E

The Iowa State University Press

A M E S

JOHN C. GREENE is professor of history at the University of Connecticut

---

© 1984 The Iowa State University Press. All Rights reserved
Composed and printed by The Iowa State University Press, Ames, Iowa 50010
First edition, 1984

Library of Congress Cataloging in Publication Data

Greene, John C.
    American science in the age of Jefferson.

    Includes bibliographical references and index.
    1. Science—United States—History.   2.   Jefferson, Thomas, 1743–1826.   I.   Title.
Q127.U6G69 1984   509.73   83–8513
ISBN 0–8138–0101–X
ISBN 0–8138–0102–8 (pbk.)

For Ellen, with love

CONTENTS

LIST OF ILLUSTRATIONS

P R E F A C E

HISTORIANS OF SCIENCE have paid relatively little attention to American science in the early national period, largely because it was overshadowed and dominated by European scientific developments to which Americans made only modest contributions. The Jeffersonian era was nevertheless a formative one for American science. This was a time in which basic institutions and deep-seated attitudes toward science and its relation to the rest of American culture took shape, institutions and attitudes that were to guide the subsequent course of American scientific development. Of the existing literature about late eighteenth and early nineteenth century American science, a considerable part concerns Thomas Jefferson's scientific interests, but hitherto no one has provided a general account of American science in its European context against which Jefferson's activities as a promoter and practitioner of science could be evaluated justly. To provide such an account and portray Jefferson's role therein is the purpose of this book. In writing it, I have ignored the long-standing controversy between the "internalists" and the "externalists" among historians of science, preferring to depict both the social setting of science and its substantive achievements. The style is narrative, not analytic. Instead of generalizing about American science and illustrating the generalizations with examples, I have described what American scientists did and said, quoting extensively from their works and the writings of those who observed them and relying on the narrative itself to give the reader a true conception of the ideas and motives that animated them and the handicaps under which they labored. The misspellings and other peculiarities in the quotations are all *sic,* as in the case of Jefferson's "knoledge." In speaking of the human race, I have adopted the terms "man" and "mankind" in conformity with the usage of Jefferson and his contemporaries. As for the "men of science," the scientists of this period were indeed men, at least in America.

Readers looking for a new interpretation of American science in the early national period will look in vain, although they will find some criticisms of existing interpretations in the last chapter. My own view of the developments in this period is a rather traditional one. It emphasizes the gradual construction of an institutional base for science and the achievements, however modest, of individual scientists working under difficult circumstances. It gives considerable prominence to the ideas and activities of Thomas Jefferson, not because he was a great scientist, which he was not, but because he participated in one way or another in nearly every field of scientific inquiry, stimulating his compatriots with his ideas and researches and inspiring them with the knowledge that their efforts were appreciated at the highest level of government.

The list of persons and funding agencies to whom I am indebted for assistance and support in bringing this work to completion is a long one. For constructive criticism of one or more of the chapters on institutional developments I am grateful to Brooke Hindle, Whitfield Bell, Jr., Silvio Bedini, Henry May, Clark Elliott, Hunter Dupree, Joseph Waring, and Henry Shapiro; for the substantive scientific chapters to James Montgomery, Gerald Holton, Brooke Hindle, Silvio Bedini, Aaron Ihde, John Allen, John Wells, George White, John Doskey, Norman Gray, Joseph Ewan, Emmanuel Rudolph, Ronald Stuckey, Keir Sterling, William Laughlin, Waldo Wedel, T. Dale Stewart, and George Aubin. For financial support of my research and writing I am deeply indebted to the National Science Foundation, the John Simon Guggenheim Memorial Foundation, the National Museum of American History, the Institute for Advanced Study, the American Philosophical Society, and the University of Connecticut Research Foundation. Nor can I forget the assistance I have received from the directors and librarians of scientific and learned institutions from Bowdoin College in Brunswick, Maine as far south as Savannah, Georgia, and west to Columbus and Cincinnati, Ohio, and Lexington, Kentucky. Finally I should like to thank the staff of the Iowa State University Press for the skill and care with which they have prepared this book for publication, and especially Judith Conlin Johnston, whose editorial suggestions produced substantial improvements in the structure, coherence, and readability of the narrative.

To my wife, who bore with me and aided me through the long years of scholarly travail, I owe more than I can say. This book is for her.

Storrs, Connecticut
Oct. 1, 1983

American Science in the Age of Jefferson

1

The American Context

THIS is a book about beginnings: the beginnings of national independence, of scientific institutions and publications, of new relations with European science — in short, of American as distinguished from British colonial science. It is not a story of great scientific discoveries and theoretical breakthroughs but of scientific exploration of the New World: naming, classifying, and describing its plants, animals, and minerals; studying its geological structure; determining the latitude and longitude of its towns and cities; researching and speculating about its aborigines and antiquities; founding botanical gardens, museums, herbaria, and scientific societies; and transplanting to America the theories, techniques, and systems of classification and nomenclature of Western science. The drama of the story is to be found, not in the brilliance of the scientific achievements (although there is drama aplenty in the Lewis and Clark expedition, the exhuming of the first mastodon skeletons, and the surprises contained in Indian languages), but in what individual scientists managed to accomplish in the face of enormous difficulties. From the point of view of the history of Western science their achievements may hardly seem worth notice, but the nation they loved and whose science they worked to establish was eventually to develop into the greatest scientific power on earth. In the years from 1780 to 1830 American scientists ceased to be mere purveyors of the raw materials of science to Europe and became junior partners in the Western scientific enterprise. In the succeeding century and a half they have become leaders in that endeavor. The story of American science in the Jeffersonian era thus has a double interest — that inherent in all difficult beginnings and that arising from the ultimate consequences of those early strivings.

This book attempts to give a full account of American science from the achievement of national independence to the 1820s, omitting only the medical sciences. There was little purely medical research in this period — medical men worked mostly in chemistry, mineralogy, botany, zoology,

3

anthropology, and like subjects; the general history of medicine as a practical art would require a separate volume. The same is true of the history of technology, which though interesting in itself, reveals few links to science in this era despite the lip service paid to "useful knowledge" by Jefferson and his contemporaries. Even Dirk Struik, who lumped science and technology together in his *Yankee Science in the Making,* had to concede that the early national period produced very little that could properly be called applied science. "Science and technology had not yet found each other," writes Struik, "and would find it difficult to see eye to eye for a long time to come."[1]

But, even when medicine and technology are excluded, it is no easy matter to give a connected account of the multifarious activities of American scientists and promoters of science in the first half-century of the nation's existence. The continued dependence of America on European science, the absence of major scientific figures and theoretical developments, the variety of sciences involved, the decentralized character of American science, and the many false starts and aborted beginnings make it difficult to present the picture of steady progress in scientific discovery that historians have led us to expect. There was progress in American science in the Jeffersonian era — in establishing an institutional base, assimilating European developments, and exploring a continent — but it took place in a context of flamboyant patriotism, political and religious controversy, and practical concern with commerce and industry that alternately inspired and distorted scientific development.

THE SCIENTIFIC CONTEXT

The closing decades of the eighteenth century and the opening ones of the nineteenth formed a brilliant period in the history of Western science. The science of mechanics inaugurated by Sir Isaac Newton's *Mathematical Principles of Natural Philosophy* (1687) was brought to a triumphant climax in the *Mécanique analytique* of Joseph Lagrange (1787) and the *Mécanique céleste* of Pierre S. Laplace. In the latter work, published between 1799 and 1825, Laplace undertook to demonstrate the motions of the bodies of the solar system in accordance with Newton's gravitational law, showing the periodic nature of their apparent irregularities and the consequent stability of the system as a whole. The result was a brilliant vindication of Newton's famous law. In observational astronomy William Herschel, a German musician turned Englishman and astronomer, startled the world with his discovery of the planet Uranus in 1781 and his explorations of stellar astronomy with greatly improved reflecting telescopes of his own devising. In experimental physics Thomas Young and Augustin Fresnel revived the wave theory of light and succeeded in measuring the wavelengths of various colors.

In chemistry Antoine Lavoisier and his colleagues launched the new oxygen chemistry and the system of nomenclature and classification based upon it, John Dalton revived the atomic theory and made it the basis of a theory of chemical reactions, Humphry Davy pioneered the new field of electrochemistry, and Jöns J. Berzelius and others established analytical chemistry on a firm basis. In mineralogy Abraham G. Werner developed a system of classification based on the external characters of minerals, while Berzelius, Martin H. Klaproth, and others applied the new chemistry to their analysis and the Abbé René J. Haüy demonstrated how to classify them in terms of crystal structure. In geology Werner laid the foundations of the systematic study of rock structures, and James Hutton and John Playfair showed how to interpret geological phenomena as the products of uniform processes of erosion, deposition, consolidation by heat and pressure, and subsequent uplift—a cycle of processes exhibiting "no vestige of a beginning, no prospect of an end."

In geography Alexander von Humboldt paved the way for geophysics and plant geography in the publications resulting from his travels in Latin America. In botany Antoine L. de Jussieu proposed a new system of classification (based primarily on the characters of the embryo and which gradually superseded the Linnaean system based on the number, figure, situation, and proportion of the stamens and pistils). At the same time Jean B. de Lamarck, Georges Cuvier, and other French zoologists completely revised the Linnaean classification of animals; and Cuvier opened the field of vertebrate paleontology by applying his knowledge of comparative anatomy to reconstruction of extinct quadrupeds whose bones and teeth had been turning up in Siberia, Europe, and America. Meanwhile, Erasmus Darwin and Lamarck propounded their highly subversive theories of organic evolution; Johann F. Blumenbach, James C. Prichard, and others mounted a scientific attack on the problem of the origin of human races; antiquaries were busy exploring mounds and earthworks scattered across the north temperate regions from Siberia to Scotland, Ireland, and North America; and students of language, stimulated by William Jones's discovery of the linguistic science of India and the affinities between Sanskrit, Greek, Latin, and various European languages, had begun to erect a science of comparative philology. From the study of the stars to the study of man it was an age of scientific advance worthy of comparison with the scientific revolution of the seventeenth century.

THE CONTEXT OF NATIONHOOD

To Americans living in this period of exploding scientific inquiry, the fundamental fact conditioning every thought and deed was the consciousness that they were now an independent nation. With respect to science this meant two things: as the example par excellence of useful knowledge, science must be cultivated to promote the interests, prosperity, and power of the rising American nation; and as the supreme example of the powers of the human mind, the successes of science challenged Americans to prove to the world that republican institutions were as favorable to intellectual achievement as they were to liberty.

The first of these challenges presented no difficulty. It required only that the utilitarian spirit of the age be given patriotic form and substance in societies for the promotion of useful knowledge, and this was soon done. "Knowledge," declared the preface to the first volume of *Transactions* of the American Philosophical Society, "is of little use, when confined to mere speculation. But when speculative truths are reduced to practice; when theories, grounded upon experiments, are applied to the common purposes of life; and when, by these, agriculture is improved, trade enlarged, the arts of living made more easy and comfortable, and, of course, the increase and happiness of mankind promoted; knowledge then becomes really useful."[2] To this end the members of the society proposed to devote their energies chiefly to "such subjects as tend to the improvement of their country, and advancement of its interest and prosperity." Among such subjects were listed (in addition to physics, mechanics, astronomy, and mathematics) improvements in agriculture; preservation of timber for shipbuilding and other purposes; improvement of materia medica; discovery and introduction of "new vegetable juices"; improvements in the manufacture of wine, cider, and vinegar; and methods of constructing highways, causeways, and bridges and of linking rivers and improving inland navigation.

In Boston the founders of the American Academy of Arts and Sciences declared their intention to pursue such inquiries as would "enrich and aggrandize these confederated States" and "advance the interest, honor, dignity and happiness of a free, independent and virtuous people." Thus the astronomical papers in the first volume of the academy's *Memoirs* were "chiefly of the practical kind . . . such observations and deductions, as are subservient to the cause of geography and navigation, the improvement of which is of great importance to this country." Looking to the future, the academy recommended to its members "useful experiments and improvements, whereby the interest and happiness of the rising empire may be essentially advanced."[3]

Not to be outdone by Boston and Philadelphia, the citizens of New York founded a Society for the Promotion of Agriculture, Arts, and Manufactures, subsequently renamed the Society for the Promotion of Useful Arts. Here again the emphasis was on practical utility for the sake of national improvement, prosperity, and power. "Our business generally," Simeon De Witt reminded the members, "is to collect all the improvements within our reach that may be made in agriculture, in manufactures and in the arts of whatsoever kind they may be." Science was to be a concern of the society, but only in a severely practical context. In practice, however, most of these early scientific societies gave more lip service than actual attention to utilitarian concerns.[4]

The other challenge facing patriotic Americans, i.e., to show that their institutions were more conducive to high intellectual attainment than those of monarchical states, was a more difficult matter. Patriotism dictated that republican institutions must favor the utmost development of the human mind, and patriots did not hesitate to assert that this was true. "The introduction and progress of *freedom*," declared John Gardiner in a Fourth of July oration in Boston in 1785, "have generally attended the introduction and progress of *letters and science*. In despotick governments the people are mostly illiterate, rude, and uncivilized; but in states where CIVIL LIBERTY hath been cherished, the human mind hath generally proceeded in improvement, learning and knowledge have prevailed, and the arts and sciences have flourished."[5] But when the champions of republican virtue put aside the orator's prophetic mantle and turned to a realistic assessment of the prospects for science in the new nation, they soon realized they had their work cut out for them. Political independence was one thing, cultural independence quite another.

In science and literature the American republic was still a British colony, as Samuel Miller acknowledged in his *Brief Retrospect of the Eighteenth Century,* published in New York in 1803. In a chapter entitled "Nations Lately Become Literary," Miller conceded that American achievements in science and literature fell far below the European standard and undertook to explain why. The answer, he insisted, did not lie in the soil or climate of the New World, as some writers had asserted, but in the absence of generously funded colleges and universities; lack of leisure; dominance of the commercial spirit; paucity of books and booksellers; absence of inducements to scientific and literary careers in the form of academic chairs, rich fellowships, ecclesiastical benefices, and the like; and the long-standing habit of dependence on Great Britain in matters of taste and culture.[6]

With the winning of political independence, this cultural dependence on the Old World, especially Great Britain, became galling to Americans. Yet in the realm of science the change of attitude is not easy to describe. Certainly there was no general hostility in America toward British science as such. Gilbert Chinard has shown clearly that the American Philosophical Society, the nation's oldest scientific society, exhibited no pro-French or anti-British bias in the election of foreign members to its ranks in the years following the revolutionary war. "Out of sixty-nine foreign members, elected between 1786 and 1793," writes Chinard, "twenty-one were French, eight lived in Germany, three in Sweden, three in Russia, three in Italy, two in Spain, one in Austria, and one in the Netherlands, while Great Britain could claim twenty-seven. The intellectual life of the country had resumed its natural and traditional course: Great Britain was decidedly the most favored country, while France came a good second."[7]

An examination of the *Memoirs* of the American Academy of Arts and Sciences in Boston indicates the same pattern. New England became increasingly Federalist and pro-British from the outbreak of the French Revolution to the end of the Napoleonic regime. "It is true beyond all controversy that there is a dangerous British faction in the heart of some of these New England states," wrote Benjamin Waterhouse to John C. Lettsom in 1810, "and I am fearful of the conflict that will attend the extirpation of this foreign influence. This party [the extreme Federalists] are zealous that we should form an alliance, offensive and defensive, with Great Britain; but the great body of us are averse to any entangling alliances with any European power whatsoever."[8] But Waterhouse, a staunch Jeffersonian Republican, did not speak for the medical profession in Boston; his colleagues were solidly Federalist. *The New England Journal of Medicine and Surgery,* established in 1812, took no notice of the war that broke out between Britain and the United States in that year except to complain that "the slow and precarious receipt of European publications" constituted an impediment to the diffusion of knowledge.

If there was no animus against British science in American scientific circles in the years following independence, neither was there any immediate change in the structure of personal relations established between British and American scientists in prerevolutionary times. These had always been more extensive in medicine and natural history than in natural philosophy, and the deaths of the astronomer John Winthrop IV in 1779 and Benjamin Franklin in 1790 increased this imbalance. The Philadelphia astronomer David Rittenhouse tried to establish a foreign correspondence in his declining years, but without success. Joseph Priestley's emigration to the United States from England in 1794 brought the chemical controversy between the phlogistonites and the champions of oxygen to America (see Chap. 7), but Priestley remained too isolated in his retreat at Northumberland, Pa., to serve as a major link between British and American science. His fellow refugee Thomas Cooper was more active, but his contacts with British science were not extensive. As late as 1817, the Philadelphia chemist Robert Hare was sufficiently despondent about the state of natural philosophy in America to urge his friend Benjamin Silliman to send scientific articles to British journals instead of publishing them at home. "It really seems bad policy to publish any thing in this country upon science especially in the first instance," wrote Hare. "It is rarely attended to in England and we are so low in capacity at home that few appreciate any thing which is done here unless it is sanctioned abroad."[9]

As in the colonial period, the medical profession formed the main link between American and British science in the postrevolutionary era. Medical schools at the University of Pennsylvania, Columbia, Harvard, and Yale soon attracted American students, some of whom might otherwise have gone to Europe for their medical education. But proud though the Americans were to be independent of Europe, they realized that a medical degree from Philadelphia, Boston, or New York was not the equivalent of training at Edinburgh, London, or Paris. The leading figures in the American medical profession, which formed the backbone of the American scientific community in everything except natural philosophy, were trained abroad, chiefly at Edinburgh and London. Silliman found twenty-five Americans studying at Edinburgh in the winter of 1805–1806. These students were learning science as well as medicine. The study of human anatomy was connected with that of comparative anatomy, paleontology, and physical anthropology; chemistry with mineralogy and geology; materia medica with botany; and the human animal frame with zoology. The personal associations formed during medical study abroad were often continued by correspondence when students returned to America. During his studies abroad David Hosack formed an attachment to James E. Smith, president of the Linnean Society of London, Benjamin S. Barton established connections with the British zoologist Thomas Pennant, and Archibald Bruce met many of the leading mineralogists of Europe.

The patronage of British physicians and scientists with whom Americans became acquainted had been and continued to be an important factor in the development of American science. Among the physicians Lettsom was the most generous in support of American efforts in medicine and science. Like his teacher and predecessor in medical practice, John Fothergill, Lettsom extended his patronage to institutions as well as individuals. The Pennsylvania Hospital sought his aid and advice in buying medical works. He sent a donation of books to Carlisle College in Pennsylvania, including a set of Fothergill's works. To Harvard College, where his friend Benjamin Waterhouse (Fothergill's cousin and former pupil) was teaching, Lettsom gave a sizable collection of minerals, birds, and quadrupeds. In 1796, when Nathan Smith went to London to buy books and instruments for a projected medical school at Dartmouth College, Lettsom provided the advice and financial assistance without which Smith's mission could scarcely have succeeded.[10]

American gratitude for these benefactions and services found expression in Lettsom's election to the American Academy of Arts and Sciences, the American Philosophical Society, and the medical societies of New York and New Haven. In 1790 Harvard conferred the degree of Doctor of Laws upon him. His correspondence with Benjamin Rush, Benjamin

Waterhouse, Samuel L. Mitchill, David Hosack, and others continued unabated until the Non-Intercourse Acts imposed on foreign commerce during the Madison administration cut it off a few years before Lettsom's death in 1815.

British botanists continued to seek the cooperation of Americans in the botanical exploration of the New World. Joseph Banks, president of the Royal Society of London, acquired plants for Kew Gardens from Philadelphia gardener Humphry Marshall. James E. Smith and Aylmer B. Lambert, president and vice-president of the Linnean Society of London, were in touch with several American naturalists, including David Hosack in New York, Jacob Bigelow in Boston, Henry Muhlenberg in Lancaster, Pa., and William Baldwin in Wilmington, Del. "In the present imperfect state of science in this country," Bigelow wrote to Smith, "I am every moment sensible of the necessity of an European correspondence, to supply the defect of books and advantages which it is impossible here to command."[11] American botanists depended heavily on British and European colleagues for assistance in identifying American plants (see Chap. 10).

Thus if we judge only by the working relationships between British and American scientists, no great change was brought about by the break with the mother country. But this conclusion overlooks one important fact—American patriotism. Science transcends national boundaries and rivalries, but scientists are citizens and patriots as well as co-workers in the pursuit of truth. The sense of nationality, developed slowly for a century and a half by the special circumstances of American life, had been raised to fever pitch by the struggle for independence. It dominated every aspect of the nation's existence. Americans felt themselves to be a chosen people with a sacred duty to prove that republican institutions were at least as favorable to letters and science as monarchical ones. To do less would be to betray the goddess of liberty in whose name they had fought.

In scientific matters American patriotism displayed itself in many ways, some of them slightly ludicrous in retrospect. Jefferson won the everlasting gratitude of his fellow citizens by refuting the Count de Buffon's aspersions on the animals and aborigines of America. James Mease, a Philadelphia physician, seconded Jefferson's cause and indignantly repelled the suggestion that the New World was of more recent origin than the Old World. Everything connected with America was defended vehemently against foreign jibe or slur. Even the diseases of the New World, such as yellow fever, were claimed as the special preserve of American medical men.

But deeper than mere solicitude for the reputation of Americans and things American was an anxious concern that the discovery and description of the country's scientific treasures—its rocks and minerals, plants and animals, Indian cultures—should be published to the world by Americans. No complaint was commoner among American botanists than that the botanical wealth of North America was being exploited by foreigners far more than by its natives. "While Americans have neglected the botanical examination of this country," wrote Jacob Green, "foreigners have immortalized themselves by doing it."[12] It was particularly galling that the plants collected on the Lewis and Clark expedition, designed by Jefferson as a means whereby American scientists would repay their long-standing scientific debt to their "elder brethren" in Europe, should have been described and classified by a visiting German botanist, Frederick Pursh (see Chap. 10). In 1814 Pursh included these plants in his *Flora Americae Septentrionalis,* published in London much to the vexation of American botanists. One American, Caleb Cushing of Boston, even found it annoying that Pursh should call his treatise a flora of North America when all the plants in it, including those Pursh had collected in Canada, could be found in the United States and hence should be "rightly appropriated to the states and territories of our republic, stretching, as they now do, from the lakes to the gulf of Mexico in one direction, and in the other from the Atlantic to the Pacific ocean. . . ."[13] Thus with one grand flourish of his pen, Cushing annexed the entire region west of the Louisiana Purchase to the United States and staked out an American claim to its botanical resources!

In retrospect we can see that the patriots of the Jeffersonian era were unduly sensitive about their dependence on Europe. Science needs much for its successful prosecution; it demands not only talent but also instrument makers, libraries, museums, herbaria, publications, and patronage and requires that these be within easy reach of each other. Americans had depended on Europe for nearly all these things before 1776; they could not be expected to supply their own needs overnight. On the contrary, the conditions of life in America militated against early development of the institutions requisite for maintenance of a high level of scientific activity. Communication among the regional centers where science developed was slow; most were in closer contact with London than with each other during the early decades of independence. Patronage for science was limited: the federal government was restrained by constitutional scruples, and state governments had not yet learned to look to science for economic improvement. Private individuals gave some support to science, but the United States lacked the kind of leisured class that contributed so much to science in Britain, both in patronage and in actual scientific achievement. The regulars of American science were professional or practical men, tied to a daily routine. "Philosophy," wrote Rush to Lettsom, "does not here, as in England, walk abroad in silver slippers; the physicians (who are the most general repositories of science) are chained down by the drudgery of their professions: so as to be precluded from exploring our woods and mountains. Besides, there are not men of learning enough in America as yet, to furnish the stimulus of literary fame to difficult and laborious literary pursuits."[14] Nor had the development of scientific institutions reached the point where a few strategically located men of science could assemble the researches of others less favorably located and work them into comprehensive treatises.

It was inevitable that the garnering of the first harvest in American natural history should have been done by outsiders, who, unhampered by the daily demands of a practical calling, were free to indulge their passion for exploring the varied productions of nature in a new world. For some reason American society did not produce scientific mavericks of this kind. They came from Europe: André and François André Michaux, Constantine Rafinesque, and John J. Audubon from France; Frederick Pursh and Aloysius Enslen from the Germanies; William Maclure, Alexander Wilson, and William Bradbury from Scotland; Thomas Nuttall from England. Some like the Michaux and Enslen came as scientific emissaries of European governments. One at least, Maclure, was a gentleman of fortune. Most, like Rafinesque, Wilson, Pursh, Nuttall, and Audubon, were scientific adventurers who lived from hand to mouth while they prospected for scientific gold in the virgin continent that lay before them. If some returned to Europe to publish their researches, it was because there they could find the libraries, herbaria, museums, and patronage that were lacking in the United States. By 1820, however, the tide had begun to turn. Americans now had their own scientific societies, journals, and books. American science was still heavily dependent on Europe but had passed from its former colonial status to a position of honorable partnership in the Western scientific enterprise.

THE RELIGIOUS CONTEXT

If science was viewed through the prism of patriotism by the first generation of American citizens, it was also refracted through another medium, that of religion. In this case, however, Americans did not see eye to eye. Abandoning belief in the divinity of Jesus and the inspiration of the Bible and retaining only the traditional faith in the existence of an omnipotent, omniscient, and benevolent Creator, deists like Jefferson, Franklin, and Ethan Allen, Vermont patriot and author of *Reason, the Only Oracle of Man,* viewed science as a God-given instrument for rationalizing human institutions in behalf of individual liberty and social progress. "Reason" and "nature" were their slogans. Science was the supreme example of reason in action, yielding useful knowledge. Nature, in their view, was not the pleasant landscape admired by the painter or wanderer in the woods; it was the atomistic world of invisible particles of matter moving inexorably in accordance with wisely ordained laws, a world epitomized in Newton's mathematical model of the solar system. Thus conceived, nature was a perfect symbol for those who viewed society as a collection of individuals seeking their own enlightened self-interest in a divinely ordained "system of natural liberty," as Adam Smith called it. Out of the competition of the marketplace the wealth of nations would result through operation of the law of supply and demand. Out of the competition of ideas and opinions Truth would emerge. Through the pooling of individual wills the general will would make itself manifest. All this would come about if individuals were freed from artificial institutional restraints and left free to follow "reason" and "nature."

Natural religion, or deism, was an integral part of the campaign to rationalize human institutions in behalf of liberty and progress. Religion was to be made a private affair outside the authority and concern of the state. Its doctrinal content was to be reduced to a minimum, thereby ending theological controversy and freeing human energies for more constructive pursuits. Anxiety about the next world would be replaced by concern for the improvement of human life in this one. But this would come about only if reason supplanted the biblical revelation as the basis of religion. Science, said Thomas Paine, had shown that revelations were absurd; the fundamental doctrines of religion were "deducible by the action of reason upon the things that compose the system of the universe." Ethan Allen agreed. Nature and its laws were perfect. Not even God could interfere with their operation.[15]

This was going too far for Jefferson. He conceded that God *could* suspend or manipulate the laws of nature but declared himself confident that God *would not* do so. Having provided man with everything needed to work out his salvation by reason and nature, God would be content to preserve the perfect order and regularity of his creation. And man, if he had the sense to put aside priestly notions of revelation and supernatural intervention, would be free to undertake the reformation of social and political institutions in keeping with the progress of science. Science would dispel the clouds of ignorance and superstition that cloaked the impostures and oppressions of priests and kings and secure for mankind the dominion over nature God had intended. Jefferson predicted after the election of 1800 that the New England states would be the last to come over to republicanism and science, "on account of the clergy, who had got a smell of union between Church and State. . . ."

> If, indeed they could have prevailed on us to view all advances in science as dangerous innovations, and to look back to the opinions and practices of our forefathers, instead of looking forward, for improvement, a promising ground-work would have been laid. But I

am in hopes . . . that they will find their interest in acquiescing in the liberty and science of their country, and that the Christian religion, when divested of the rags in which they have enveloped it, and brought to the original purity and simplicity of its benevolent institutor, is a religion most friendly to liberty, science, and the freest expression of the human mind.[16]

Such was Jefferson's view of the relations between science, religion, and republican government, but the picture he painted of Christian, and especially clerical, attitudes toward science was grossly inaccurate. The New England clergy—men like Jeremy Belknap, Manasseh Cutler, Joseph Willard, John Prince, Jedidiah Morse, Ezra Stiles, and Timothy Dwight—were not against science and secular improvement. Dwight, for example, worked his way through Newton's *Principia* several times as a student at Yale. On succeeding Stiles as president of Yale in 1795, he persuaded the Connecticut legislature to establish four new professorships, including the one in chemistry and natural history to which he appointed young Benjamin Silliman. In subsequent years Dwight supported Silliman unstintingly and played an important role in the Connecticut Academy of Arts and Sciences. His *Travels in New England and New York,* published in London several years after his death in 1815, rivaled Jefferson's *Notes on the State of Virginia* in its attention to the climate, soil, geography, geology, flora and fauna, and aborigines of the region described. Indeed, if one had to choose a Christian-Federalist counterpart to Jefferson as a champion of science and its potential benefits to society, Timothy Dwight would rank high on the list of candidates. He was, said Silliman, "truly a great man," one who, at a time when physical science had not yet proved its social utility, "saw with a telescopic view both its intrinsic importance and its practical relations to the wants of man and to the progress of society."[17]

Dwight's ideas about science are best conveyed in the course of sermons he delivered annually to the junior and senior classes at Yale, in which he presented science as the study of God's works. Like the deists, he portrayed nature as a stable framework of wisely designed structures fitted as a stage for the activities of intelligent beings. Nature was a perfect means to perfect ends, designed to minister to the physical, intellectual, moral, and spiritual needs of the creatures who inhabited it. The solar system with its inhabitants was but one of thousands of such systems circling the fixed stars, "all . . . sustained, regulated, and moved by the hand of the Almighty Being, who created them, and *whose kingdom ruleth* over all."[18] This glorious system of worlds without end was governed by divinely ordained laws, but the God who had established them could suspend them or modify their operation at will to accomplish his purposes.

Thus Christian natural theology was indistinguishable from that of the deists, which was borrowed from the Christian tradition. What separated the "friends of revelation," as the Christians liked to call themselves, from the deists with respect to nature and natural science was not natural theology but belief in the divine inspiration of the Bible and the view of human nature and human history derived from it. To the deists, confident in man's ability to know and do the right and preoccupied with advancing the cause of liberty and social progress here and now, nature and reason seemed to reveal all that was necessary and practical about man's duty, destiny, and relation to the Creator. To Dwight, however, it was clear that man was out of harmony with nature and its Creator, "possessed in the humblest degree of rational attributes; the subject of extreme weakness, sluggishness, and ignorance; hastening with a rapid tendency to decay, old age, and death; without love to God, or his fellowmen; depraved throughout with sin; and voluntarily yielding by himself to final perdition."[19] In this sorry condition, only God's direct revelation of himself as recorded in the Bible could save mankind from self-destruction. Deists might argue that nature itself was sufficient revelation of the existence and attributes of God, but they forgot it was the Bible that had taught them to see the wisdom of God in nature.

That the Bible was indeed the inspired Word of God was evident, said Dwight, from internal evidence—its lofty language, spiritual profundity, and exalted moral precepts—and from external evidence, especially the miracles and prophecies recorded in Scripture and attested by numerous reliable witnesses. Since the Bible was from God, written under the influence and direction of the Holy Spirit, everything in it was infallibly true if properly interpreted. In view of its divine origin, it was to be expected that many of its doctrines should transcend human reason. In interpreting the Bible, therefore, the proper method was to accept the words in their plain and obvious sense unless such a meaning was clearly at variance with reason and observation, in which case other interpretations might be considered, but always with a due sense of the fallibility of human reason and the divine source of the biblical revelation. The Bible was the Word of God; nature was his work, hence there could never be any real conflict between Scripture and science. Any apparent conflict must be the result of either faulty science or an incorrect interpretation of Scripture and hence be capable of resolution by the further progress of science and biblical exegesis.

From this point of view, the enemy of religion was not science but "science falsely so-called," as the Scriptures termed it. The difference between the true and the false scientist was a matter of attitude, not method. The true scientist approached nature reverently, seeking to discover in God's handiwork the evidences of his power and wisdom. Whatever wonders the searcher found, whatever uses flowed from discoveries, were attributed to God's design and providence. If his studies yielded results that appeared to contradict the Bible, he preferred to trust God's Word rather than his own feeble reason until such time as the apparent inconsistency was resolved. In moments of success he remembered that man's finite mind can penetrate but a little way into the unsearchable mysteries of God's creation; the ultimate truths of nature, except as they were revealed in the Bible, were too high for human understanding. Valuing natural knowledge for its own sake and its utility, he kept it in due subordination to the most important and useful kind of knowledge, the knowledge of eternal salvation vouchsafed by God himself in his Word.

The false scientist, on the contrary, attempted to reduce nature to "the blind action of stupid matter." He was more concerned with studying the laws of nature than with God's purposes in establishing them and, indeed, often forgot to attribute them to God. For all he cared, the universe and its laws were eternal and all the beauty and contrivance of creation the product of chance. The infidel philosopher had little reverence for God's works and none at all for his Word. He put the speculations of human reason above the truths of revelation and even denied its possibility. Not content to describe the harmony of the created world, he presumptuously inquired into the origin of things, substituting his own crude theories for God's inspired narrative. In short, false science openly contradicted God's revealed Word or the conceptions of man, nature, history, and salvation contained therein. It denied the divine origin of nature, its appointment as a theater for the moral development of rational beings, its wise contrivance and divine superintendence, the moral significance of natural events, man's fall from original perfection to a state of moral probation, and the consequent necessity of a second revelation for the redemption of mankind and the restoration of the harmony of nature.

Within this framework of ideas Christians could be as enthusiastic about the progress of science and its practical applications as any scoffer at revelation. Discussing the question of human perfectibility with a senior class at Yale, Dwight called attention to the sudden burst of progress in Western Europe, beginning in the sixteenth century, and added:

> The Millennium is not to come all at once, but is to alleviate the evils of mankind by degrees. . . . Improvements are making on so large a scale, as to outrun every thing which had been known before. . . .
>
> See the improvements we have. If we invent a way to turn a thousand spindles we pay an hundred hands. We have found that the small-pox may be prevented by a milder disease arising from the udders of cows. . . . These inventions we commonly attribute to human ingenuity: but the author is God. Knowledge will increase, and great improvements will still be made. I believe it will not be more than one or two centuries before there will be such a restraint on human passions . . . that it will make the world appear as if all the good men

Two champions of the Christian view of science and its social role: Timothy Dwight (by John Trumbull) (left), President of Yale College 1795–1817, and Samuel Miller (by Thomas Sully) (right), Presbyterian clergyman and author of A Brief Retrospect of the Eighteenth Century. *Courtesy of the Yale University Art Gallery (Dwight) and the Frick Art Reference Library (Miller), the latter with the permission of Mrs. Arnold A. Willcox*

in it were brought together. [In a sermon before the American Board of Commissioners for Foreign Missions in 1813 Dwight set the probable date of the millennium, "in the full and perfect sense," at about A.D. 2000!]. All this will take place, but when we have done, it will be as far from the perfectibility which fools talk of, as the mite from a man, and an oyster from a whale.[20]

If further proof was needed that the clergy, at least in denominations insistent upon a learned ministry, were not opposed to science and its applications, it could be found in Samuel Miller's *Brief Retrospect of the Eighteenth Century,* cited earlier. Miller, a young Presbyterian clergyman in New York City, was the brother of Edward Miller, coeditor with Samuel L. Mitchill of the *Medical Repository,* the nearest thing to a general scientific journal in the United States at that time. Having undertaken to give a sermon commemorating the dawn of the nineteenth century, Miller chose as his theme the revolutions and improvements in science, arts, and literature in the previous century. Eventually this sermon was expanded into a two-volume work, the *Brief Retrospect,* characterized by impressive erudition and a felicitous style.

Miller defined the sciences very broadly. But although he included theology, morals, and politics among the sciences, he decided to postpone consideration of those branches of knowledge to a later time, devoting his first volume to the sciences of nature and their applications in navigation, agriculture, and the mechanic arts and his second to the philosophy of the human mind, educational theory, literature, and the growth of scientific and literary societies and publications. Throughout the work, he explained, he would be concerned to show the perfect harmony between the Christian religion and all sound science and learning, but he would not fail to describe the scientific and literary achievements of infidel writers, however false and misleading the philosophical and religious conclusions they drew from those achievements. "It is proper," said Miller, "to commiserate the mistakes of such persons, to abhor their blasphemy, and to warn men against their fatal delusions; but it is surely difficult to see either the justice or utility of withholding from them that praise of genius or of learning to which they are fairly entitled."[21] The true Christian had nothing to fear from sound science. On the contrary, he should take pride in the achievements of the human mind, confident that true science would triumph over false and contribute to the progress of mankind in accordance with the wise dispensations of an overruling Providence.

True to his promise, Miller surveyed the scientific and literary developments of the eighteenth century comprehensively, but always with an eye to their bearing on Christian principles and dogmas, especially the doctrine of the inspiration of the Bible. Various theories of the earth, from those of Thomas Burnet and William Whiston in the seventeenth century to those of James Hutton and Richard Kirwan in Miller's day, were described in some detail; and Kirwan's authority was invoked in support of the historicity of Noah's flood and its importance in explaining geological phenomena such as the presence of elephant and rhinoceros bones and teeth in northern regions far from their apparent native habitats. The geographical discoveries of Captain Cook, Vitus Bering, and others were also cited as providing confirmation of the truth of the Bible, especially in its account of the common origin and subsequent dispersion of the human race. In the field of linguistics and Oriental studies, the writings of William Jones provided the requisite demonstration of the harmony of science and Scripture. "In Massachusetts," noted a reviewer of Miller's book in the *Monthly Anthology and Boston Review,* "the *Retrospect* has become a fashionable book, and will be more so, as it is introduced to the social libraries. . . . Religious people, especially, will be fond of reading a book which makes every event and every work illustrative of the belief of the scriptures. . . ."[22]

The Age of Enlightenment, Miller concluded, had been in many respects a wonderful one. Those who had lived in that era might well consider themselves a favored generation.

> Though they have been pained with the sight of some degrading retrocessions in human knowledge, and almost stunned with the noisy pretensions of false philosophy, they have seen at the same time, improvements in science, which their fathers, a century ago, would have anticipated with astonishment, or pronounced altogether impossible. They have seen a larger portion of human society enlightened, polished, and comfortable, than ever before greeted the eye of benevolence. They have in a word, witnessed, on the one hand, the accession of honours to science, which it could boast of in no former period; and, on the other, a degree of usefulness reflected from science to economy and art, no less conspicuous and unrivalled.[23]

But these triumphs of science and its applications gave no support to deistic dreams of unlimited future progress through the accumulation and dissemination of secular knowledge, Miller warned. Although the achievements of one generation in science and technology were handed down to the next, progress in morality, religion, and civic virtue was not. Every generation had to learn again the lessons taught by the wisdom of the ages. Moreover, as Thomas Malthus had shown in his *Essay on Population* (1798), the tendency of human population constantly to outrun the food supply forever doomed to ineffectiveness utopian visions of increasing plenty for all mankind. Also, the doctrine of man's perfectibility through his own efforts was blatantly in conflict with the revealed Word of God, which represented him as a fallen, selfish creature incapable of doing good without the aid of divine grace. A millennium would undoubtedly come, but it would be a divinely wrought millennium of benevolence, peace, purity, and universal holiness, not the millennium of knowledge without wisdom, benevolence without piety, and purity and happiness without virtue that was depicted by infidel philosophers.

Thus the battle lines were drawn, not between the friends and enemies of science and progress, but between those who conceived the progress of knowledge and society in Christian terms and those who conceived it as a consequence of the inevitable triumph of reason and nature over arbitrary and oppressive institutions of church and state that found their ultimate sanction in supernatural revelation. Not all Jeffersonian Republicans were deists or atheists, nor were all Christians anti-Jeffersonian in politics — far from it. Jefferson had the political support of many God-fearing Presbyterians, Methodists, Baptists, and even some Congregationalists. But whether the progress of science and technology was to be conceived in secular terms without reference to the Bible or understood in the light of the biblical conception of man, nature, and history was a profound question with important consequences for the future.

Although public interest in science was aroused from time to time by its bearing on religious belief or by dramatic scientific revelations such as William Herschel's discovery of a new planet, the comet and meteor shower of 1807, and Charles Willson Peale's exhumation of two mastodon skeletons, it could not be depended on to sustain a vigorous scientific enterprise in the new nation. Quite to the contrary, if we can believe the American scientists of this period, struggling as they were to create a place for science in the thoughts of their fellow citizens. "I was compelled to clear the ground to prepare it before I sowed the seed," said Benjamin Waterhouse. "I found I must first excite a taste, and then try to gratify it."[24] The same was true of all the sciences. "Chemistry," wrote Benjamin Silliman in 1810, "is here almost a new pursuit, and hence it is not uncommon to find even intelligent men manifesting an entire ignorance of its nature and utility."[25] Popular interest in science was something to be wished for and worked for, not something to be taken for granted.

The United States was not greatly different from European countries in this respect. The period was a brilliant one in the history of Western science, but it was not a great age of popular science. The earlier enthusiasm for popularizing Newtonian physics and astronomy was on the wane, and the day of the mechanics' institute, the lecture circuit, and the popular scientific press had not yet arrived on either side of the Atlantic. In France one could point to the public lectures at the Museum of Natural History, in Britain to Humphry Davy's lectures at the Royal Institution, Frederick Accum's chemical school in London, attendance of the working class at lectures on experimental physics at the University of Glasgow, and the gradually increasing number of introductory works on natural philosophy, some frankly popular in character.[26] But these were only the beginnings of a movement that was not to come into its own until the second quarter of the nineteenth century, by which time the progress of urbanization and industrialization, the spread of public education, and the growth of a strong will for self-improvement among the artisan class had created the conditions requisite for large-scale popularization of science.

The flowering of popular science requires not only a potential audience but also some degree of maturity in science itself as well as a sense of security and leisure on the part of the scientist. In these respects conditions in the United States were distinctly less favorable than those in Europe. "Our elder brethren in Europe know not the difficulties that the first settlers in science have to encounter," wrote Waterhouse in 1811. What were some of these difficulties? First, very few Americans were in a position to devote themselves wholeheartedly to scientific research. Science was an avocation for most who loved it, and even those who taught natural philosophy or natural history in the colleges were hard pressed with other duties. Benjamin S. Barton, Samuel L. Mitchill, and Benjamin Waterhouse were practicing physicians as well as teachers of botany and natural history. Mitchill was also active in politics. Astronomers David Rittenhouse and Andrew Ellicott were kept busy with surveying, public office, and other activities to support their families. Henry Muhlenberg, James Madison, and Manasseh Cutler were clergymen. It was much to be regretted, declared Barton, "that the principal cultivators of natural science, in the United States, are professional characters, who cannot, without essentially injuring their best interests, devote to these subjects that sedulous attention which they demand."

When time for scientific research was available, the necessary tools were frequently lacking. "The study of natural history in this country," wrote Cutler to a European correspondent in 1799, "is in its infancy."

> This deficiency has been in part, owing to the great scarcity of books on natural history. . . . We have many public libraries, consisting of large and well chosen collections of books from Europe, excepting on the different branches of natural history, in which there are few, and those mostly ancient authors who wrote before the Linnaean system was formed, and our booksellers import no books on this subject. . . . We have no cabinets of natural history in America, excepting one in Philadelphia and another in Boston.[27]

In a similar vein a decade later Silliman lamented the scarcity of chemical apparatus in the United States and the inability of American industry to supply the necessary instruments. "Under such circumstances," he observed, "it is easy to see, that the practice of chemistry, even for philosophical [scientific] purposes, must be attended with difficulties of no small magnitude. . . ."[28] The energies of American scientists could not be devoted to popularization except insofar as the creation of an intelligent appreciation of science was necessary to its very existence.

The place of efforts at popularization in the larger struggle to put American science on its feet may be illustrated in various ways. *The Catalogue of All the Books Printed in the United States,* issued by the Boston booksellers in January 1804, provides a convenient starting point. Of the 1,338 titles listed, not more than twenty can be considered works of science if medical titles are excepted. Judging from these titles and the notices in the *Medical Repository* in subsequent years, chemistry was the most popular science of the day. "Elementary books on chemistry are become almost as common as books on spelling and arithmetic," the editors of the *Repository* observed in 1808. "The compilations of Parkinson, Thompson, Henry, Accum, Jacobs, Murray, Lagrange, Spalding, Ewell, and several others which have been published within a few years, evince the prevailing taste for this interesting science." As the list of authors suggests, most of these books were reprints of foreign works. Few were truly popular in character. The two that ran to several American editions before 1820, James Parkinson's *Chemical Pocket-Book* and William Henry's *Epitome of Experimental Chemistry,* were technical manuals for the beginning student. Works of a more popular character, such as Madame Marcet's *Conversations on Chemistry* or Thomas Ewell's *Plain Discourses on the Laws or Properties of Matter,* were few and far between. Jefferson's complaint to Ewell that the chemists "seem to write only for one another" was echoed by Bushrod Washington:

> I have not met with a single treatise which has not appeared unnecessarily obscured by technical terms, which only scholars can understand. They have been more generally addressed to the comprehension of professional and learned men, than to those of the humble walks of life, for whose use this science might be made most essentially to contribute, by adapting it to their capacities, and by pointing out the way in which its principles may be applied to the more common arts, in which they are daily employed.[29]

Natural history was in a similar position, except that books in this field were less numerous. One of the few popular works in botany was Waterhouse's *The Botanist,* which was published serially in the *Monthly Anthology and Boston Review* before appearing in book form in 1811. It had been his original idea, Waterhouse explained in the "Advertisement," to include lectures on other branches of natural history, but he had relinquished that plan on the plea of the editors "that mineralogy would be less popular than botany; and therefore less adapted to such a monthly magazine of knowledge and pleasure, as the Anthology was meant to be; and less likely to attract the attention and patronage of readers of both sexes." Waterhouse was also influenced by the knowledge that the merchants of Boston were being asked to contribute to establishment of a botanical garden at Harvard, although few of them knew enough of botany to understand the purposes that such a garden might serve. These facts were set forth, he said, so that the European disciples of Linnaeus might "see the reason, and therefore excuse the popular dress, in which BOTANY, that beautiful handmaid of Medicine, has been introduced to the inhabitants of a region characteristically called by the English a century ago, THE WILDERNESS."[30]

The other botanical works that issued from the American press in this period (see Chap. 10) could scarcely be described as popular. Benjamin S. Barton's *Elements of Botany* was a textbook designed for his students at the University of Pennsylvania. Jacob Bigelow's *Florula Bostoniensis* was produced in response to popular acclaim of his botanical lectures in Boston, but it was little more than a catalogue of plants in the vicinity of Boston. For his Harvard students, Bigelow undertook to edit an American edition of James E. Smith's *Introduction to Physiological and Systematic Botany*. "Our Botanists," he wrote to Smith, "are not yet sufficiently numerous to induce the booksellers to publish large works; but as the country grows I hope the taste for science will increase." The works of Linnaeus were in demand among American botanists, but none appears to have been reprinted in the United States. Nor do the more popular works of the Count de Buffon, although available in English translation, seem to have attracted American booksellers. Erasmus Darwin's poetical renditions of Linnaean botany, as in his *Botanic Garden,* were reprinted in the United States and considered appropriate reading for young women, but the work from Darwin's pen that ran through more American editions than any other, his *Zoonomia,* was a medical work of little interest to the general public. Apparently the American market for scientific books in this period was largely among serious students of the sciences. The potential audience was probably fairly well described by the anonymous authors of *An Epitome of Electricity and Galvanism,* published in Philadelphia in 1809, when they wrote: "Our views will be fully answered, if [our work] shall be found well adapted to assist youth in their academical and philosophical studies, and at the same time, to afford amusement to men of learning, and some useful information to gentlemen of leisure."[31]

Scientific periodicals were also serious in tone and content. The various scientific societies, notably the American Philosophical Society and the American Academy of Arts and Sciences, published transactions from time to time, but these publications had very limited circulation. The short-lived *Emporium of Arts and Sciences,* published by John R. Coxe and Thomas Cooper, was devoted to the technological applications of science, but it was not a popular science magazine in the modern sense. "I will not," Cooper announced on taking over the editorship in 1813, "condescend to make this a work of mere amusement, for the purpose of sale — one that shall suffice merely, under the show of science, to enable the reader to trifle away an hour, and to skim the surface of a great many subjects for the purpose of a superficial and conversation knowledge. Many pages of this work to a general reader will be very dull; but it will be my fault if they are not useful to those who read for improvement."[32] Apparently the number of such readers was not large, for the magazine was soon discontinued.

The only thing approaching a general scientific magazine was the *Medical Repository,* published in New York by Mitchill and his associates. In its early years this was by no means simply a medical journal. The preface to the first volume, issued in 1798, declared the journal to be "a depositary of facts and reasoning relative to Natural History, Agriculture, and Medicine." In subsequent prefaces the editors announced their intention "to notice every leading fact, and every important improvement in the progress of the physical sciences both in Europe and America." They were pleased, they said, to find the scope and circulation of the publication increasing, but they had made no money in the enterprise, "and every prospect of that kind is at present so remote and uncertain, as to offer no encouragement of such expectations." Instead, their reward must be "the consoling and animating reflection that the importance of the objects we pursue is not limited to the place in which we reside, nor to the times in which we live."

True to their word, the editors of the *Repository* posted its readers on scientific developments on both sides of the Atlantic. From its pages they learned not only of Cuvier's work on fossil quadrupeds and Davy's researches in chemistry but also of Maclure's geological explorations in the United States, Peale's exhumation of the mastodon, the appearance of the successive volumes of Alexander Wilson's *American Ornithology,* the founding of new American scientific publications, and many other undertakings and accomplishments. All these were reported with great enthusiasm for science, especially American. "Too long," declared the editors, "has it been fashionable for our people to seek scientific news from transatlantic regions, while they neglected the manifold novelties by which they were surrounded at home. But latterly a more correct opinion has prevailed among them . . . to turn their backs to the east, and direct their views to the inviting and productive regions of the interior of America." The day was at hand, said the editors, when the businessman, the gentleman, and the finished scholar must all pay greater attention to natural history: "How boorish . . . is it to be ignorant of the general history of the elephant!" With such exhortations did Mitchill and his colleagues seek to promote the cause, but their approach was scholarly, their audience limited, and their reward uncertain.[33]

The demand for public lectures on science, though not totally absent, was not sufficient to support sustained enterprises of this kind in most cities. The only person who earned a substantial part of his living by popular scientific lectures during this period was John Griscom. About 1795 while teaching school in New Jersey, Griscom became interested in chemistry through reading a translation of Lavoisier's *Elements of Chemistry*. He went to Philadelphia to attend James Woodhouse's chemical lectures, ordered apparatus from London, and soon was teaching chemistry in Burlington, N.J. Upon moving to New York in 1800, he tried his hand at lectures by public subscription. They were an immediate success, so much so that Griscom rented part of the Friends' graveyard on Liberty Street where he built a substantial brick building to house his lectures. He continued there until the New York City authorities donated rooms in the Old Alms House for use by Griscom and other scientific and cultural enterprises (see Chap. 4). Griscom's lectures thus became a familiar institution in New York, attended by humble mechanics as well as society women, noticed from time to time in the newspapers, and mentioned with respect in Mitchill's *Medical Repository*. The lectures of 1807–1808 were attended, said the *Repository,* by "upwards of one hundred persons of both sexes, and have obtained a distinguished degree of approbation." Several of Griscom's auditors that winter volunteered to inhale nitrous oxide, "laughing gas," to demonstrate its effects. One person fainted, but persons in good health, especially those inclined to corpulency, exhibited "the highest degree of exhilaration and rapture."[34]

Griscom did not confine himself to chemistry. In 1811 he announced that his chemical lectures would be preceded by a full course on natural philosophy, including astronomy. The topics discussed and the fashionable character of his audience are suggested in Fitz-Green Halleck's poem "Fanny," in which the heroine is pictured as one of the "bright-eyed maids and matrons" who attended "Griscom's conversations," and there learned:

> Words to the witches in Macbeth unknown,—
> Hydraulics, Hydrostatics, and Pneumatics,
> Dioptrics, Optics, Katoptrics, Carbon,
> Chlorine, and Iodine, and Aerostatics;
> Also,—why frogs, for want of air, expire;
> And how to set the Tappan sea on fire.

Griscom's career continued well into the age of popularization in the second quarter of the nineteenth century. Along with Benjamin Silliman he could claim the honor, as Silliman put it, "of inaugurating a system which has since become almost universal in the United States."

But Griscom was the exception rather than the rule in the opening decades of the century. Most public lectures on science before 1815 were one-season performances before small audiences, often without recompense to the speaker. In the winter of 1807–1808 Boston society turned out to hear a visiting Frenchman, Silvain Godon, lecture on mineralogy. In the spring of 1809 chemistry was the topic, the lecturer John Gorham, recently returned from studies in Europe. "Although the subject was a novel one, and though few persons in this metropolis had cultivated a taste for it," reported the *Monthly Anthology and Boston Review,* "yet his lectures were as fully and constantly attended as the scale upon which his modesty had induced him to commence them could admit."[35] The response to the botanical lectures of Benjamin Waterhouse and Jacob Bigelow was equally gratifying.

Albany reported a course of chemical lectures attended by about sixty persons, including many women. Philadelphia was much taken with the botanical lectures of Joseph Correa da Serra, Portuguese minister to the United States and a naturalist of international reputation. In Wilmington, Del., William Baldwin found great enthusiasm for botanical instruction among the "ladies of the city," and John Vaughan complimented them for their faithful attendance at his lectures on natural philosophy. In Charleston, S.C., Stephen Elliott lectured on natural history to the Literary and Philosophical Society.[36] These various public lectures evinced a growing interest in the sciences, compounded of native curiosity and civic pride, but they were small affairs compared to Silliman's grand performances at the Lowell Institute a quarter of a century later.

A final illustration of the effort to create and exploit popular interest in science is to be found in the story of Peale's Museum, established in Philadelphia in 1786–1809 by the well-known portrait painter Charles W. Peale (see Chap. 2). Of all the museum entrepreneurs of the period, and there were many, Peale alone seems to have had a vision of the service a popular museum of natural history could render to the progress of science and the scientific education of the public. His ideal was that of a great museum publicly supported as part of a state or national university, containing within its walls a "world in miniature," its exhibits open alike to the student of science and the public, its program embracing the display of nature's productions and the instruction of the public by lectures and guided tours and the linking of science to the arts of music and painting. Such a museum would be a veritable "Temple of Wisdom," a place where men and women would learn to imitate the benevolence of the Creator by observing the order and beauty of his creation.[37]

If the reality fell short of the ideal, it was not for lack of effort on Peale's part. He labored constantly to extend and improve his collections and arrange them scientifically. He studied the methods of European museums and exchanged specimens with them. He undertook to produce a catalogue of the exhibits in his museum and to lecture on natural history. He tried again and again to secure public support for the institution; however, in this he failed. The Jeffersonian party to which he gave allegiance was too much attached to the principles of economy in government and strict construction of the Constitution to approve federal expenditures for a national museum, and the Pennsylvania legislature confined itself to donating the upper rooms of the old State House (Independence Hall) to house the collection. Like most of the visitors who paid their twenty-five cents to see the exhibits, the legislators regarded the museum as a kind of show; they were impressed by Peale's financial success but took it as proof that his museum did not need public support.

After Peale's retirement from the management of the museum in 1809, the precarious balance between scientific integrity and pecuniary profit tilted dangerously in the practical direction. In the hands of his son Rubens Peale the "Temple of Wisdom" became increasingly a business operation, hard pressed by competition from the numerous rivals to which its financial success had given rise. Charles W. Peale's vision of the museum as a great instrument for the promotion and popularization of science was not to be realized in his day nor for many decades to come. Jefferson probably appraised the situation correctly when in 1807, in response to a request for his subscription to a museum venture in Williamsburg, Va., he wrote:

> In the particular enterprises for museums, we have seen the populous and wealthy cities of Boston and New York unable to found or maintain such an institution. The feeble condition of that in each of these places sufficiently proves this. In Philadelphia alone, has this attempt succeeded to a good degree. It has been owing there to a measure of zeal and perseverance in an individual rarely equalled; to a population, crowded, wealthy and more than usually addicted to the pursuit of knowledge.[38]

It appears, then, that popular interest in science in this period, though by no means completely absent, was inadequate to provide solid support for scientific enterprises and that such public interest as existed was due in no small part to the efforts of the devotees of science to convince their fellow citizens of the importance and utility of scientific research.

THE JEFFERSONIAN CONTEXT

In most of its leading characteristics — patriotism; utilitarianism; antitheoretical bent; fascination with the geography, flora, and fauna of the North American continent; and interest in the relations of science, politics, and religion — American science found an appropriate spokesman and symbol in Thomas Jefferson. Like Francis Bacon, whom he admired greatly, Jefferson took all knowledge for his province. His interest in science went well beyond acquiring the smattering expected of a gentleman in those days. While he was in Paris as American minister to France, he visited private cabinets and royal collections and sought out such promi-

N O T E S

ON THE

STATE OF VIRGINIA,

WRITTEN BY

THOMAS JEFFERSON.

ILLUSTRATED WITH

A MAP, including the States of VIRGINIA, MARY-
LAND, DELAWARE and PENNSYLVANIA.

L O N D O N:

PRINTED FOR JOHN STOCKDALE, OPPOSITE
BURLINGTON-HOUSE, PICCADILLY.

M.DCC.LXXXVII.

Title page of the London edition of Jefferson's only book, which set the tone and agenda for much of American science in Jefferson's lifetime. Courtesy of the American Philosophical Society

A comparative View of the Quadrupeds of Europe and of America.

I. *Aboriginals of both.*

	Europe.	America.
	lb.	lb.
Mammoth		
Buffalo. Bison		*1800
White bear. Ours blanc		
Caribou. Renne		
Bear. Ours	153.7	*410
Elk. Elan. Original, palmated		
Red deer. Cerf	288.8	*273
Fallow deer. Daim	167.8	
Wolf. Loup	69.8	
Roe. Chevreuil	56.7	
Glutton. Glouton. Carcajou		
Wild cat. Chat sauvage		†30
Lynx. Loup cervier	25.	
Beaver. Castor	18.5	*45
Badger. Blaireau	13.6	
Red fox. Renard	13.5	
Grey fox. Isatis		
Otter. Loutre	8.9	†12
Monax. Marmotte	6.5	
Vison. Fouine	2.8	
Hedgehog. Herisson	2.2	
Marten. Marte	1.9	†6
	oz.	
Water rat. Rat d'eau	7.5	
Wesel. Belette	2.2	oz.
Flying squirrel. Polatouche	2.2	†4
Shrew mouse. Musaraigne	1.	

"A Comparative View of the Quadrupeds . . ." from Jefferson's Notes on the State of Virginia, *in refutation of Buffon's aspersions on the size and vigor of New World animals. Courtesy of the American Philosophical Society*

nent figures in natural history as the Count de Buffon, the anatomist
J. M. L. Daubenton, and the botanist André Thouin, with all of whom
he subsequently corresponded. In his outings along the Potomac River
"he would climb rocks, or wade through swamps to obtain any plant he
discovered or desired and seldom returned from these excursions without
a variety of specimens."[39] His knowledge of the various departments of
natural history, especially botany and zoology, was, said Benjamin S.
Barton, "equalled by that of few persons in the United States." His acquaintance with astronomy and natural philosophy was such that Andrew
Ellicott deemed him "more capable than any other gentleman of my acquaintance" of judging the merit of Ellicott's astronomical and
geographical observations during the survey of the southern boundary of
the United States. Jefferson was his own weather bureau before such
facilities existed, keeping daily records of temperature, wind, and rainfall. He investigated and collected the plants of Virginia and compiled
his own list of its birds, using the works of Linnaeus, Mark Catesby, and
Buffon. He also gathered information about the Indian tribes of North
America, with special reference to their languages, and excavated an Indian
mound near Monticello.

In 1780–1781 Jefferson brought this and much other information
together in his one and only book, *Notes on the State of Virginia, written . . . in answer to certain queries . . . respecting its boundaries, rivers,
sea ports, mountains, cascades and caverns, productions, mineral,
vegetable and animal, climate, population, military force, marine force,
aborigines, counties and towns, constitution, laws, college, buildings and
roads, proceedings as to Tories, religion, manners, manufactures, subjects of commerce, weights, measures and money, public revenue and expenses, histories, memorials, and state papers,* which set the tone and foreshadowed the content of much of American science for the next three
decades. It provided a preview of things to come. It will be worthwhile,
therefore, to examine some of the arguments of the *Notes,* comparing them
with the characteristics of early American science already described and
with the ideas and attitudes reflected in Jefferson's other writings.

The patriotic sensitivity of Americans with respect to everything connected with the New World is nowhere better illustrated than in Jefferson's refutation in his *Notes* of the confident assertions about the
degeneracy of nature's productions on the American continent contained
in the thirty-six volume *Natural History, General and Particular* of the
celebrated Count de Buffon, Intendant of the Jardin des Plantes and
Keeper of the Royal Cabinet of Natural History in Paris. Buffon was a
man of wide knowledge and great scientific imagination, one of the early
foreshadowers of the theory of organic evolution. In particular, he was
interested in the nature, stability, and distribution of animal species. As

specimens of the quadrupeds of the Old and New Worlds accumulated at the Royal Cabinet of Natural History and were dissected by Buffon's anatomist, Jean Marie L. Daubenton, Buffon compared the animals of the eastern and western hemispheres. He attempted to explain their similarities and differences by supposing that the New World was geologically younger than the Old World and that animals from the latter had migrated to America and had there degenerated in size and vigor owing to the influence of a different climate, topography, and diet. The European inhabitants of the New World were infuriated by this invidious comparison and tried in every way to refute it.

Among the various champions of the honor of American animals, none was better equipped to meet Buffon's challenge than Thomas Jefferson. He had an ardent sense of patriotism, a passion for collecting facts, and a mania for measurement. He measured all kinds of things: the trunks of trees, the height of mules, the weight of peas and strawberries grown on his plantation, the time it took a workman to fill a wheelbarrow and roll it thirty yards, the time it took a stone dropped in the fountain at Nîmes to reach the bottom of the pool, the dimensions of the arches in the amphitheater at Arles, the size of the Roman bricks in the Bordeaux circus, how long it took to pass through the locks at Bézieres, even the time it took to dig the grave of his deceased friend Dabney Carter.

Small wonder, then, that Jefferson was able to refute Buffon's allegations about the size of American animals. Presenting in tabular form his "Comparative View of the Quadrupeds of Europe and America," Jefferson was able to show that, whether one considered the quadrupeds common to both continents or those peculiar to each, the American animals came off very well both as to numbers and weight.[40] On being appointed minister to France, Jefferson carried the argument directly to the aged Buffon in Paris, taking with him an unusually large panther skin he had bought in a hatter's shop in Philadelphia. On receiving this gift and a copy of Jefferson's *Notes,* Buffon invited Jefferson to enjoy his hospitality at the Jardin des Plantes and later at his country estate at Montbard, where Jefferson found him "a man of extraordinary powers in conversation." Jefferson was then in his forties, a tall vigorous man known throughout Europe as the author of the bold and eloquent Declaration of Independence. Buffon was in his seventies, "a handsome figure, noble and calm," with quick black eyes beneath white hair curled in ringlets and tied at the back. The count was properly impressed by Jefferson's proofs of nature's power and vigor in the New World, which included the skin and bones of a moose Jefferson had procured from New Hampshire at great personal expense, and promised to make his apologies to the partisans of the American continent in the next volume of his *Natural History,* but he died before he could do so.[41]

Two representative figures in the science of the Age of Enlightenment: the Count de Buffon (by P. Sauvage, 1785) and Thomas Jefferson (by Charles Willson Peale, 1791), depicted about the time of their scientific interchange in Paris. Courtesy of Paul Farber, Oregon State University (Buffon) and the Independence National Historical Park Collection (Jefferson)

The antitheoretical bent of Jefferson's mind and of American science generally in his day is reflected in his *Notes*. Consider, for example, his discussion of the problem of explaining the presence of marine fossils in strata high in the mountains and far from the ocean. If these fossils were genuine organic remains (he was half inclined to believe that they had been formed by "calcareous juices"), Jefferson was sure that they had not been placed there by Noah's flood. He likewise rejected the theory that the fossil-bearing strata had been formed at the bottom of the ocean and had then been elevated to their present locations. No known forces of nature, he declared, could have raised the Andes from the sea in one mighty upheaval. Since none of the available theories could account for the facts, it was best to have no theory at all: "It is always better to have no ideas than false ones."[42]

In general, Jefferson had no use for the "theories of the earth" that proliferated in the eighteenth century and eventually gave rise to scientific geology. Mineralogy was a science of practical use in exploring and exploiting nature's mineral resources, he acknowledged, but "the dreams about the modes of creation, inquiries whether our globe has been formed by the agency of fire and water, how many millions of years it has cost Vulcan or Neptune to produce what the fiat of the Creator would effect by a single act of will, is too idle to be worth a single hour of any man's life."[43] This in 1826 when the science of geology was firmly established!

Jefferson was equally conservative in his views about the fixity of species, a scientific dogma that had been challenged by Buffon and was soon to be entirely rejected by Erasmus Darwin and J. B. de Lamarck. Jefferson scoffed at Buffon's suggestion that the similarity with a difference between the quadrupeds of the Old World and those of the New World could be explained by migration and gradual organic modification through the influence of climate, diet, and other environmental factors. "All the manna of heaven cannot change the mouse into a mammoth," Jefferson declared. In the same way he dismissed the idea that any species could ever become extinct. Like most of his contemporaries, Jefferson thought the extinction of species inconceivable: "For if one link in nature's chain might be lost, another and another might be lost, till this whole system of things would vanish piecemeal."[44] In his *Notes* Jefferson applied this reasoning to the giant creature (the American mastodon) whose bones and teeth had been found at Big Bone Lick in Kentucky, and in 1799 he invoked the same argument with reference to the huge clawed animal whose remains he had procured from a cave in western Virginia (see Chap. 11). Unfortunately for Jefferson, 1799 was the year in which Cuvier began publishing a series of memoirs on extinct quadrupeds, proving beyond doubt that the earth had undergone a series of revolutions in its structure and animal inhabitants.

Jefferson's conservatism in scientific matters also extended to the innovations in nomenclature and classification that were being proposed in various sciences in the late eighteenth and early nineteenth centuries. In Paris he lived virtually next door to innovators of this kind—Antoine Lavoisier in chemistry, A. L. de Jussieu in botany, and the Abbé Haüy in mineralogy—but he opposed all these innovations. Concerning the new chemical nomenclature advocated by Lavoisier and his colleagues, Jefferson wrote home in July 1788:

> It [chemistry] is yet indeed a mere embryon. Its principles are so minute as to escape our senses; and their result too fallacious to satisfy the mind. . . . The attempt therefore of Lavoisier to reform the chemical nomenclature is premature. One single experiment may destroy the whole filiation of his terms, and his string of sulfates, sulfites, and

sulfures may have served no other end than to have retarded the progress of science by a jargon from the confusion of which time will be requisite to extricate us. Accordingly it is not likely to be admitted generally.[45]

The same situation existed with the revisions of nomenclature and classification undertaken by Cuvier and Johann F. Blumenbach in zoology, by Jussieu in botany, and by Haüy in mineralogy. "Nature," Jefferson wrote to a friend in 1814, "has produced units only through all her work." All systems of nomenclature and classification, Jefferson believed, were conventional and arbitrary, hence the main thing was to get scientists to agree in adopting one system. This had been fairly well accomplished by the Swedish naturalist Linnaeus, hence it was unwise to propose innovations in the Linnaean nomenclature and classificatory system.[46]

Underlying Jefferson's antipathy to scientific theorizing and innovations in nomenclature and classification was the spirit of utilitarianism so typical of eighteenth-century minds and American minds in particular. Dismissing geology as "too idle to be worth an hour of any man's life," Jefferson assigned botany a high rank among the sciences because plants furnished "the principal subsistence of life to man and beast, delicious varieties for our tables, refreshments from our orchards, the adornments of our flower-borders, shade and perfume for our groves, materials for our buildings, or medicaments for our bodies."[47] In chemistry and medicine, too, Jefferson valued the practical over the theoretical. "The common herd of philosophers," he complained to the author of a treatise on the uses of chemistry, "seem to write only for one another. The chemists have filled volumes on the composition of a thousand substances of no sort of importance to the purposes of life." Instead, they should apply their science "to domestic objects, to malting, for instance, brewing, making cider, to fermentation and distillation generally."[48] As for medicine, Jefferson was more impressed by Edward Jenner's discovery of the cowpox vaccine for smallpox than by William Harvey's demonstration of the circulation of the blood. The latter, he conceded, was "a beautiful addition to our knowledge of the animal economy," but what had it done to improve the practice of medicine?

But if Jefferson was unimpressive as a scientist and unimaginative in his attitude toward innovation in science, he was unrivaled as a promoter of science; no other high public official in American history has been so strongly identified with that cause. After serving as vice-president, Jefferson was elected president of the American Philosophical Society in 1797 and continued in that office until 1815, contributing to the *Transactions*, serving on its committees, and enriching its collections with the bones of the mastodon and other extinct animals, specimens from the Lewis and Clark expedition, and historic documents such as the journals of that expedition and a draft of the Declaration of Independence.

Just as Theodore Roosevelt dramatized the importance of conservation for the American public and John F. Kennedy the importance of the arts, so Jefferson dramatized the advancement of science. During his term as president distinguished foreign scientists came to visit him, among them Alexander von Humboldt and his botanical colleague Aimé Bonpland fresh from their scientific explorations in Latin America. The unfinished East Room of the White House was turned into a cabinet of natural history in which were displayed fossil bones from Big Bone Lick in Kentucky. For a few months after the return of Zebulon Pike's expedition, two young grizzly bears could be seen on the White House lawn; they were later shipped to Peale's Museum in Philadelphia.

So pronounced was Jefferson's interest in science that his political opponents tried to use it against him, linking this zeal to his sympathy for the French Revolution and his "infidel" religious views. Filled with Federalist indignation against Jefferson and all his works, the young poet William C. Bryant (later a political liberal and defender of Andrew Jackson) apostrophized the "infidel" president as follows:

> Go, wretch, resign the Presidential chair,
> Disclose thy secret measures, foul or fair —
> Go, search with curious eye for hornèd frogs
> 'Mid the wild wastes of Louisiana bogs;
> Or where the Ohio rolls his turbid stream,
> Dig for huge bones, thy glory and thy theme.[49]

When else in American history could a president of the United States have been criticized for showing too much interest in science?

But Jefferson's services to science were not limited to giving it public recognition. He was ever active in collecting scientific information and getting it into the right hands. Having procured and described the remains of a hitherto unknown animal, the huge, clawed megalonyx, he deposited them with the American Philosophical Society and sent casts to Georges Cuvier in Paris. A few years later he commissioned William Clark to collect fossil bones at Big Bone Lick and divided the resulting collection between the society and the Museum of Natural History in Paris (see Chap. 11). He collected Indian vocabularies and made them available to scholars interested in comparative linguistics (see Chap. 14) and gave a detailed account of his excavation of an Indian mound located near Monticello (see Chap. 13). When Benjamin Waterhouse appealed to him for help in introducing Jenner's technique of vaccination for smallpox, Jefferson sent for some of the virus, persuaded Edward Gantt of Washington, D.C. to try it, and when it did not work, suggested to Waterhouse that he try shipping the next bottle of virus enclosed in a larger bottle of water. This batch proved effective, and Jefferson then tried the vaccine on his relatives and slaves at Monticello and encouraged other planters to do likewise. He even persuaded Chief Little Turtle and some of his braves to undergo vaccination and gave them some vaccine to take back to their tribe. From Monticello he shipped samples of the virus to Georgetown, Washington, Richmond, St. Petersburg, and Philadelphia. Later he provided some vaccine for the Lewis and Clark expedition.[50]

Nor should we forget Jefferson's splendid library, rich in works of science, which became the nucleus of the Library of Congress. His collection of Linnaeus's works was, says the botanist-historian Joseph Ewan, "surely the largest private collection in America by 1815."[51]

Jefferson also furthered the progress of science by lending some government assistance in scientific enterprises. His role in this respect was ambivalent, however, for his principles of strict construction acted as a barrier to rapid development of government support for science. Jefferson opposed establishment of a national university, and when Peale invited the United States government to take charge of his famous museum, Jefferson replied that this would exceed the powers of the new government.

But Jefferson managed to do a good deal for science through the auspices of government despite his strict-construction principles. When Peale mounted an expedition to exhume a skeleton of the so-called mammoth, Jefferson offered the use of some army tents and a pump from a navy frigate. When the talented Swiss geodesist Ferdinand Hassler stressed the importance of establishing a coast survey, Jefferson found a place for him in the Treasury Department and supported his project in Congress. Above all, Jefferson conceived and engineered the Lewis and Clark expedition, the most dramatic American exploration to precede the Apollo 11 moon shot and one that set the pattern for American exploring expeditions for a century to come. As the scientific results of the expedition became available, Jefferson made very effort to place the seeds, plants, minerals, and animals where they would be of maximum use to science (see Chap. 8). Thus, although Jefferson's contributions to the literature were modest and largely of the fact-gathering sort, he was keenly interested in all the sciences, highly knowledgeable about developments in most of them, convinced of their importance for the American experiment in republican government and human progress generally, and extraordinarily active in promoting them in every way he could, so much so that he became a national symbol of interest and faith in science. It seems appropriate, therefore, to designate his epoch in American science "The Age of Jefferson."

EPILOGUE

There was still another way in which Jefferson influenced the development of early American science, however unintentionally. A staunch advocate of the southern interest in the political struggles of the 1790s, he helped to engineer the political compromise that resulted in moving the national capital from Philadelphia to Washington, D.C. With this transfer it became apparent that the United States was not to have a capital in the European style, a city that was the political, economic, and cultural center of the nation. Philadelphia was on its way to becoming such a capital in the years before 1800 (see Chap. 2). But Washington, D.C., a raw new city on the banks of the Potomac, had no such potential. Although the city's population reached 10,000 by 1816, its residents were mostly diplomats and government employees, few of whom had more than a passing interest in science. Economy in government and strict construction of the Constitution were the watchwords of the Jeffersonian Republican administrations that guided the nation's destinies in the first quarter of the nineteenth century; proposals for establishing a national university, botanical garden, or national observatory at public expense received short shrift in the halls of Congress. Private efforts to found institutions for promoting science and its applications bore little fruit until after the War of 1812, when they gave rise to the Columbian Institute for the Promotion of Arts and Sciences, the Washington Botanical Society (interested in a botanic garden), and the Medical Society of the District of Columbia. Congress granted the Columbian Institute a twenty-year charter and a small tract of land at the east end of the Mall for a botanical garden, but no funds to implement the project. The institute continued in existence until 1837, but it was unable to realize any of its ambitious plans for collecting and distributing information on plants, minerals, analyses of mineral waters, agricultural improvements, and topographical and statistical data. Not until the revival of the Coast Survey in the 1830s and the founding of the Smithsonian Institution and the United States Naval Observatory in the 1840s was the national capital to acquire institutions capable of supporting sustained scientific activity.[52]

Lacking the energizing influence of a capital city on the model of London or Paris, American science developed regionally. Philadelphia continued as the leading scientific center for many years, but without hope of dominating the scientific scene as it had before 1800. In Boston, New Haven, New York, Charleston, Cincinnati, and Lexington the Philadelphia pattern of scientific institutions was copied with varying degrees of success. To what extent the champions of science in the nascent urban centers of the new republic succeeded in establishing a solid institutional base for scientific research will be seen in Chapters 2–5.

2

The Philadelphia Pattern

PHILADELPHIA, the Quaker city on the Delaware, set the pattern of institutional development for the sciences in the urban centers of the new nation. From the early eighteenth century Philadelphia's central location, flourishing overseas trade, tolerant Quaker tradition, and enterprising medical community, combined with the scientific genius and organizing talents of Benjamin Franklin, had set the stage for the city's emergence as a scientific and medical center. In rapid succession Franklin and his associates had organized the Library Company of Philadelphia, the American Philosophical Society, the Pennsylvania Hospital, and the College of Philadelphia with its medical school. Meanwhile, John Bartram established his botanical garden on the banks of the Schuylkill River and collected specimens for the naturalists of Europe. Before the quarrel with the mother country erupted into revolution, Franklin had won the Royal Society's Copley Medal for his researches on electricity, Bartram had been pensioned by King George III for his contributions to natural history, the medical school had acquired a brilliant faculty and the beginnings of a library, and the Philosophical Society had published a first volume of transactions containing valuable papers on the transit of Venus in 1769. Small wonder, then, that John Adams called Philadelphia "the pineal gland of the republic," alluding to René Descartes's notion that the pineal gland was the meeting place of the incommensurable worlds of mind and matter.[1]

Before describing the postrevolutionary development and expansion of this pattern of institutions supportive of science, however, let us take a tour of the Quaker City as it appeared to Manasseh Cutler, a New England clergyman with broad scientific interests, when he visited in the summer of 1787. On Friday, July 13, Cutler set off with his Philadelphia friend Gerardus Clarkson to see the city and its environs, traveling in a handsome phaeton drawn by two spirited horses. Passing the Pennsylvania Hospital on the outskirts of the city, they drove along the Schuylkill River

37

*View of the State House Garden, showing the State House
(now Independence Hall) on the left and Philosophical Hall,
rear center. The sign "Museum" may be seen on the door of
the hall. In 1802 Peale was given the use of the upper floors of
the state House rent-free for his collections. Courtesy of the
American Philosophical Society*

to view the country seats of financier Robert Morris and other prominent
Philadelphians. Returning, they stopped to see the celebrated Benjamin
Rush, America's leading physician and a signer of the Declaration of
Independence, visited Charles Peale's museum and portrait gallery at Third
and Lombard streets, and then went to the State House. The Supreme
Court of Pennsylvania was in session in the west room. In the chamber
above the east room the Constitutional Convention was busy drafting a
new instrument of government for the United States, with sentries posted
in front of closed doors. Cutler and his guide went through the back door
to view the public garden, recently landscaped under the direction of
Samuel Vaughan, a Jamaica planter who was a leading spirit in the
Philosophical Society.

> Here is a fine display of rural fancy and elegance [Cutler noted in his diary]. . . . The trees are yet small, but most judiciously arranged. The artificial mounds of earth, and depressions, and small groves in the squares have a most delightful effect. The numerous walks are well graveled and rolled hard; they are all in a serpentine direction, which heightens the beauty, and affords constant variety. . . . On one part [of the square containing the garden] the Philosophical Society are erecting a large building for holding their meetings and depositing their Library and Cabinet. This building is begun, and, on another part, a County Court-house is now going up.[2]

The touring party then inspected the University of Pennsylvania and several churches and called on John Vaughan, son of Samuel Vaughan and destined to serve the Philosophical Society as treasurer and librarian for several decades. About five o'clock Cutler, John Vaughan, and Elbridge Gerry, one of the Massachusetts delegates to the Constitutional Convention, went to visit Benjamin Franklin. They found him in his garden, "sitting upon a grass plat under a very large Mulberry, with several other gentlemen and two or three ladies."

> I saw a short, fat, trunched old man, in a plain Quaker dress, bald pate, and short white locks, sitting without his hat under the tree, and, as Mr. Gerry introduced me, rose from his chair, took me by the hand, expressed joy to see me, welcomed me to the city, and begged me to seat myself close to him. His voice was low, but his countenance open, frank, and pleasing. . . . The tea-table was spread under the tree, and Mrs. Bache, . . . who is the only daughter of the Doctor and lives with him, served it out to the company. She had three of her children about her . . . who appeared to be excessively fond of their Grandpapa.[3]

After dark, Franklin took Cutler into his library-study, "a very large chamber, and high studded." Books lined the walls and overflowed into four large alcoves extending two-thirds of the length of the chamber. There was also a glass machine for exhibiting the circulation of the blood, a rolling press for making copies of letters and documents, and a "great armed chair, with rockers, and a large fan placed over it, with which he fans himself, keeps off flies, etc., while he sits reading, with only a small motion of his foot; and many other curiosities and inventions, all his own, but of lesser note."

Since Cutler was a botanist, Franklin brought out a huge illustrated edition of Linnaeus's *Systema Vegetabilium,* so large he had difficulty lifting it to the table on which it was examined. Cutler was delighted with it. He and Franklin spent "a couple of hours" going through it while the other gentlemen discussed politics. To Cutler's surprise, Franklin displayed a keen interest in natural history.

An engraving of Manasseh Cutler, clergyman, botanist, congressman, and promoter of the Ohio Company and its settlement at Marietta. Courtesy of the Massachusetts Historical Society

I was highly delighted with the extensive knowledge he appeared to have of every subject, the brightness of his memory, and clearness and vivacity of all his mental faculties. Notwithstanding his age (eighty-four), his manners are perfectly easy, and every thing about him seems to diffuse an unrestrained freedom and happiness. He has an incessant vein of humor, accompanied with an uncommon vivacity, which seems as natural and involuntary as his breathing.[4]

About ten o'clock Cutler took his leave of the famous "Doctor" and returned to his lodgings at the Indian Queen, where many of the members of the Constitutional Convention were also staying.

THE AMERICAN PHILOSOPHICAL SOCIETY

Benjamin Franklin's return to the United States from France in 1785 gave a much needed impetus to the American Philosophical Society, whose president he had been since 1769. The society, so vigorous before the Revolution, had fallen on hard times. Its platform for astronomical observations in the State House yard had fallen into disrepair, and its meetings were poorly attended: "scarce ten [members] can be got together, unless upon some *very special occasion,*" a visiting New Englander reported. Samuel Vaughan, Francis Hopkinson, and the astronomer David Rittenhouse had tried to reinvigorate the society, but with little success.

With Franklin's return, the society's prospects began to improve. At a meeting at his house in September 1785 Franklin rallied the spirits of his colleagues by giving precedence to the reading of two scientific papers, after which the society voted to proceed with the construction of Philosophical Hall "with all convenient dispatch." Two years later Franklin resolved the society's financial problems by donating £100 and lending an additional £500 in support of the building project. Slowly but surely Philosophical Hall took shape on a lot next to the State House while the nation debated and ratified the federal Constitution that had, with Franklin's aid, been drafted during the summer of 1787.

In November 1789, seven months after George Washington took office as head of the new government established by the Constitution, the society held its first meeting in Philosophical Hall. There were no festivities in connection with this event, for Franklin had entered on his last illness and the society ordinarily met at his house in consideration of his feeble health. The man who more than any other had taken the lead in founding Philadelphia's scientific and learned institutions, who more than any other symbolized American democracy and American aspiration and achievement in science, was approaching the end of his days. Before he died, however, he received a visit from an American destined to succeed him as president of the Philosophical Society and as an international symbol of American political and scientific ideals. In March 1790 Thomas Jefferson, recently returned from France, paid his respects to the ailing Franklin and gave him news of friends in Paris and the revolutionary events taking place there. It was the last meeting of these two men, both champions of American liberty, American science, and the American Philosophical Society. Franklin died on April 17. Four days later, it is recorded in the minutes of the society he helped to found, twenty-three members "went in procession to the funeral of their late illustrious President, Dr. Benjamin Franklin."

In the decade following Franklin's death the Philosophical Society slowly gathered strength and prestige. Its new president was David Rittenhouse, to whom Franklin had bequeathed his reflecting telescope "for the use of his observatory" and who now took Franklin's place as the leading American natural philosopher. From the observatory near his house at Seventh and Mulberry streets Rittenhouse made regular observations, which, added to those he had made earlier on the transit of Venus in 1769, won him election to the Royal Society of London in 1795 and the favorable notice of Continental astronomers like Joseph Jérôme de Lalande and Franz von Zach. Charles W. Peale's portrait of Rittenhouse, done for the society in 1791, shows the fifty-nine-year-old astronomer seated at his desk with pen in hand, his thin gray hair parted down the middle, his direct gaze and firmly set lips suggesting the precision, integrity, and reserve of

Three presidents of the American Philosophical Society—Benjamin Franklin (by Charles Willson Peale, 1787) (upper left), David Rittenhouse (by C. W. Peale) (upper right), Robert Patterson (by Rembrandt Peale) (lower left)—and the treasurer and librarian for several decades, John Vaughan (by Thomas Sully) (lower right). Courtesy of the Pennsylvania Academy of the Fine Arts (Franklin) and the American Philosophical Society (Rittenhouse, Patterson, Vaughan)

his personality. Although ill health and his duties as technical expert for the Philadelphia-Lancaster road survey and the Delaware-Schuylkill canal project made it impossible for him to attend society meetings regularly, he nevertheless gave prestige and direction to its activities.

Philosophical Hall provided plenty of space for society gatherings and collections. Regular meetings were held at six o'clock in the evening on the first and third Fridays of each month from October through May and at seven o'clock on the third Friday in other months. On other Friday evenings the rooms were open for conversation and reading, guests could be invited, and the "philosophical apparatus" was available for purposes of experiment. The society's collections had grown rather haphazardly since 1769, when three curators had been appointed "to take care of, and preserve, all *Specimens of natural Productions,* whether of the ANIMAL, VEGETABLE, or FOSSIL Kingdom; all Models of Machines and Instruments, and all other matters and things belonging to the Society, which shall be committed to them; and to class and arrange them in their proper order, and to keep an exact list of them, with the names of the respective donors, in a book provided for that purpose. . . ."[5] In the years following, collections accumulated in all these categories, but little was done to order or display them until in 1794 Peale leased the unused parts of Philosophical Hall for his family and his museum, agreeing to act as "*depository* of the Models, drawings, plans, natural and artificial curiosities, and all their other property; and the same preserve in order, and exhibit at proper times, under the direction of the Curators."

Peale discharged these functions admirably. He continued to occupy a substantial part of Philosophical Hall until 1811. When Peale and his wife could not agree on a name for their youngest son, born in the hall in 1796, he presented the four-month-old infant to the society at their meeting on February 19, 1796, and asked the members to name him. They lost no time in naming the child Franklin after their departed president.

Peale was but one of the society's core members, men linked to Rittenhouse as much by their Jeffersonian-Republican leanings in politics as by their enthusiasm for science. Among the others were the astronomer-surveyor Andrew Ellicott, Francis Hopkinson, Rittenhouse's nephews William Barton and Benjamin S. Barton, James Hutchinson, and Jefferson himself, now secretary of state in the Washington administration and vice-president of the society. Jefferson was an admiring friend of Rittenhouse and a frequent guest at his house, where Jefferson could seek advice in preparing his Report on Weights and Measures for Congress, designing the coinage, reviewing applications for patents, and designing his moldboard of least resistance to improve the efficiency of plows. The two men were intensely devoted to the affairs of the society, and it was no surprise to anyone when, on Rittenhouse's death in June 1796, Jefferson was elected to succeed him.

The early minutes of the society and a description by a visiting foreigner, M.L.E. Moreau de St. Méry, of its place and manner of meeting enable us to reconstruct the scene that took place shortly after Jefferson's election as its president. On Friday evening, March 10, 1797, Jefferson called the meeting to order. The members were ranged around a long table in a room on the first floor of Philosophical Hall. Under Franklin's portrait on the south side of the table Jefferson sat alone in the shabby old armchair Franklin had sat in. Twenty-one members occupied the other three sides of the table, including Professors Barton, Wistar, and Patterson from the University of Pennsylvania; Joseph Priestley; the Duc de Liancourt; Ambroise Palisot de Beauvois; John Vaughan; George Turner from the Northwest Territory; and Hugh Williamson of North Carolina and his guest Governor Martin of North Carolina. Two papers (one dealing with archaeological finds in southeastern Ohio, the other with calculations relating to grist and sawmills) and two prize essays submitted in the society's competition for the best discourse on "The Best System of Liberal Education and Literary Instruction" were referred to appropriate committees. A deed of gift from Mrs. Rittenhouse and her daughters relating to the observatory of her husband was presented for inspection by the members. The group was then ready for the main business of the evening, the reading of "A Memoir on the Discovery of Certain Bones of a Quadruped of the Clawed Kind in the Western Parts of Virginia," submitted to the society by their new president and read for him by one of the secretaries.

One could wish to have been present at this historic meeting when Jefferson's description of the remains of the megalonyx was first presented to the world of science and discussed, no doubt with considerable animation, by Priestley, Wistar, Liancourt, and the rest of the distinguished company. Did Jefferson then verbally assert, as he did in the published memoir, that the great clawed animal must still be alive in some remote part of the earth? Did Wistar take exception to Jefferson's guess that the newly found *incognitum* was a huge lion, as he did in his remarks appended to the published account? The minutes record only that the memoir was referred to the Committee of Selection and Publication with an authorization to hire someone to delineate the megalonyx bones and that Mr. Peale was requested to "cause those bones to be put in the best order, for the Society's use."[6]

As the presence of Liancourt suggests, Philadelphia attracted a good many refugees from the French Revolution in this period, and Jefferson's role in the society and the federal government did much to secure the interest and support of foreign visitors. Many were elected to the

Philosophical Society, and several, including Liancourt, Beauvois, and St. Méry, whose bookshop and printing establishment at the corner of Front and Walnut streets became a meeting place for displaced Frenchmen, attended meetings frequently.[7]

However cordial Jefferson and his Philadelphia colleagues may have been in welcoming these visitors from abroad, the Federalist administration of President John Adams remained suspicious of them, fearing they would spread the radical principles of the French Revolution in America and strengthen the political opposition forming around Jefferson. To prevent this, Congress passed the Alien and Sedition Acts authorizing deportation of suspicious foreigners, and soon after several sympathizers with the French Revolution departed for Europe. Jefferson was victorious in the presidential election of 1800, but this meant that he must leave Philadelphia for the new capital on the banks of the Potomac. The Philosophical Society was left with a president in absentia.

Despite the move to Washington, D.C., Jefferson continued to be active in the society, relying for the conduct of its ordinary business on a cadre of regulars including Caspar Wistar, James Woodhouse, Benjamin S. Barton, and John R. Coxe from the University of Pennsylvania medical school; Robert Patterson, professor of natural philosophy at the university and director of the United States Mint; Andrew Ellicott, who laid out the District of Columbia and ran the boundary line between the American and the Spanish possessions in North America; Robert Hare, son of a well-to-do Philadelphia brewer; Adam Seybert, trained in chemistry and mineralogy at the *Écoles des Mines* in Paris and subsequently a prosperous merchant and manufacturer in Philadelphia; Charles W. Peale, artist and museum entrepreneur; and the ever-present and indefatigable John Vaughan, treasurer of the society.

In 1803 Vaughan was given additional duties as librarian and began a program of systematic acquisition of books and journals. Up to that time the society's library had grown at random from individual donations, including those of Franklin, the Vaughans, and various foreign members. Now it was put in good order and new acquisitions were made by exchange, purchase, and donation. Among the most important purchases were those made at the sale of Franklin's library in 1803. Vaughan himself donated hundreds of volumes and in 1826 sold his private library, which had been housed in Philosophical Hall, to the society.[8]

Although Philadelphia was no longer the national capital, it continued to attract distinguished scientists from abroad, many of whom played an important role in the society. Among these were the Abbé Correa da Serra, Portuguese diplomat and naturalist; José J. de Ferrer, a Spanish naval officer; Silvain Godon and Gerard Troost, mineralogists; naturalists Alexander Wilson and Thomas Nuttall; Charles A. Lesueur, an able French naturalist who came to Philadelphia in 1816; and Charles L. Bonaparte, nephew of Napoleon and an excellent ornithologist. The contributions of these men to American science were of great importance.

The main business of the society, "Held at Philadelphia for Promoting Useful Knowledge," was publication of its *Transactions,* which were published from time to time as the supply of scientific communications and financial resources permitted. The first volume had been published in 1771, and succeeding volumes followed in 1786, 1793, 1799, 1802, and 1809, after which there was a nine-year pause before a new series was begun in 1818. Compared to the *Philosophical Transactions* of the Royal Society of London the *Transactions* of the Philadelphia society were infrequent in appearance and inferior in quality and importance, yet the papers contained in its pages were of major significance in the development of American science and useful contributions to science generally. The sixth volume (1809), for example, contained the Silliman-Kingsley memoir on the chemistry of the Weston meteor, William Maclure's pathbreaking account of the geology of the United States, Joseph Cloud's experiments on palladium found combined with pure gold, and observations on the eclipse of the sun in 1806 by several American observers. By this time the number of memoirs on applied science and technology had declined. Articles of this sort continued to appear from time to time, but the early emphasis on "useful knowledge" had given way to an interest in science and learning for their own sake.

THE UNIVERSITY OF PENNSYLVANIA AND ITS MEDICAL SCHOOL

No less important than the American Philosophical Society in the scientific life of Philadelphia was the medical school of the University of Pennsylvania, an institution formed in 1791 by legislative merger of the College of Philadelphia (organized in 1751 by Franklin and other leading Philadelphia citizens) and the newly established University of the State of Pennsylvania. The undergraduate college did not amount to much for the next forty years, but the medical school, which had been added to the College of Philadelphia in 1765, prospered greatly after the merger. Despite the ravages of yellow fever, which carried away one tenth of the city's 50,000 inhabitants and ten of its physicians in 1793 and more in succeeding years, the number of medical students rose rapidly from about 60 to 400 in 1810 and continued at a high level. They came from all over the United States, especially from the South, and occasionally from the West Indies, South America, Canada, and Europe. Normally they attended classes for two years during the four winter months and had clinical lectures and demonstrations at the Pennsylvania Hospital or the Philadelphia Almshouse. Although there was no formal connection between the hospital and the university, most of the leading professors were physicians at the hospital and students accompanied them on their rounds.[9]

The medical school was important not only from a medical point of view but also for the sciences generally. Supported by the fees students paid to attend courses and by private practice as well, the professors extended their researches in many directions. Some like Benjamin Rush and William Shippen, Jr. confined themselves entirely to medicine, but others like Caspar Wistar, James Woodhouse, and Benjamin S. Barton ranged far beyond the bounds of medicine. Wistar was drawn into paleontology through anatomy and his association with Jefferson, whom he eventually succeeded as president of the Philosophical Society. Woodhouse devoted himself heart and soul to chemical research and teaching after his appointment to the chemical chair in 1795, extending his researches to combustion, respiration, plant chemistry, metallurgy, coal, and industrial processes. On a hot summer day, wrote one of his students, he might be found in his laboratory in Surgeons' Hall "stripped to his shirt and summer pantaloons, his collar unbottoned, his sleeves rolled up above his elbows, the sweat streaming copiously down his face and person, and his whole vesture drippingly wet with the same fluid." Founder of the Chemical Society of Philadelphia, Woodhouse was also one of the secretaries of the Philosophical Society.

Most versatile of all the professors was Benjamin S. Barton, professor of natural history, botany, and materia medica; editor of the short-lived *Philadelphia Medical and Physical Journal* (1804–1809); founder of the Linnaean Society of Philadelphia; subsidizer of botanical researches by Thomas Nuttall and Frederick Pursh; member of innumerable foreign societies; and author of books, pamphlets, and articles on botany and materia medica, comparative linguistics, and the animals, fossils, and antiquities of North America. A tall, handsome man with strong features and a face "lit up by eyes inordinately large, black and full of fire," he was known for his irascibility, occasioned perhaps by the fits of gout to which he was subject. In his botanical lectures, his students reported, he displayed "an earnest and exciting enthusiasm, by which he induced his pupils to engage in the study of the science with a corresponding earnestness, accompanied by a resolution to teach themselves." Visitors to his house on Chestnut Street found him "surrounded with books, bottles of insects, the bones of the mammoth, and other evidences of his ruling passion. . . ." In addition to his professorial duties Barton was a physician at the Pennsylvania Hospital and vice-president of the Philosophical Society.[10]

Surgeons' Hall (with cupola), containing the anatomy room and James Woodhouse's chemical laboratory, and the building of the Library Company of Philadelphia. Across Fifth Street was Philosophical Hall and the State House Garden. Courtesy of the American Philosophical Society, whose present library is a replica externally of the one shown here

THE NATURAL HISTORY ESTABLISHMENTS

Botanical Gardens. Philadelphia had better facilities than any other American city for the study of natural history. Botany, in particular, had many devotees, and there were several gardens and nurseries where they could gather to exchange specimens and information. The oldest of these was Bartram's Garden, established in 1730 on the west bank of the Schuylkill River a mile or two below the lower ferry by John Bartram and continued after his death in 1777 by his two sons John and William. In the year of his father's death William returned from explorations in the Carolinas and Florida, settled down to write his delightful and highly informative *Travels,* and joined with his brother in the business of selling plants and seedlings. In 1807 the Bartrams published a *Catalogue of Trees, Shrubs and Herbaceous Plants,* announcing that their gardens were "the

Seminary of American vegetables, from whence they were distributed to Europe, and other regions of the civilized world," and adding that they deserved to be called "the Botanical Academy of Pennsylvania, since, being near Philadelphia, the Professors of Botany, Chemistry, and Materia Medica, attended by their youthful train of pupils, annually assemble here during the floral season." These claims were not without foundation. In the eight acres of the garden could be found the bald cypress, the buckeye, the persimmon, the oak-leaved hydrangia, the long-leaved pine, several kinds of magnolias, the horse chestnut, and other southern trees, many of them shipped to Philadelphia by John Lamboll and his wife of Charleston, with whom John and William Bartram stayed during their southern tours. There, too, could be seen the *Franklinia alatamaha,* a beautiful tree with fragrant flowers the Bartrams had seen growing wild in Georgia and had saved for posterity by transplanting it to their garden and naming it after their friend Benjamin Franklin.

Because of its many attractions, Bartram's Garden had many visitors. President Washington came there twice during the summer of 1787 and Thomas Jefferson, who lived across the river from the Bartrams for several years and was an ardent gardener himself, was a frequent visitor. Manasseh Cutler visited the garden with several members of the Constitutional Convention during his brief stay in Philadelphia.

> This is a very ancient garden [Cutler noted in his diary], and the collection is large indeed, but is made principally from the Middle and Southern States. It is finely situated, as it partakes of every kind of soil, has a fine stream of water, and an artificial pond, where he [Bartram] has a good collection of aquatic plants. . . . From the house is a walk to the river, between two rows of large, lofty trees, all of different kinds, at the bottom of which is a summer-house on the bank, which is here a ledge of rocks, and so situated as to be convenient for fishing in the river. . . .[11]

But Bartram's Garden was more than a nursery and a tourist attraction. It was, as the Bartrams themselves said, "the Botanical Academy of Pennsylvania." There Benjamin S. Barton's students William P. C. Barton, William Darlington, and William Baldwin learned to recognize plants. There Alexander Wilson acquired his love of natural history and lived from time to time while he wrote the last volumes of his *American Ornithology.* Thomas Nuttall had a room set aside there for his work on *The Genera of North American Plants.* There Thomas Say found encouragement for his early interest in entomology and conchology. Wandering naturalists like Johann D. Schoepf, André and François A. Michaux, Frederick Pursh, Constantine Rafinesque, and John E. Le Conte were assured a cordial welcome and free access to William Bartram's ample fund of botanical and zoological knowledge. "Mr. William Bartram has a Library within himself," wrote Zacchaeus Collins in 1813. "Some great and famous Botanists have robbed of his Honey, without thanking him for his Trouble."[12]

A mile or two up the Schuylkill River from the Bartrams' house was another famous Philadelphia garden, William Hamilton's estate "The Woodlands." Designed in the style of the Romantic gardens Hamilton had visited in England shortly after the revolutionary war, this garden impressed Thomas Jefferson as "the only rival I have known in America to what may be seen in England." In 1802 the gardens were put under the care of an English nurseryman, John Lyon, who did much to improve them. Lyon was succeeded in 1803 by Frederick Pursh, a botanist from the Royal Botanical Garden in Dresden. Manasseh Cutler was overwhelmed by the magnificence of the estate when he visited there with Timothy Pickering en route to Washington in 1803. In every direction, Cutler noted, there were walks with borders of flowering shrubs and trees collected from all parts of the world. The greenhouses, occupying "a prodigious space," were "crowded with trees and plants from the hot climates, and such as I had never seen, all the spices, the tea-plant in full perfection; in short, he [Hamilton] assured us there was not a rare plant in Europe, Asia or Africa, many from China and the islands in the South Seas, none, of which he had not procured." On retiring to the house, Cutler was treated to an equally dazzling display of botanical books. Hamilton himself proved to be excellent botanist. His enthusiasm for his books and plants knew no bounds. Gardeners were sent with torches to fetch specimens of rare plants from the greenhouses, and Cutler, who had fractured two ribs in an accident before leaving Massachusetts, was torn between ecstasy and agony as he examined the never-ending succession of botanical treasures. It was one o'clock in the morning before he was permitted to go to bed.[13]

Three other gardens in the environs of Philadelphia were destined to play a role in early American science. At Marshallton, Pa., at the forks of Brandywine Creek was the greenhouse and garden of Humphry Marshall, author of *Arbustum Americanum: The American Grove,* the first indigenous botanical book published in the United States; it was dedicated to the Philosophical Society of which Marshall was a member. Like John Bartram, whom he knew and admired, Marshall had built his house with his own hands, equipping it with a hothouse and an observatory. He began a garden and entered the business of selling seeds and plants at home and abroad, carrying on a botanical commerce and correspondence with a great variety of European naturalists and plant dealers, including Joseph Banks, John C. Lettsom, John Reichert (gardener to the Duke of Saxe-Weimar), Jean Descemet in Paris, and others. He was aided by his nephew Moses Marshall, who made trips from time to time at Banks's request, collecting not only plants and seeds but also shells, tortoises, frogs, bird eggs, and the like.[14]

Although Marshall's garden did not long survive his death in 1801, its place was taken by a nursery for useful and ornamental plants begun in the following year by Bernard M'Mahon, an Irish gardener who had come to Philadelphia in 1796. M'Mahon improved his knowledge of botany by attending Barton's lectures and in 1806 published *The American Gardener's Calendar: Adapted to the Climate and Seasons of the United States,* the first book on gardening in the new nation. Copies of the *Calendar* were sent to the Philosophical Society and to Thomas Jefferson, with whom M'Mahon had struck up a friendly correspondence on gardening. In December 1808 Jefferson wrote M'Mahon that André Thouin at the Jardin des Plantes in Paris had sent him 700 different seeds from countries all over the world. It was to M'Mahon and William Hamilton that Jefferson entrusted the precious cargo of seeds and plants from the Lewis and Clark expedition for safekeeping and cultivation, assigning to Benjamin S. Barton the task of writing the descriptions (see Chap. 8).

M'Mahon had a seed store on Second Street where devotees of botany and horticulture met to exchange information and ideas and browse among his books and plants. In 1808 he purchased twenty acres for a new nursery and botanical garden just north of Philadelphia on the Germantown road, naming it "Upsal," presumably in honor of Linnaeus's garden at Upsala. M'Mahon died in 1815, but his business was carried on by his wife for many years. His *Calendar* eventually underwent eleven editions, and his friend Thomas Nuttall named the genus *Mahonia* in his honor.

A third garden of special interest to Philadelphia botanists was that of David Landreth, whose seed house was the first such establishment in the United States. In Landreth's nursery John Lyon, formerly in charge of Hamilton's garden, grew the plants he collected on his botanical expeditions in the South. In November 1809 Henry Muhlenberg of Lancaster, Pa., noted that Lyon had just passed through that town on his way back to Philadelphia from a southern expedition and had shown him plants collected in Virginia, the Carolinas, and Tennessee.

> Mr. Lyons is an indefatigable collector of mountain plants and deserves every encouragement. I am now preparing for him a collection of grasses and cryptogamous plants which he has neglected hitherto. In my next journey to Philadelphia I hope to see his numerous living plants. Your new shrub he had last spring as likewise the Pontederia lancifolia, the Thalia coerulea and many other southern plants.[15]

Henry Muhlenberg, youngest of three distinguished sons of the patriarch of the Lutheran Church in Pennsylvania, must be counted as one of the Philadelphia botanists, though he lived sixty miles away in the town of Lancaster. Educated at Halle in Germany, Muhlenberg developed a keen interest in botany during the revolutionary war and was soon in correspondence with leading European botanists, thanks in part to the German physician Johann D. Schoepf, who discovered him on a visit to Lancaster during travels after the war. "The most important thing for me at Lancaster," Schoepf wrote in his *Travels,* "was the very agreeable acquaintance which I had the pleasure of making with the pastor of the Lutheran congregation there (and now Principal of the new college), Mr. Heinrich Muhlenberg. This excellent man, through his own diligence, has gained a very considerable knowledge of natural history and is unwearied in the study of the animals, plants, and minerals of his region."[16]

Muhlenberg was in close touch with the Philadelphia botanists. When his duties as pastor and college president permitted, he journeyed to Philadelphia to see his fellow naturalists, making the rounds to Barton's house, M'Mahon's store, Bartram's and Hamilton's gardens, and Landreth's nursery in search of new plants and information. Between visits he kept in touch with his Philadelphia friends by correspondence. In 1810 he wrote to William Bartram: "Hardly a Day passes but I am in Spirit with You and wander with You Hand in Hand through Your Garden. . . ."[17]

Peale's Museum. On the zoological side of natural history, Peale's Museum was the counterpart of the city's botanical gardens. This museum was the single-handed creation of Charles W. Peale, portrait painter and amateur naturalist. The project began to take shape in his mind when on a summer day in 1784 a Hessian officer brought in a collection of mastodon bones and asked Peale to make sketches of them. These bones, lying about in the portrait gallery adjoining Peale's house, caught the eye of his brother-in-law who suggested there were many people like himself "who would prefer seeing such articles of curiosity than any paintings whatsoever." About this time Robert Patterson contributed a paddlefish caught in the Allegheny River, and Peale began to entertain the idea of developing a natural history collection in connection with his portrait gallery.[18]

Peale was a man of many projects. When an idea struck him that promised to redound to the public good, he could not rest until he had put it into execution, whether it concerned a natural history museum, improvements in dentistry, better farming methods, construction of a polygraph, or invention of a better stove. He had, moreover, a genuine interest in nature, and his practical ingenuity was challenged by the problems involved in creating a museum. He began to experiment with ways of preserving animals. Spirits of turpentine proved damaging to the plumage of his birds. He then tried an arsenic solution, which was hard on his hands and lungs but kept the insects away. Peale labored on, inspired by "the idea of handing down to posterity a work [which] if judiciously managed might become equal to any undertaking of the like kind in Europe." The museum would not only help to provide for his growing family but would "diffuse a general knowledge of the wonderful works of creation."[19]

Manasseh Cutler visited Peale's Museum with some friends in 1787 when it was housed in the gallery adjoining Peale's house at Third and Lombard streets. The exhibits, including portraits, historical paintings, and "natural curiosities," were displayed in a long narrow room two or three stories high and lighted from above by adjustable openings in the roof. At one end of the room General Washington, "at full length and nearly as large as the life," presided over an assembly of revolutionary war heroes. At the opposite end under a small gallery the natural history exhibits were arranged "in a most romantic and amusing manner" on or around a mound of earth adorned with trees filled with birds; an artificial pond with an assortment of stuffed fish, turtles, frogs, lizards, and waterfowl; a small thicket with game birds, bears, deer, feline species, rabbits, raccoons, and the like; a rock garden with appropriate snakes; and excavations designed to display various minerals and soils. "Mr. Peale was very complaisant," Cutler recorded, "and gave us every information we desired. He requested me to favor him with any of the animals and fossils from [New England], not already in his museum, which it might be in my power to collect."[20]

In general, however, Peale was his own collector. Every year, especially in the spring when the birds were in their finest plumage, he sallied forth to collect specimens, some to be added to his own exhibits, others to be used in foreign exchange. In 1795 while visiting his daughter and family in Baltimore, he made the acquaintance of the Reverend Mr. Kirtz, pastor of the German Lutheran church and an enthusiastic collector of beetles. Peale had already done something with butterflies and moths, but he had never collected beetles or insects. Under the tutelage of Kirtz, he learned to look for beetles under stones, rags, dung, and carrion and for insects in the flowers and small bushes along the edges of meadows.

In 1798 Peale and his family made one of their many trips to New York to visit Mrs. Peale's relatives, the De Peysters and the Staggs, who were engaged in commerce there. He returned with an entomologist named Jotham Fenton, whom he had met at Gardiner Baker's museum in New York. Largely self-taught, Fenton was an excellent mechanic and optician and very useful to Peale in arranging the entomological collections and making microscopes for patron use. Peale also collected a good many specimens of fish at the New York fish market, where a much greater variety could be obtained than in Philadelphia.[21]

Peale augmented his collections by foreign exchange. Duplicates of birds were sent to the Museum of Natural History in Paris, the Royal Academy at Stockholm, and various provincial societies in Europe. Paris and Stockholm sent specimens in exchange, although this commerce was interrupted by the Napoleonic wars. In London, Peale obtained many valuable specimens from a commercial dealer, John Hall. He complained of the high exchange rate exacted by Hall but could not resist the opportunity to obtain a specimen of *Menura superba,* one of the Australian lyrebirds, and of the platypus, "a Quadruped with a ducklike bill from New Holland."

Peale moved his collections to Philosophical Hall in 1794, hiring men and boys to carry the exhibits from his house on Lombard Street in a grand parade, "at the head of which was carried on men's shoulders the American Buffalo, then . . . the Panthers, Tyger Catts and a long string of Animals of small size carried by the boys." Once established in Philosophical Hall, Peale secured permission from the governor of Pennsylvania to fence in part of the State House grounds as a yard for young living animals, which were being maintained until full grown and ready to be stuffed and mounted for display. The menagerie included a bald eagle kept at the top of Philosophical Hall in a case with a sign, "Feed me daily 100 years," emblazoned on it in letters of gold. There was also a female elk staked in an open space outside the animal yard. When the elk got loose one day, Peale threw a halter with rope attached over its neck. The elk bolted, the rope jammed in Peale's pocket, and he was thrown to the ground and dragged thirty or forty feet before the pocket tore loose.

Amidst these adventures, Peale was bending every effort to make his museum a useful scientific and cultural institution. The exhibits were arranged in Linnaean order, each tagged with a number referring to a gilt-framed catalogue giving the Linnaean and common names of all the specimens. The French naturalist Palisot de Beauvois was engaged to prepare a catalogue of the museum for publication, but this project had to be abandoned for lack of money. Meanwhile, Peale was busy preparing a course of lectures on natural history, delivered in the winter of 1799 to "a small class of very respectable citizens." In all, Peale gave twenty-seven lectures from 1799 to 1802 on the exhibits in the museum, mixing natural theology with natural history as he made his way through the *Scientific and Descriptive Catalogue of the Collections.*[22]

In 1801 Peale's literary labors ended with an expedition to dig for a skeleton of the "mammoth," the giant *incognitum* whose remains had been turning up in various parts of North America from early colonial times. In this enterprise Peale displayed the same enthusiasm, tenacity, and ingenuity that characterized all his endeavors. He was rewarded with two nearly complete skeletons, one of which was mounted and exhibited in a special room in his museum, where it excited as much popular curiosity as it did scientific interest. A year later the Pennsylvania legislature voted to allow Peale the use of the upper floor of the State House free of rent, and Peale lost no time in moving his exhibits into these more spacious quarters. From that time until 1816, when the State House was deeded to the city of Philadelphia by the state legislature, the museum enjoyed a high prosperity. People came in great numbers to see the mastodon skeleton, splendid collection of birds, portrait gallery, and other exhibits. For those curious about natural philosophy there were lectures on electricity and magnetism, with experimental demonstrations. Musical concerts were presented in the evening.

Peale himself described the attractions of his museum in a brochure, "A Walk with a Friend through the Museum." Entering through the front door of the State House, over which was inscribed "Museum, Great School of Nature," patrons proceeded up the great stair at the rear of the central hall to the turnstile, where a chime of bells announced their arrival. After paying twenty-five cents for admission, the visitors passed through the lobby, where their guide pointed out a large machine for generating static electricity, "the subtle fluid so actively associated with Life, matter and motion." This machine, Peale told his customers, was "frequently resorted to by Professional advice, in various complaints."[23]

The first display room, forty feet in length, was devoted to quadrupeds exhibited in lifelike poses against naturalistic backgrounds painted by Peale. "The Lama of South America is raring up, in the act of spitting through the fissure of his upper lip, which he used to do when he was alive in the Museum." The orangutan stood "erect and holding a staff in its right hand." Then came the baboons and monkeys. "That Rib-nosed Baboon was alive at the Museum several years — his sagacity in drawing fruit that was thrown to him, if out of reach of his hands, by making use of a stick (which if he found too short, he would change for another), plainly demonstrates that he was not devoid of reflection." Next in line were the sloth, the anteater, and the armadillo followed by the order of *Ferae,* including the grizzly bear, a raccoon, and a badger presented to the museum by President Jefferson. Lastly there were the opossum, mole, shrew, hedgehog, porcupine, cavy, and the American buffalo, "but alas! they are destroyed wantonly! and, in a little while must be extinct."

From the quadrupeds Peale led his visitors to the Long Room, his pride and joy, stretching the length of the building. Along the inside wall were ranged the cases of birds, with two rows of portraits above them. Opposite the birdcages were nine windows divided by partitions holding cases of insects, minerals, and fossils. Over the center window was a "neat and well-toned" organ for the use of "such Visitors as understand Musick." There were also microscopes for examining small insects and, at the lower end of the room, a physiognotrace for making profiles. At that end also

Charles Willson Peale's self-portrait, "The Artist in His Museum," shows him in the Long Room on the upper floor of the State House. His portraits are ranged above the bird cases. Note the mastodon tooth at the lower right. Courtesy of the Pennsylvania Academy of the Fine Arts, of which Peale was the principal founder was a painting of a flight of stairs, showing his two sons Raphaelle and Titian positioned on the stairs in so lifelike a manner that visitors mistook them for real persons (George Washington is said to have tipped his hat to them). Finally, on the payment of an additional fifty cents, Peale took his visitors to another building to see the skeleton of the mastodon and various models, wax statues, Indian costumes and artifacts, and the like.

The Academy of Natural Sciences. Peale's Museum and the various gardens and nurseries in or near Philadelphia were valuable assets to the study of natural history, but the city still lacked a society devoted specifically to these branches of science. Benjamin S. Barton had attempted to provide one when he organized the Linnaean Society of Philadelphia in 1806, but that venture lasted only a few years. Then in 1812 a new society, calling itself the Academy of Natural Sciences of Philadelphia, was begun and was to endure to the present day. The impetus for the new venture came from a small group of devotees of mineralogy, geology, and zoology: John Speakman, a Quaker apothecary; Jacob Gilliams, a well-known Philadelphia dentist; John Shinn, Jr., a manufacturer of chemicals; Nicholas Parmentier, a native of France engaged in the manufacture of spermaceti and cordials; Gerard Troost, a Dutch scientist trained in chemistry, medicine, pharmacy, and mineralogy at Amsterdam, Leyden, and Paris who had arrived in Philadelphia on his way to Java in 1810, settled there, and opened a pharmaceutical and chemical laboratory; William Maclure, the roving Scottish merchant-geologist-reformer who had already made a name for himself with his account of the geology of the eastern United States; Thomas Say, Speakman's business partner and a passionate devotee of entomology and conchology; and Camillus McMann, about whom little is known.

In January 1812 Speakman, Gilliams, Shinn, Parmentier, McMann, and Troost met at Speakman's apothecary shop at the northwest corner of Second and Market streets and organized the new academy. Troost was elected president, and Say was put in charge of the books and collections. In 1815 the academy rented a three-story building in Gilliams Court on Arch Street between Front and Second. The upper floors were used for the museum and library and for meetings and lectures, the ground floor for laboratory purposes. In 1816 the Pennsylvania legislature granted a charter of incorporation, and in the following year the first numbers of the academy's *Journal* appeared.[24]

These auspicious beginnings would scarcely have been possible without the strong support, financial and otherwise, of Maclure, who became president of the academy in 1817 and was reelected annually until his death in 1840. In 1819 Maclure presented the academy with 1,500 volumes on natural history, antiquities, fine arts, and voyages, which he had collected on his travels. His contributions in money, books, and geological and mineralogical specimens were continued in subsequent years; his frequent trips to Europe, especially to Paris, kept the academy in touch with the latest developments in European science. He also organized and paid for natural history expeditions to various places, often taking academy members with him. In the winter of 1817–1818 he set off for the coast of Georgia and Florida accompanied by Say, George Ord, and Titian Peale. In 1819–1820 Titian Peale, Say, and Augustus Jessup accompanied Major Long's expedition to the Rocky Mountains. Say published his researches on this expedition in the *Journal*.

Once established, the academy attracted to its ranks nearly all the leading naturalists of Philadelphia, including foreign naturalists who found their way to the Quaker City in their scientific wanderings. Thomas Nuttall found it a convenient place to work on his *Genera of North American Plants*. Charles A. Lesueur, arriving in Philadelphia in 1816 with Maclure at the end of a series of natural history explorations in the West Indies, New York, and Ohio, found congenial company and encouragement for his researches at the academy, to whose *Journal* he was an early and frequent contributor. Among the native Philadelphians, Say, son of a physician-apothecary and great-nephew of William Bartram, was the most faithful member of the academy, inhabiting its rooms, cooking simple meals there, and caring for its collections, to which he added substantially through his own researches.

At the weekly meetings one might also encounter merchant-naturalists Reuben Haines, Charles J. Wister, and Zacchaeus Collins; Solomon W. Conrad, bookseller, printer, botanist, and mineralogist; physician-botanist John Barnes; Richard Harlan, anatomist and budding paleontologist; George Ord, the retired chandler and ropemaker who completed the last two volumes of Alexander Wilson's *American Ornithology* and acted as his literary executor and champion against Audubon's claims to fame; Thomas Cooper, now a professor in the newly established Department of Physical Sciences and Rural Economy at the University of Pennsylvania; Adam Seybert, the chemist-manufacturer-congressman whose mineral cabinet formed the nucleus of the academy's fast-growing mineralogical collections; and a group of younger geologist-mineralogists trained at the School of Mines and the Museum of Natural History in Paris, among whom were Henry Seybert, Isaac Lea, Lardner Vanuxem, and William Keating.

In addition to its regular Saturday night meetings, the academy sponsored popular lectures. Troost spoke on crystallography; Shinn on chemistry; Say on entomology; and John Barnes, John Waterhouse, and Thomas Nuttall on botany. Through lectures like these and the rapid increase in the size and diversity of its collections, the treasurer reported in December 1815 that the academy had "risen into public notice, . . . attracted the attention of strangers, and . . . established for itself a character and reputation, far exceeding our most sanguine expectations."[25]

CONCLUSION

With an appropriateness seldom found in historical events the institutional development of science in Philadelphia returned at the end of the Jeffersonian era to the spirit of Benjamin Franklin, the founder or cofounder of so many of the city's institutions. At a public meeting on February 5, 1824, Samuel V. Merrick (nephew of John Vaughan) and his associates launched the Franklin Institute of Pennsylvania, dedicated to "the promotion and encouragement of manufactures and the mechanic and useful arts by the establishment of popular lectures on the sciences connected with them, by the formation of a cabinet of models and minerals and a library, by offering premiums on all objects deemed worthy of encouragement, by examining all new inventions submitted to them, and by such other measures as they may judge expedient."[26]

Franklin would have felt quite at home in the new society, whose aims were utilitarian like his own, and whose membership included a sampling of merchants, manufacturers, engravers, jewelers, carpenters, hatters, dentists, metalworkers, druggists, lawyers, coachmakers, tailors, accountants, and instrument makers as well as some of Philadelphia's leading scientists. The Franklin Institute was formed too late to influence science and technology in the Jeffersonian era, but it rounded out the pattern of Philadelphia institutions related to science. The city's gardens and its natural history museum would not long outlast their founders, but the Institute, the American Philosophical Society, the Academy of Natural Sciences, and the University of Pennsylvania would provide a solid institutional base for the cultivation of science and learning for many years to come.

3

Scientific Centers in New England

PHILADELPHIA'S pretensions to national leadership in science and culture did not long go unchallenged. New England had always considered itself the stronghold of learning, and Timothy Dwight, president of Yale College, was not slow to press its claims to this honor in his *Travels*. What other region of the United States was so well endowed with colleges, each with its library, museum, and "philosophical apparatus?"

THE TRADITION OF LEARNING – YALE

As of about 1810 Yale had three "colleges," a chapel with rooms for the philosophical apparatus, and a building known as the Connecticut Lyceum containing the college library of 7,000 volumes. Harvard, "now styled the University in Cambridge," had four colleges, a chapel, Harvard Hall (containing the library of 15,000 volumes, a "rich and ample" philosophical apparatus, a museum, and a dining hall), and a botanical garden with an endowed professorship. Brown University, with only one college building, had a library of about 3,000 volumes, a philosophical apparatus, and a museum "containing a number of natural and artificial curiosities." Dartmouth College, begun as a school for the education of Indians, was now a liberal arts college with the same attractions as Brown but with a chapel as well. Bowdoin College and Williams College had two dormitories each, with libraries of 1,500 and 1,000 volumes respectively and the usual chapel, museum, and philosophical apparatus. Middlebury College had but one collegiate building when Dwight visited it in 1811, but its 110 students were "probably as virtuous a collection of youths as can be found in any seminary in the world."[1]

Dwight was especially proud of Yale, with its young aggressive professor of chemistry and natural history Benjamin Silliman, and of the Connecticut Academy of Arts and Sciences, organized in 1799 by himself, Noah Webster, Jeremiah Wadsworth, and other Connecticut literati. From 1810 to 1818 the academy published a very creditable first series of *Memoirs,* including Silliman's account of his experiments on refractory substances with an oxyhydrogen blowpipe, the Silliman-Kingsley memoir on the Weston meteor of 1807, Jeremiah Day's "Observations on the Comet of 1811," Dwight's "Observations on Language" and his letter "On Light," and Noah Webster's essay entitled "Origin of Mythology."[2]

In 1811 Yale acquired on loan the splendid collections of over 12,000 minerals purchased in Europe by George Gibbs of Newport, R.I., and Silliman persuaded President Dwight to allocate the second floor of South Middle Hall for use as a mineral gallery, which soon became a major attraction for sightseers passing through New Haven.[3] Silliman's lectures on chemistry, mineralogy, and geology were the delight of all who heard them, including Dwight who was especially pleased with the geological lectures demonstrating the perfect harmony of science and religion.

In 1813 Yale added a medical school to its complement of scientific institutions. In 1818 Silliman launched the *American Journal of Science and Arts,* destined to be the leading American scientific journal for the next several decades. In 1819 he and Gibbs founded the American Geological Society, with headquarters in New Haven. Thus, thanks to Dwight, Silliman, and their colleagues, New Haven became a scientific center of some importance.

THE HARVARD CONNECTION

But whatever Dwight and Silliman may have wished, the intellectual capital of New England was not New Haven but the Boston area, dominated intellectually by Harvard College, whose influence radiated throughout eastern Massachusetts and into New Hampshire and the District of Maine. Boston, Cambridge, Beverly, Ipswich, Newburyport, Byfield, Plymouth, Quincy, and a score of other Massachusetts towns had their communities of Harvard graduates (lawyers, physicians, clergy, merchants, politicians) with wide-ranging intellectual interests and a strong disposition to foster the growth of their alma mater.

Manasseh Cutler, himself a Connecticut Yankee and graduate of Yale, discovered such a community when he accepted a call to a congregation in Ipswich Hamlet (now Hamilton), Mass., in 1771. A man of varied scientific interests (astronomy, physics, botany, zoology, and medicine), Cutler soon found kindred spirits among the Harvard graduates in Ipswich and surrounding towns. In Beverly there was Joshua Fisher, a physician whose passion for natural history led him to endow the Fisher Professorship of

Botany and Natural History at Harvard in 1833, and the clergyman-astronomer Joseph Willard, who in 1780 was elected president of Harvard. In Salem, Cutler became acquainted with William Bentley, diarist and polymath; Edward A. Holyoke, son of a former Harvard president; Joseph Orne, physician and poet; and John Prince, a Boston mechanic's son who became pastor of the First Church of Salem after graduating from Harvard. Prince employed his mechanical talents in devising an improved air pump. His friends and visitors, his biographer reports, "were introduced through his admirable apparatus and specimens, to all the wonders of Astronomy, Optics, Pneumatics, Botany, Mineralogy, Chemistry, and Entomology. . . . As an experimental lecturer and operator, in his own parlor and surrounded by his private friends, he was never surpassed by any public professor of science."[4]

Cutler, Prince, Willard, Fisher, Holyoke, and Orne were all members of the Philosophical Club formed in 1761 to purchase the part of the library of the Irish chemist Richard Kirwan that had been captured at sea by a Salem privateer. "Rode to Beverly, and went to Mr. Willard's with Mr. Barnard, Mr. Prince, and Mr. Fisher," Cutler noted in his journal for June 19, 1781. "The Proprietors of the Philosophical Library we lately purchased . . . met at Mr. Willard's to overhaul the books and repair them, which we did, and established the regulations of the Proprietary. We valued a share at thirty hard dollars. Chose Mr. Willard, Librarian; and Mr. Prince, Clerk."[5] After Willard's departure to assume the presidency of Harvard in 1780, the library was moved to Salem. In 1810 it was joined to the social library to form the Salem Athenaeum. Kirwan was offered indemnification for his unintentional contribution to the cultural life of Salem, but he refused the offer, "expressing his satisfaction that his valuable library had found so useful a destination."[6]

Newburyport, not many miles north of Salem, was a somewhat longer ride for Cutler. There, too, he found a community of Harvard graduates with scientific interests. The most important of these was Theophilus Parsons, a lawyer-jurist who moved to Boston in 1800 and played an important role in the development of Harvard and other literary and scientific institutions in the Boston area. Parsons's ruling passion, according to his son, was "a universal and ardent desire for knowledge." Surrounded by his astronomical, optical, electrical, and chemical instruments and his library of several thousand books, he devoted his leisure hours to scientific studies, especially mathematics and natural philosophy. He was especially intimate with Prince, whose visits to Newburyport were anticipated eagerly by Parsons's son: "I was sure, whenever he made his appearance, that some new instrument would come out to be explained and tried, or some old ones would be taken, and new trials made."[7] Cutler applied to Parsons in August 1782 for the loan of a sextant for use in observing a conjunction of Saturn and Jupiter: "I have been desired to make observations of their approach to each other," Cutler explained, "in order to determine the true time of their conjunction, but find myself unable to make the necessary observations without a sextant."[8]

After moving to Boston in 1800, Parsons became a close friend of the instrument maker and self-taught astronomer William C. Bond, whose shop on Washington (then Cornhill) Street was near Parsons's house on Bromfield Lane, and corresponded on problems of celestial mechanics with Nathaniel Bowditch, author of the *American Practical Navigator* and translator of Laplace's *Mécanique céleste*. Botany, mineralogy, and agriculture also interested Parsons, though not to the same extent as natural philosophy. A tall powerfully built man of slow firm step and high-strung temperament, wearing a wig to conceal his baldness, Parsons cut an impressive figure appropriate to a member of the politically conservative Essex Junto. He eventually became chief justice of the Supreme Court of Massachusetts.

John Lowell, another Newburyport lawyer and patron of science and letters, had moved to Boston in 1776 and became judge of the Federal District Court for Massachusetts. His scientific interests were less extensive than Parsons's, but he and his son, the second Judge John Lowell, were deeply involved in development of science in Cambridge and Boston, especially at Harvard. Both served on the Corporation of Harvard College, where the younger Lowell became associated with John Davis, his father's successor as district court judge and also a Harvard graduate with scientific interests. In no other part of the United States were members of the legal profession so active in the institutional development of science.

Boston-Cambridge was the heart and center of the network of Harvard-bred literati of which Cutler's friends formed an integral part. Reaching into every part of eastern Massachusetts and northward into the District of Maine, the Harvard connection formed the basis for the founding of a variety of literary and scientific institutions and for the strengthening of Harvard College. The American Academy of Arts and Sciences, Massachusetts Historical Society, American Antiquarian Society, Massachusetts Society for Promoting Agriculture, Boston Athenaeum, Harvard School of Medicine, Massachusetts General Hospital, *Monthly Anthology and Boston Review, North American Review,* and Linnaean Society of New England formed a set of interlocking institutions with overlapping personnel and purposes. In the rosters of their officers and members the same names appear again and again: Lowell, Davis, Adams, Quincy, Warren, Jackson, Dexter, Kirkland, Pickering, Sullivan, Winthrop, Vaughan, Bowdoin, Higginson, Parsons. John Lowell the younger, like his father, served as president of the Agricultural Society and was a member of the Harvard Corporation and a Fellow of the American

Academy. When the Boston Athenaeum was formed in 1807, he became
a member. John Davis and Josiah Quincy were active in all these institu-
tions and the Massachusetts Historical Society as well. Davis was a found-
ing member and president of the Linnaean Society of New England. John
C. Warren followed his father as professor of anatomy and surgery at
Harvard. He and his colleague James Jackson were active in the Anthology
Club, the American Academy, the Athenaeum, the Massachusetts Medical
Society, and the founding of the *New England Journal of Medicine and
Surgery* and the Massachusetts General Hospital. John Kirkland, pastor
of the New South Church, belonged to the American Academy, the
Agricultural Society, the Historical Society, and the Athenaeum. His
parishioner Theophilus Parsons was especially influential in securing his
election to the presidency of Harvard in 1810. Nowhere else in the United
States was the network of scientific and literary institutions so tightly interwoven by ties of family, friendship, and loyalty to a common alma mater
as in the Boston area.

THE AMERICAN ACADEMY OF ARTS AND SCIENCES

The first and most comprehensive effort to organize the intellectual
resources of eastern Massachusetts was instigated by John Adams, who
had become acquainted with the American Philosophical Society during
his trips to Philadelphia as a delegate to the Continental Congress. The
society, he wrote his wife in August 1776 in a letter describing the at-
tainments of the Philadelphia literati, "excites a scientific emulation, and
propagates their fame."

> If ever I get through this scene of politics and war, I will spend the
> rest of my days in endeavoring to instruct my countrymen in the art
> of making the most of their abilities and virtues; an art which they
> have hitherto too much neglected. A philosophical society shall be
> established in Boston, if I have wit and address enough to accomplish
> it, sometime or other.[9]

Adams's determination to pursue this project was raised to an even
higher pitch when he went to France in 1778, visited French scientific institutions, witnessed Franklin's enormous prestige and popularity as a
"philosopher," and received repeated inquiries from French savants about
the American Philosophical Society. Returning to Boston in 1779 in company with the French ambassador to the United States, Adams attended
a dinner for the ambassador held in the Philosophy Chamber of Harvard
and there outlined to Samuel Cooper, pastor of the Brattle Street Church
and a Fellow of Harvard College, his project for establishing a philosoph-

ical society in Boston.[10] Cooper was won over, but only after Adams had persuaded him that the new organization would reinforce rather than undermine Harvard, and Adams himself drafted the section of the constitution of Massachusetts that declared it to be the duty of legislators and magistrates "to cherish the interests of literature and the sciences, and all seminaries of them" and "to encourage private societies and public institutions, rewards and immunities for the promotion of agriculture, arts, sciences, commerce, trades, manufactures, and a natural history of the country."

In accordance with this injunction and in response to the petition of Cooper and his associates, the Massachusetts legislature granted a charter on May 4, 1780, incorporating John Adams, Samuel Cooper, and sixty other citizens of Massachusetts into "a Body Politic and Corporate, by the name of The American Academy of Arts and Sciences," organized "to promote and encourage the knowledge of the antiquities of *America,* and of the natural history of the country, and to determine the uses to which the various natural productions of the country may be applied; to promote and encourage medical discoveries, mathematical disquisitions, philosophical inquiries and experiments; astronomical, meteorological, and geographical observations, and improvements in agriculture, arts, manufactures, and commerce; and, in fine, to cultivate every art and science which may tend to advance the interest, honor, dignity, and happiness of a free, independent, and virtuous people."

The list of sixty-two charter members of the academy read like a *Who's Who* of Boston's political and cultural elite. Among the "Honorables" were John and Samuel Adams, John Hancock, James Bowdoin, Levi Lincoln, James Sullivan, Robert T. Paine, William Cushing, and Francis Dana. Among the physicians were Edward Holyoke, son of a former Harvard president; Cotton Tufts; and Joseph Orne. Charles Chauncy and Samuel Cooper led the long list of "reverends." Harvard was represented by President Samuel Langdon; his successor-to-be Joseph Willard; Edward Wigglesworth, Hollis Professor of Divinity; Samuel Williams, Hollis Professor of Mathematics and Natural Philosophy; Stephen Sewall, Hancock Professor of Hebrew; tutors Benjamin Guild and Caleb Gannett; librarian James Winthrop, and such present or future members of the Harvard Corporation as John Lowell, and Thomas Cushing.

On May 30, 1780, a "fair, pleasant day" immediately preceding election day, the American Academy of Arts and Sciences met for the first time in the Philosophy Chamber. On June 17 a committee of the academy met at James Bowdoin's house to draw up rules for the conduct of business, and on August 30 the academy met to adopt a code of statutes and elect officers. Bowdoin, Franklin's friend and correspondent on scientific subjects, was elected president. Willard became vice-president and corresponding secretary, Gannett was elected recording secretary and keeper of the cabinet. On November 8, "somewhat cloudy, cold and chilly," the academy gathered at the county courthouse in Boston and proceeded to Cooper's meetinghouse to hear the inaugural "Philosophical Discourse" of their newly elected president. After describing the wide field of investigation opened to the "sons of literature," Bowdoin turned his attention to the subjects of civil and natural history, stressing the value of a thorough knowledge of the natural history of the United States for the development of its agriculture, commerce, and manufactures. With a prescient glance into the future, he stressed the central role that Harvard would play in the activities of the academy. Most of the members, he observed, were sons of Harvard, as would most of its future members be. He concluded with a glowing but less clairvoyant prophecy of the achievements that would be recorded by their historian at the hundreth anniversary in 1880. The entire academy then went to the president's house for dinner.[11]

Although it was inspired by the example of the American Philosophical Society, the American Academy of Arts and Sciences was somewhat different from the older group in its character and operations. It had no building of its own and its members were scattered over a wide area, making frequent meetings difficult. Instead of meeting every other week, the academy met four times a year, convening in the Philosophy Chamber in the summer and fall and in Boston wherever accommodations could be found in winter and spring. Beginning at the county courthouse, the Boston meetings moved successively to the Manufactory House on Tremont Street, to a room in the State House, to the Boston Marine Insurance Company building, until in 1817 the academy found quarters with the Boston Athenaeum in its house on Tremont Street. The library, which up to that time had been deposited at Harvard, was then moved to the Athenaeum, along with the cabinet of minerals. In general, meetings in Boston were better attended than those in Cambridge. Scheduling meetings about the time of the Harvard commencement and the annual elections in Boston made it easier for out-of-town members to attend.

The early records of the academy are preserved at the Boston Athenaeum.[12] Fortunately, the rather terse account of activities contained in the official minutes can be supplemented for the early years from the journal kept by Manasseh Cutler. Cutler was elected to the academy on January 31, 1781. On May 29 he and Joseph Willard rode to Boston to attend a meeting of the group at the county courthouse.

> About twenty-two members present [Cutler noted]. Several communications. I communicated a meteorological journal of the weather, from July, 1780, with the diseases most prevalent in Ipswich, Beverly, and Salem. I also presented the Society a sample of sheet-lint, from

Dr. Spofford, who has contrived a machine for scraping it with great dexterity. It was much admired.[13]

In August of the same year and again in November Cutler went to Cambridge for meetings in the Philosophy Chamber. Few members were present at the November meeting, but Cutler seems to have enjoyed himself. He had dinner with Samuel Williams and borrowed Catesby's *Natural History of the Carolinas* from the college library.

The Boston meetings were made pleasant for the members by President Bowdoin's custom of inviting them to dine. Cutler's diary for January 29, 1783, gives some idea of how activities were organized:

> The Academy met in Boston at Concert Hall in the forenoon. I dined with the Hon. James Bowdoin, Esq. The Academy adjourned to his house in the afternoon. A Committee was chosen for two years, responsible for some communication once a year. The business of this Committee was divided into three general heads, and three members chosen to each; first, Mathematics, Geography, and Astronomy; second, Natural Philosophy, Natural History, etc. including many other branches; third, Physic. The gentlemen appointed were, on the first, President Willard, Professor Williams and Mr. Gennet; on the second, Theophilus Parsons, Esq., General Lincoln, and M. Cutler; on the third, Dr. Holyoke, Dr. Warren, and Dr. Tufts. . . . This Committee are, among other matters, to examine the communications that have been made to the Academy, and if they find materials, prepare for publishing a volume as soon as may be.[14]

On March 25 Edward Holyoke of Salem rode with Cutler to Cambridge for a meeting of this steering committee at Willard's house. Two days later the academy met at the courthouse, dined with President Bowdoin, and in the afternoon met for the first time in the "new hall" in the Manufactory House, "which is assigned by the General Court for the use of the American Academy and Medical Society."

Cutler was also appointed to the academy's Committee on Printing, charged with supervising the production of the first volume of *Memoirs,* and to the Committee of Agriculture, with the result that he made several trips to Boston or Cambridge each year. On March 12, 1784: "Went to Boston in my sleigh. Mr. Prince rode with me." Apparently the secretaries of the academy were not as punctual in the performance of their duties as Cutler wished, for in September 1789 he sent Winthrop Sargent, secretary of the Northwest Territory, his certificate of election to the academy, explaining that otherwise it might not reach him for two or three years owing to the *vis inertiae* of the secretaries.[15]

Cutler's diary for January 26, 1791, noted the death of the academy's first president. "We walked in procession from the Hall of the Bank to Mr. Thatcher's Meeting House, where the Hon. Mr. [John] Lowell delivered a most elegant oration in honor of our late President, Governor Bowdoin. Very full assembly. We returned in procession."[16]

Bowdoin was succeeded by John Adams, the original instigator of the academy. Adams's long tenure as president, from 1791 to 1814, exceeded that of Jefferson as president of the American Philosophical Society. Willard served as vice-president and corresponding secretary until his death in 1804. Unfortunately, he did not bother to preserve copies of his official correspondence, but his son had access to some of the letters Willard received in performance of his duties. These included several from Joseph Priestley, who sent the academy the five volumes of his *Observations on Air* in care of Benjamin Vaughan as he left England for his estate in Hallowell, Maine. The Count de Chastelleux was also a frequent correspondent. In November 1782 he visited Cambridge, "a little town, inhabited only by students, professors, and the small number of servants and workmen they employ." He had words of high praise for Willard: "He unites with great understanding and literary acquirements a knowledge of the abstruse sciences, and particularly astronomy."[17]

Like the American Philosophical Society, the academy was launched in a burst of utilitarian enthusiasm, with a heavy emphasis on the promotion of agriculture, manufactures, commerce, and the practical arts generally. The astronomical papers in the first volume of the *Memoirs* were advertised as "chiefly of the practical kind." Natural history was recommended as an aid to agriculture, medicine, and the development of natural resources. Physics and chemistry were praised for their practical applications, and the mechanical arts, manufactures, and commerce for their power to "enrich and aggrandize these confederated States." The utility of medical studies was too obvious to require elaboration.

In keeping with these objectives, the academy established a Committee of Agriculture, out of which developed the Massachusetts Society for Promoting Agriculture. The first two volumes of the *Memoirs* contained articles on the preparation of yellow pigments; culture of Smyrna wheat, Indian corn, and apple trees; grafting of fruit trees; manufacture of steel and potash; extraction of salt from seawater; softening of hard water; and improvements on the fire engine and the air pump. Benjamin Waterhouse submitted a proposal for collecting material on the history of epidemics; Edward Wigglesworth described improved methods of forming mortality tables. In the 1790s the academy urged the state legislature to revise and expand its plans for conducting surveys for an accurate map of the state.

In 1795, in response to a request from Secretary of the Treasury Alexander Hamilton, the academy appointed committees in the various regions of eastern Massachusetts to "enquire into the state of new Arts and manufactures which have been introduced, the time of the introduction of each, and the means which may be had for the improvement of the same." And in 1806 Loammi Baldwin, a Woburn engineer, drafted a resolution asking the academy to correspond with other learned societies "upon the practicability and expediency of forming an universal language, standards of measures, weights, &c."[18]

As time went on, the utilitarian spirit waned, and the *Memoirs* of the academy, like the *Transactions* of the Philosophical Society, became largely a vehicle for publication of scientific articles. Of these a disproportionate share were astronomical and meteorological papers, the former supplied chiefly by Joseph Willard and after 1800 by Nathaniel Bowditch. In 1796 Count Rumford, who in his youthful days as Benjamin Thompson had walked regularly from Woburn to Cambridge with Loammi Baldwin to attend John Winthrop's lectures on natural philosophy, gave the academy $5,000 in stock to support a biennial award for outstanding American contributions to the study of heat and light; the academy could find no worthy recipient until 1839 when Robert Hare of Philadelphia was belatedly given the first Rumford Medal for his invention of the oxyhydrogen blowpipe in 1801 (see Chap. 7).[19]

Meeting infrequently and without a building of its own, the academy could not provide a stimulating environment for scientific discussion and experiment. In the years from 1780 to 1820 only four volumes of *Memoirs* were published. About 1812 John C. Warren, distressed at the purely business character of the gatherings, attempted to arrange special meetings for scientific discussion. A few were held, but the effort soon collapsed.[20] In 1817 Benjamin Vaughan suggested a plan for making the academy the center and moving force of a network of scientific societies in New England. Massachusetts, he complained, had done little to assert intellectual leadership in the United States.

> Its American Academy is asleep perhaps in the sleep of death. . . . Instead of being the organ of New England's learning, its own journals and its own petty societies carry away what ought to belong to it; and Connecticut has instituted a rival Society. The American Academy by the suspension or slow rate of its publications, and by its want of zeal in soliciting pieces, must receive the chief share of censure. . . . An annual publication would satisfy the impatience of authors and the public; and thus prevent many contributions from being diverted, . . . and seeking the public notice through other publications. . . . I perceive no real objection to collecting the published pieces of New England . . . and annexing them . . . to the Massachusetts Volume: but if that be thought degrading let pieces which can not otherwise be obtained, be first presented to the Academy, then

published elsewhere with the *leave* of the Academy . . . and be resumed by the Academy at the end of the year, with the corrections of the author and the remarks of others.[21]

Nothing ever came of Vaughan's suggestion. The academy resigned itself to remaining a local organization for publishing scientific memoirs and recognizing scientific achievement by the bestowal of its membership and prizes.

AUXILIARY SOCIETIES

Except for the American Academy of Arts and Sciences and the Linnaean Society of New England, none of the Boston institutions formed in the surge of organizing activity after the Revolution was predominantly scientific in character, but a surprising number showed some interest in science, especially natural history.

Massachusetts Society for Promoting Agriculture. The Massachusetts Society for Promoting Agriculture took shape in response to a suggestion of Manasseh Cutler, himself a successful farmer, that the American Academy do something to further the cause of scientific agriculture. In November 1785 the academy appointed a committee "whose special Business shall be to attend to the several branches of husbandry; to make experiments, and to endeavor improvements therein; to pursue such methods as they shall judge proper to engage others in promoting the same designs; to connect any gentlemen, whether Fellows or not, with themselves, in prosecuting and perfecting the end of their appointment; to publish from time to time such observations as they may determine to be beneficial to the community at large; and from time to time report their proceedings."[22] The outcome of this resolution was the formation of the Massachusetts Society for Promoting Agriculture, incorporated by the state legislature in the spring of 1792. Very few of the founding members were practical farmers. Many, like the first president, Thomas Russell, were successful merchants. A few, like John Lowell and James Sullivan, were lawyers; others, like Aaron Dexter and Cotton Tufts, were physicians. In due time the clergy came to be represented by Manasseh Cutler, John T. Kirkland, William Emerson, and Joseph S. Buckminster.

In 1793 the society offered a prize of $50 for "the most satisfactory account of the natural history of the canker-worm," which had been causing considerable damage to crops. The prize was won by William D. Peck, a Harvard graduate who, after an unsuccessful venture in business, retired to a small farm in Kittery, Maine, and devoted himself to natural history.

In 1797 Peck won another prize for an essay "on the natural history of the worm that has lately infested cherry, pear, quince and plum trees, called the snail or slug worm." Both essays were published by the society, and Peck became a candidate for the professorship of natural history proposed to be established at Harvard along with a botanical garden. In 1801 the society appropriated $500 for these purposes, and the Massachusetts legislature was persuaded to grant a township six miles square in the District of Maine, the society stipulating that an acre of land in the botanical garden should be devoted to raising seeds of culinary vegetables and producing specimens of new and useful grains and grasses. Another township in Maine was granted to the society by the legislature in 1809. The trustees, along with the presidents of Harvard, the American Academy, and the Massachusetts Medical Society, were made ex officio Visitors of the new professorship and charged with the care and superintendence of the garden. The recording secretary of the Agricultural Society was Kirkland, soon to become president of Harvard. The president, John Lowell, was a member of the Harvard Corporation. In 1814 the legislature granted an annual allowance of $1,000, and the society in turn voted an annual grant to Harvard in support of the garden. Not until 1830 when the legislature withdrew its support was the connection between the society and the natural history establishment at Harvard terminated.[23]

Massachusetts Historical Society. The Massachusetts Historical Society included natural history within its scope. Organized in 1791 in response to the initiatives of Jeremy Belknap, its declared purpose was "the collection of observations and descriptions in natural history and topography, together with specimens of natural and artificial curiosities, and a selection of everything which can improve and promote the historical knowledge of our country, either in a physical or political view." For nearly forty years thereafter the society received natural history specimens and kept them in its quarters on the third floor of the central pavilion in Charles Bulfinch's handsome Tontine Crescent Building on Franklin Street. To promote the gathering of these items, the society published in its *Collections* for 1795 some directions drafted by Manasseh Cutler and William D. Peck for preserving whole animals and their parts, along with further instructions for preserving and labeling plants, marine productions, minerals, and petrifactions. In 1798 the society reprinted Peck's "Natural History of the Slug-Worm," published by the Agricultural Society in the *Massachusetts Mercury*. The Historical Society continued to collect and display natural history specimens until 1833, when the collections were turned over to the newly formed Boston Society of Natural History.[24] A similar fate befell the natural history collections of the American Antiquarian Society (organized in 1812) (see Chap. 13).

Boston Athenaeum. The Boston Athenaeum also had a scientific side in its early years. It grew out of the Anthology Club organized in 1804 to undertake publication of the *Monthly Anthology and Boston Review,* a general magazine that paid considerable attention to scientific as well as literary developments in Europe and America. John C. Warren and James Jackson were assigned responsibility for scientific intelligence. They performed this duty conscientiously, reporting weather data, diseases, and vital statistics regularly and noticing European developments in chemistry and mineralogy from time to time. Many of the magazine's scientific notices were copied from the *Medical Repository,* however.

In 1806 the Anthology Club issued a prospectus for establishing a reading room by public subscription. Sixteen hundred dollars were subscribed by one hundred participants, among whom were Theophilus Parsons; John Lowell; Robert H. Gardiner, president of the Anthology Club; Joseph S. Buckminster; and Obadiah Rich, a wealthy bookseller who collected natural history specimens. On February 13, 1807, the Boston Athenaeum was incorporated and a memoir was circulated to potential subscribers.

> The Reading-room and Library being considered leading objects and chief departments of the Athenaeum, it is proposed . . . to join to the foundation a museum or cabinet, which shall contain specimens from the three kingdoms of nature, scientifically arranged; natural and artificial curiosities, antiques, coins, medals, vases, gems, and intaglios; also . . . a repository of arts, in which shall be placed for inspection models of new and useful machines; likewise drawings, designs, paintings, engravings, statues, . . . and especially the productions of our native artists. Lastly, the plan of the Athenaeum includes a laboratory, and an apparatus for experiments in chemistry and natural philosophy, for astronomical observations, and geographical improvements, to be used under the direction of the corporation. . . . The *Laboratory* and *Apparatus* may be used . . . for the purpose of lectures on chemistry, natural philosophy, and astronomy.[25]

Among the publications listed as being received regularly were the *Annales du Muséum d'Histoire Naturelle, Annales de Chimie, Journal Polytechnique, Annals of Philosophy,* Aikin's *Annual Review,* Tilloch's and Nicholson's philosophical magazines, the *Medical and Physical Journal,* the *Medical Repository,* and Coxe's *Medical Museum.*

The Athenaeum's scientific apparatus was augmented considerably by the purchase in 1807 of instruments belonging to the Society for Study of Natural Philosophy that had been organized a few years earlier by John T. Kirkland, John Davis, Josiah Quincy, John Lowell, John C. Warren,

James Jackson, and others "for the purpose of cultivating, among its members, the science of Natural Philosophy." The members bought shares in the club, the money being used to buy books and scientific apparatus. At each meeting, parts of some work on natural philosophy were read and discussed. The members also delivered lectures upon topics of their choice. Kirkland agreed to give six lectures on astronomy. Davis chose botany; Jackson, chemistry; Warren, human and comparative anatomy and physiology; Stephen Higginson, mechanics.[26] Some idea of the club's activities is given in Jackson's letter of January 10, 1802, to John Pickering, lawyer and linguist and apparently also a student of the sciences.

> It will be my business in the course of the present week to take up the subject of electricity, to make a digest thereof for our society. For this purpose I want to look over Adams on Electricity, which I believe you have. If you have also Cavall's or any other works not exceeding half a dozen volumes, I will thank you to let me have them. . . . The philosophers of Europe must look well to it that they may not soon be eclipsed by the new light that is here springing up.[27]

In 1806 Jackson and Warren were appointed as a committee to observe the solar eclipse predicted for June 16. By the summer of 1807, however, the society had lost its élan. The group was dissolved and the scientific instruments were sold to the Athenaeum, except the chemical apparatus, which went to John Gorham, a newly elected member who had recently been appointed adjunct professor of chemistry at Harvard.

In 1809 the Athenaeum purchased a house on Tremont Street to accommodate its library, natural history specimens, scientific apparatus, and other collections. Three years later Obadiah Rich deposited his natural history collection in the Athenaeum. The trustees authorized the expenditure of $200 to prepare a room for the reception of the cabinet and appointed Rich its curator. By 1822 the Athenaeum had become a place of deposit for the books of the Agricultural Society, the Boston Medical Library, the American Academy, and the private library of John Quincy Adams. By this time, however, the proprietors began to realize that neither their funds nor their new building on Pearl Street, donated by James Perkins in 1822, were adequate to sustain the great variety of collections and activities they had hitherto undertaken. They appointed a committee to dispose of the books and apparatus they had purchased from the Society for the Study of Natural Philosophy and subsequently declined to receive the burgeoning collections of the Linnaean Society of New England when these were offered to them in 1822. The Boston Athenaeum became what it continued to be thereafter, a gentlemen's library with reading rooms and picture and sculpture galleries.

Linnaean Society of New England. The Athenaeum, the Historical Society, and the Agricultural Society made useful contributions to the development of natural history in the Boston area, but a society devoted exclusively to investigation of that field of science was needed. A first step in this direction was taken in December 1814 when a group of interested persons gathered at the house of Jacob Bigelow to discuss formation of a society dedicated to the study of natural history. Besides Bigelow, the group included John Davis, Walter Channing, George Hayward, John W. Webster, and John Ware; James Freeman Dana, a recent graduate of the Harvard Medical School; William S. Shaw, librarian of the Boston Athenaeum; and such other Boston gentlemen as Octavius Pickering, Ezekiel D. Cushing, La Fayette Perkins, and Nathaniel Tucker. A constitution was adopted at the next meeting, providing for weekly meetings and the formation of a natural history cabinet. The name agreed upon was the Linnaean Society of New England. Davis was elected president; Shaw, vice-president; and Webster, keeper of the cabinet.[28]

The enthusiasm of the members was truly remarkable. The collections grew rapidly, and arrangements were made to house them in a room over the Boylston Market in Boston and to open them to the public on Saturday afternoons. The members contributed their own collections, and a circular containing instructions for caring for specimens was printed and distributed to masters of sailing vessels and others, with the result that two live tigers were received from Captain Stewart of the U.S. frigate Constitution and a live bear from Comm. Isaac Chauncy. A collection of African birds and several cases of insects, minerals, shells, and corals were among other donations. Channing was asked to prepare a box of minerals for shipment to France in exchange for French specimens, and Shaw was requested to procure a moose. An artist was hired, and curators were assigned to the various departments. An expedition to the White Mountains in New Hampshire in August 1816 resulted in still further acquisitions. Within a year and a half it was announced in the *New England Journal of Medicine and Surgery* that the collections had outgrown their original apartments and had been moved to "a spacious hall over the new south market house."

> Among the quadrupeds may be mentioned the Lion, Tiger, Leopard, Catamount, Wolf, Bear, Stag, Sea-Elephant, and a great number of smaller species, principally native. The birds amount to nearly three hundred in number, . . . from the Albatross and the large Sea-Eagle of North America, to the minutest Humming birds of Cayenne. . . . The fishes are prepared in uniform half specimens fixed upon a white ground, and afford a fine display. . . . The insects and shells amounting to some thousands in number, include many rare and elegant species, . . . a fine collection of insects from China, and of

shells from the Isle of France and Calcutta. The mineralogical specimens already fill four large cabinets, and the herbarium of native plants, . . . is perhaps inferior to none for the neatness, and perfection of its specimens.

The whole collection, with the exception of the fishes . . . have been inclosed at a great expence, in mahogany cases with glass fronts. They have been arranged according to their orders and genera . . . and most of the specimens are labelled with their scientific and English names . . . [the] collection . . . being always accessible to persons properly introduced, and being opened once a week for the free reception of citizens and strangers.[29]

Exhilarated by the rapid progress of the collections, the recording secretary predicted that if they continued to increase at the same rate the museum would eventually surpass every other American institution of its kind and even rival those of Europe.

In the end, however, the enthusiasm of the society's members was its undoing. The collections grew faster than the capacity to find space and provide for their care. By 1818 the situation had become desperate. Overtures were made, first to the Boston Athenaeum and then to Harvard, to take over the museum and maintain it. Finally, in 1823 an agreement was signed with Harvard and the Board of Visitors of the Massachusetts Professorship of Natural History in which the college agreed to accept the collections, erect a building for them, and grant society members free access to them and to the botanical garden.

Unfortunately, the college did not keep its end of the bargain. The collections gradually deteriorated and the society became inactive, but the zeal for natural history that had initiated it did not. In February 1830 a group of enthusiasts, including five members of the moribund Linnaean Society, met at the house of Walter Channing to discuss the formation of a new society. Thus was born the Boston Society of Natural History, which was to have a long and distinguished career worthy of comparison with that of the Academy of Natural Sciences of Philadelphia.[30]

HARVARD COLLEGE AND ITS MEDICAL SCHOOL

Whatever the successes and failures of the various societies for promoting literature, science, and the practical arts, it was to Harvard College that all looked for intellectual leadership. But Harvard was at a low ebb when Samuel Langdon resigned the presidency in August 1780. Washington's troops had occupied the college buildings during the Revolution, and the library and scientific apparatus had been moved to Andover and Concord. The college finances had suffered under the treasurership

*The so-called "Abbot view" of Harvard College (1798), show-
ing (from left to right) Holden Chapel, Hollis Hall, Harvard
Hall, and Massachusetts Hall, the first and third of which
housed the scientific apparatus and classrooms. Courtesy of
the Fogg Art Museum, Harvard University*

of John Hancock, and the salary of the new president, Joseph Willard,
remained uncertain for several years. Willard was a modest, studious man,
thoroughly competent in astronomy. In the winter he slept in his study
and rose at five in the morning to ready himself for six o'clock chapel,
after which he returned to conduct family devotions before breakfasting
and beginning the day's work. "An engaging degree of modesty was combined with great dignity in his deportment. And when free from care a
temperate cheerfulness always beamed in his countenance."[31]

Under Willard's resolute leadership and with strong support from
John Lowell, James Bowdoin, and other leaders of the Boston community the college began to revive. It had enjoyed a high reputation in natural
philosophy during the colonial years owing to the publications of Hollis
Professor of Mathematics and Natural Philosophy John Winthrop IV in
the *Philosophical Transactions* of the Royal Society of London, to whose
company Winthrop was elected in 1766. But Winthrop's performance was
not to be equaled or surpassed for sixty years after his death in 1779. His

successor in the Hollis professorship, Samuel Williams, began promisingly enough but was forced to resign in 1788 when the Board of Overseers charged him with impropriety in his financial dealings with the college. His replacement, Samuel Webber, was a competent teacher and textbook writer but probably not as good a practical astronomer as Willard, whom he succeeded as president in 1806. According to Willard's son, Webber was a thorough but uninspired teacher, who made up for his lack of vivacity by "the evident desire of making himself understood."[32]

The level of instruction in natural philosophy was raised to a higher plane when John Farrar was appointed to the Hollis chair following Webber's elevation to the presidency. Farrar had no special training in natural philosophy, having been brought to Harvard in 1805 as a tutor in Greek, but Theophilus Parsons, now a member of the Harvard Corporation, was impressed by Farrar's enthusiasm and aptitude in mathematical studies. Parsons's confidence was eventually justified, for Farrar applied himself to these studies with great diligence and introduced the works of Lacroix, Legendre, Biot, Bézout, and other French mathematicians and physicists to the Harvard students in his own translations. Although he published little of importance, he was an excellent teacher: "the best lecturer we ever listened to in any department," one of his students testified.

> Whatever the subject, he very soon rose from prosaic details to general laws and principles, which he seemed ever to approach with blended enthusiasm and reverence, as if he were investigating and expounding divine mysteries. His face glowed with the inspiration of his theme. His voice . . . broke into a shrill falsetto; and with the first high treble notes the class began to listen with breathless stillness, so that a pin-fall could . . . have been heard throughout the room.[33]

Meanwhile, efforts were being made by friends of the college to secure the establishment of an astronomical observatory. As early as 1805 the younger John Lowell, urged on by Webber, visited the Paris Observatory on his European tour and obtained from the celebrated Jean Baptiste J. Delambre a written description of the instruments and buildings needed for a good observatory. These were transmitted to Webber, but the project lay dormant until 1815 when the president and fellows of Harvard voted "that the President, Treasurer [Davis], and Mr. Lowell, with Professor Farrar and Mr. Bowditch, be a committee to consider upon the subject of an Observatory, and report to the Corporation their opinion upon the most eligible plan for the same, and the site." Farrar and Bowditch were given special responsibility for the project. They in turn decided to enlist the aid of William C. Bond, whose observations on the great comet of 1811 had come to their attention only a few years earlier. Bond was about

John Farrar (artist unknown), professor of mathematics and natural philosophy, whose translations of French texts introduced Harvard students to Continental methods in mathematics and physics. Courtesy of Fogg Art Museum, Harvard University

to sail for Europe and agreed to acquire exact information concerning the location, instruments, foundations, buildings, costs, and the like of the Greenwich observatory and others where he could obtain similar information. He kept his promise faithfully and brought back details about the Greenwich, Glasgow, Edinburgh, Highbury, Milford-Haven, and other observatories. The Harvard Corporation was staggered at the amount of money required for a good facility. Wealthy friends of the college were approached, both in 1815 and several years later, but the necessary funds were not forthcoming. Not until the great comet of 1843 reawakened public interest in astronomy did Harvard acquire an observatory worthy of the name.[34]

The development of sciences other than natural philosophy at Harvard was closely linked to progress of the medical school. Philadelphia and New York acquired their medical schools in the 1760s, but Harvard made no move to found one until after the revolutionary war. In May 1780 Boston physicians formed the Boston Medical Society and in the winter of 1781 asked John Warren, then only twenty-five years old, to present a course of anatomical demonstrations at the military hospital on the corner of Milton and Spring streets. Warren, whose brother Joseph had immortalized himself at Bunker Hill, had taught himself anatomy while still an undergraduate at Harvard, forming an anatomical society among his classmates for mutual instruction. During the war he acquired all the practical knowledge of surgery he could wish for as a hospital surgeon, devoting his leisure hours to the study of French and devouring such books on anatomy and surgery as he could procure from the French surgeons with whom he came into contact. His course of anatomical demonstrations in 1781-1782, the first ever given in Boston, drew a wide audience, including President Willard and various members of the Harvard Corporation. As a result, he was asked to submit a plan for establishing a medical school at Harvard. In due course, the School of Medicine was launched with Warren as professor of anatomy and surgery, Aaron Dexter as professor of chemistry, and Benjamin Waterhouse as professor of the theory and practice of medicine.[35]

A beginning had been made, but the new school was a rather feeble affair during the first thirty years of its existence. Warren had a wealth of practical experience, tremendous energy, and unusual eloquence as a lecturer, but both he and Dexter lived in Boston and had to ride almost daily in winter to Cambridge to give their lectures. Even after the Charles River Bridge and West Boston Bridge were built, this was a chore. Before that time Warren had the choice of taking the Charlestown ferry or riding eight miles through Roxbury and Brookline to Cambridge. His contemporaries have left us a picture of him in powdered wig, colored coat with yellow metal buttons, colored and figured waistcoat, leather breeches, and long boots with white tops, riding furiously to visit his patients or meet his class. Toward the end of his life, says his son, "he used a chaise with a powerful horse, and drove with the utmost possible speed, avoiding with great adroitness all obstacles. He never had any attendant except in the winter, when he indulged in a good booby-hut, with a pair of horses driven by a black servant."[36]

Aaron Dexter seems to have been a man of mediocre talents, famous chiefly for his habit of telling students who had witnessed an unsuccessful experiment, "Gentlemen, the experiment has failed, — but the principle, Gentlemen, the principle remains firm as the everlasting hills." He did, however, introduce the new French chemistry at Harvard, much to the delight of the visiting Frenchman Brissot de Warville, who described him as "a man of extensive knowledge, and great modesty."[37] Partly on Dexter's account his friend William Erving endowed the Erving Professorship of Chemistry in 1790. But not until John Gorham was appointed adjunct professor of chemistry in 1806 after completing his medical and scientific studies at Edinburgh was the teaching of chemistry at Harvard brought up to date. Succeeding to the Erving professorship in 1816, Gorham lectured to both medical students and undergraduates. In 1819-1820 he brought out a two-volume textbook, *The Elements of Chemical Science,* described by Benjamin Silliman as "a perspicuous, chaste and philosophical treatise" and by William Brande of Cambridge University as "a most excellent and complete digest of everything at present known on that science."[38] He also planned the laboratory and lecture room in Holden Chapel when that building was renovated in 1814 and persuaded President Kirkland to send Gorham's pupil-assistant James Freeman Dana to London to buy new chemical apparatus from Frederick Accum during his studies with that well-known chemist and manufacturer of chemical instruments. Gorham was also interested in mineralogy, which he included in his lectures, but he soon found that the demands of his teaching and medical practice precluded active field work.

The third member of the original triumvirate at the Harvard School of Medicine, Benjamin Waterhouse, came from a Quaker family in Newport, R.I. In 1775 after a medical apprenticeship there, he was sent to live and study in London with his mother's cousin, the famous John Fothergill. He then spent nine months in Edinburgh and two years at the University of Leyden, living during the latter period with the Adams family at The Hague. After taking his medical degree at Leyden in 1780, he traveled for a year in Europe, visiting Franklin at Passy in France. Gilbert Stuart's portrait of Waterhouse about this time shows a handsome, serious-looking young man with dark hair, high forehead, wide-set eyes, straight nose, and pointed chin.[39]

On his return to the United States in 1781 with a letter of recommendation from Fothergill, Waterhouse seemed an obvious choice for the newly established professorship in the theory and practice of medicine at Harvard. Unfortunately, Waterhouse did not get on well with his medical colleagues, who resented his condescending manner, Jeffersonian principles in politics, and resolute Quakerism. At first Waterhouse planned to live and practice in Boston like Warren and Dexter, but on finding that the Boston physicians did not welcome an addition to their number, he settled in Cambridge where the opportunities for practice were much more limited. To add to his discomfiture, his colleague John Warren was paid from an endowment of $1,000 left by Ezekiel Hersey in 1770 and Aaron Dexter soon received support from William Erving's bequest in support of the chemical chair, but Waterhouse remained dependent on student fees. In 1786 Waterhouse tendered his resignation and prepared to return to Rhode Island where he had been elected professor of natural history at Rhode Island College (named Brown University in 1804), but the Harvard Corporation persuaded him to stay, offering as an inducement permission for him to give lectures on natural history as well as the theory and practice of medicine.

Having settled in Cambridge in a house facing the Common, Waterhouse devoted himself largely to working up his natural history courses and building what he called "the natural history establishment" at Harvard.

> I commenced my Natural history lectures in the autumn of 1788, by
> giving my first course *gratis,* as I found there was no disposition to
> subscribe to a course of lectures on a subject of which these young
> men had not formed . . . any previous notions. The 2d. year I opened
> my course for five hearers, at a guinea a piece. The 3d. year I had
> seven. The 4th year I allowed each to subscribe whatever he chose;
> then I had about thirty pupils, some subscribed three guineas, some
> two, some one; and some half a guinea, others clubbed together, and
> divided the half guinea and lectures between them. . . .[40]

The students seem to have enjoyed these lectures, however discouraged
Waterhouse may have been. According to President Willard's son:

> They afforded much entertainment to the students, besides the scien-
> tific instruction they imparted. In his style there was great vivacity
> and compass of expression, with the added attraction of anecdote and
> humor, all which combined made the lectures very popular. . . .[41]

Waterhouse was not much of a field naturalist, however, nor did he
stimulate his students to do field work. His surviving lecture notes are
distinguished more for their literary quality and general erudition than
for practical instruction in the study of nature.

In 1787 Waterhouse wrote to John C. Lettsom, Fothergill's successor
as patron of American science, and to Joseph Banks, president of the Royal
Society of London, describing his lectures on natural history and soliciting
their support for establishment of a botanical garden at Harvard. Banks
endorsed the idea warmly, but Lettsom advised him to undertake the less
expensive project of forming a mineral cabinet and sent him a collection
of specimens as a beginning. Inspired by this donation with "an ardor
bordering upon enthusiasm," Waterhouse began the study of mineralogy
and communicated his zeal to the French consul in Boston, who arranged
for a further gift of 200 mineral specimens from the revolutionary government in France. Fothergill sent new donations, and James Bowdoin contributed 150 specimens of volcanic lavas and Italian marbles. Impressed
by Waterhouse's success in securing specimens, the Harvard Corporation
appropriated funds for purchasing an assaying apparatus and a cabinet
for displaying the minerals. Soon the Lettsomian Cabinet of Minerals,
housed in elegant mahogany and declared by Waterhouse to be "by far
the richest and most extensive collection of minerals in the United States,"
was on display in the Philosophy Chamber of Harvard Hall cheek and
jowl with the library, the philosophical apparatus, and a gallery of portraits of patriots and benefactors of Harvard.[42]

Despite Waterhouse's achievements in acquiring mineral specimens,
he was not popular with his colleagues nor college authorities. Webber

objected to the presence of stuffed birds, minerals, and other "impedimenta" in the Philosophy Chamber, and Waterhouse complained that no accommodations whatsoever had been arranged for his medical lectures. To relieve the congestion in Harvard Hall, Holden Chapel was remodeled in the summer of 1800. Waterhouse was given a room on the ground floor for his medical lectures, as was the professor of chemistry and materia medica, Aaron Dexter. As lecturer in natural history, Waterhouse had the right to use the second-floor chamber reserved for the professor of anatomy and surgery, but only when Warren was not using it.[43]

The crowning blow to Waterhouse's wounded pride was the appointment of William D. Peck to the Massachusetts Professorship of Natural History when that was established in 1805 through the efforts of the Agricultural Society. By 1804 Peck had taught himself enough natural history to win two prizes offered by the society; to win election as a Fellow of the American Academy for his description of four new species of fishes; and to attract the attention of John Lowell, Manasseh Cutler, and others in their campaign to promote the study of natural history at Harvard. Since the Agricultural Society had raised the money for the botanical garden and the professorship of natural history, Harvard had little choice but to accept Peck as the first incumbent. Installed in 1805, Peck was sent to Europe for three years, where he made valuable acquaintances with European naturalists such as Joseph Banks, William Kirby, and James E. Smith in England; Olaf Swartz, Carl P. Thunberg, and Adam Afzelius in Sweden; Jens W. Hornemann in Denmark; Franz C. Mertens in Germany; and Pierre A. Latreille in France. The fact that several of these men corresponded with Peck after his return to the United States indicates that they were favorably impressed with his abilities as a naturalist.[44]

To Benjamin Waterhouse, however, Peck was a man "who could neither write nor read," put forward by "the high-toned Essex Junto political party" and sent off to Europe to acquire the education he lacked. Waterhouse found himself shoved aside and treated in the American Academy of Arts and Sciences "as if he had never known what a mineral, plant, or animal was." In 1809 the corporation deprived him of his posts as lecturer in natural history and keeper of the mineral cabinet. Two years later his relations with the college, embittered by prolonged quarrels with John Warren and his son and with James Jackson, the new professor of clinical medicine, were severed completely. Waterhouse blamed the political machinations of Kirkland, Davis, and the Essex Junto for his downfall, but it appears in retrospect that he had outlived his usefulness. Peck, though not a brilliant lecturer, was a highly competent practicing naturalist, and James Jackson, fresh from his studies in Europe, soon surpassed anything Waterhouse could offer as a professor of medicine.

Three important figures in the development of natural history in the Boston area: Benjamin Waterhouse (by Samuel Harris) (upper left), William Dandridge Peck (attributed to Gilles-Louis Chrétien) (upper right), and Jacob Bigelow (photograph, left). Courtesy of the Francis A. Countway Library of Medicine, Harvard University (Waterhouse and Bigelow) and of the Fogg Art Museum, Harvard University (Peck)

Meanwhile, the Boston community was astir with new interests and projects, and Harvard was gathering momentum for a strong surge forward during the presidency of John Kirkland, the Essex Junto's candidate when President Webber died in 1810. Kirkland was a member of the Agricultural Society, and its pet project, the botanical garden, moved ahead rapidly with his support. Land was purchased half a mile north of the college and was soon augmented by Andrew Craigie's donation of an adjoining piece of land, making a total of about seven acres. A handsome dwelling, known as the Garden House, was erected for Peck's occupancy; and a Yorkshireman named William Carter, whom Peck had hired in England, began work on the grounds. By 1816 the garden had become a tourist attraction.

> This spot [wrote an observer] is handsomely inclosed with a high fence and a belt of trees. . . . It contains three buildings; the professor's house, situated on an eminence on the western side, and affording a delightful view of the garden; a gardener's house on the opposite side, and an extensive green house and conservatory for the plants of warm climates in the centre. The latter building contains apartments admitting different degrees of heat, calculated for the plants of all latitudes and climates south of us. It is furnished with a rich and splendid collection of exotic trees, shrubs and plants. . . . A fountain and a pool of water near the centre furnish a place for the cultivation of aquatics, and an unfailing supply for the purpose of watering the garden.[45]

In 1818 Peck published a catalogue of the plants, noting that the garden was intended "for the cultivation of plants from various parts of the world, to facilitate the acquisition of botanical knowledge." There were a good many foreign plants, he explained, contributed by "gentlemen who have visited tropical regions in the East and West Indies, and in Africa" or grown from seeds received from botanical gardens in Europe. Domestic plants were fewer in number, owing to the lack of funds to employ a full-time collector, but efforts were being made to procure "all such indigenous trees, shrubs, and herbaceous plants, as are worthy of attention, as being useful in domestic economy, in the arts, or in medicine." Members of the Board of Visitors, original subscribers to the establishment of the garden, the governor and legislators of Massachusetts, the officers of Harvard, strangers of distinction, and clergymen were admitted to the garden without charge. Others paid twenty-five cents or bought an annual family admission ticket for five dollars. "Living exotic plants may be purchased by applying to the gardener," the "Advertisement" announced.[46]

Peck made good use of the garden in teaching his students. Among those who later distinguished themselves by work in natural history were Francis Boott, Benjamin D. Greene, James F. Dana, Thaddeus W. Harris, Charles Brooks, George B. Emerson, William Oakes, and Charles Pickering. "The first visit I made, after being established in college," Emerson recalled, "was to the Botanic Garden, to learn from Prof. Peck the name of the plants I had examined in Wells [Maine] for which I had found no name. He recognized them instantly from my description."[47]

In the medical school the impetus for change came largely from John Warren's son, John C. Warren, and his close friend and colleague James Jackson. Like his father, young Warren was a man of great energy, "of a highly excitable temperament, keen sensibility to impressions of pain or joy, prompt to impulsive action, but yet controlled by habits of reserve." His medical education was the best then available. After studying medicine and chemistry at Edinburgh, he went to Paris where he lived with family of Baron Antoine Dubois, surgeon of the Hospice de l'École de Médecine, and attended the lectures of such leading French physicians as Jean N. Corvisart, François Ribes, Hyacinthe Gavard, François Chaussier, and Guillaume Dupuytren, as well as those of Dubois. He also continued his chemical education in a small private class conducted by Louis N. Vauquelin. "By its aid," wrote Warren, "I was able to go into practical and experimental chemistry, as soon as I returned to Boston, in the Natural-Philosophy Society, which, by its apparatus, the zeal of its members, and the necessity of lecturing on different topics, gave me a very satisfactory acquaintance with the subject."[48] When the winter courses were over, Warren attended the public lectures of Fourcroy, Cuvier, Desfontaines, and Sabatier at the Jardin des Plantes.

Learning that his father had suffered an attack of paralysis, Warren sailed for Boston in October on the same ship with George Gibbs of Newport, R.I., an ardent mineralogist known to Warren from Gibbs's family connections with the Channings in Boston. Warren was soon caught up in the intellectual and professional circles that were beginning to form: the Society for Study of Natural Philosophy, the Anthology Club, the Boston Athenaeum, and the Boston Medical Society among others. He soon made friends with young James Jackson, who had also returned recently from medical studies abroad, and they set out to infuse life and scientific vigor into the Massachusetts Medical Society, which up to that time had been notable chiefly for its opposition to the development of the Harvard School of Medicine. Warren and Jackson took over and regularized the main publication of the society; composed and published the *Massachusetts Pharmacopoeia;* and in 1806 issued a report on vaccination, which had been a highly controversial subject since its introduction into Massachusetts by Waterhouse in the 1790s. About the same time, they founded the short-lived Boston Medical Society from which sprang the Boston Medical Library, which was eventually absorbed by the Athenaeum.

Warren and Jackson were an impressive team, Warren contributing "scientific ability, tireless energy, an inflexible will, and an ardent zeal for righteous causes"; Jackson, "finely disciplined patience, wise judgment, firmness in purpose, and unusual power of persuasion."[49] Their next project was the improvement of the Harvard Medical School. Warren had moved to a house on Park Street in Boston in 1805. In the same year he rented a room over White's apothecary store at 49 Marlborough Street and gave public demonstrations in anatomy for the Boston physicians, especially

the younger men. In 1806 he was appointed adjunct professor of anatomy and surgery at Harvard, and he and his father began to press hard for removal of the medical school to Boston. The Corporation agreed, despite the objections raised by Waterhouse, on condition that the courses given there should also be given at Cambridge. In 1810 the move was made, but not until 1815 did Harvard supply a building to take the place of the hall above the apothecary shop. Meanwhile, Jackson supplanted Waterhouse as professor of the theory and practice of medicine, and Warren and Jackson, aided by John Gorham, launched the *New England Journal of Medicine and Surgery* (1812), the first such publication in the Boston area.

With the erection of the Massachusetts Medical College, as the building on Mason Street not far from the Boston Common was called, the Harvard Medical School began to emerge as an important center of instruction. John C. Warren succeeded his father in the Hersey Professorship of Anatomy and Surgery. Jackson taught theory and practice of

*Four early professors of medicine at Harvard (left to right):
John Warren (after the portrait by Rembrandt Peale), John
Collins Warren (by Gilbert Stuart), James Jackson (by Spiri-
dione Gambardella), and John Gorham (lithograph). Cour-
tesy of the Francis A. Countway Library of Medicine,
Harvard University*

medicine; Gorham, chemistry and mineralogy. To these were added two
able young Bostonians, Walter Channing and Jacob Bigelow, both of
whom had studied in Philadelphia where both acquired an interest in
natural history from Benjamin S. Barton. Channing, who had continued
his medical education in Edinburgh and London before returning to
Boston, became professor of midwifery and medical jurisprudence at
Harvard. Bigelow was appointed professor of materia medica and, soon
afterward, Rumford Professor of the Application of Science to the Useful
Arts, a professorship founded by bequest of Count Rumford "to teach
by regular courses of academical and public lectures, accompanied with
proper experiments, the utility of the physical and mathematical sciences,
for the improvement of the useful arts, and for the extension of the industry, prosperity, happiness, and well-being of society."

Not content with these developments, Warren and Jackson, strongly
supported by President Kirkland, John Lowell, John Davis, and others,
launched a campaign to provide hospital facilities where medical students

The "Massachusetts Medical College," Mason Street, Boston,
seat of the Harvard Medical School, 1816–1847. Courtesy of
the Francis A. Countway Library of Medicine,
Harvard University

could receive clinical instruction. Once again the Massachusetts legislature, which had contributed handsomely to the cost of the Massachusetts Medical College, lent its support, and wealthy patrons in the Boston community came forward with donations amounting to $300,000. In September 1821 the Massachusetts General Hospital, a handsome building fronting on the Charles River, opened its doors to the public and its wards to the professors of the medical school and their students.

A pamphlet published in 1824, entitled *Some Account of the Medical School in Boston, and of the Massachusetts General Hospital,* described in loving detail the facilities available to students of medicine. On the first floor of the "Medical College" were a spacious lecture hall, an octagonal room for chemical lectures, a chemical laboratory, and a library of 3,000 books. On the second floor were the anatomical theater, large and small dissecting rooms, and an anatomical museum with more than a thousand preparations. Students, the pamphlet announced, were supplied with sets of osteological specimens for study and were admitted free to the surgical operations and clinical practice of the Massachusetts General Hospital. Those who wished to broaden their training were urged to take advantage of the lectures on natural philosophy, botany, anatomy, mineralogy, and chemistry at Harvard and of the botanical garden, the philosophical

The Massachusetts General Hospital, founded 1821, provided Harvard medical professors and students with opportunities for clinical instruction. Courtesy of the Francis A. Countway Library of Medicine, Harvard University

and chemical apparatus, the mineral cabinet, and the wax models available to students in the medical courses. "Probably no place in the United States has collected so many of the useful aids of science, as this ancient and distinguished seminary," the account concluded.[50]

CONCLUSION

Boston had indeed made important progress since 1780 in establishing institutions for the pursuit of science. Instruction in natural philosophy and astronomy at Harvard was equal if not superior to that in any other collegiate institution in the United States. In chemistry, mineralogy, and geology it was surpassed by Yale, but its mineral collections were increasing steadily by purchase and private donations. In natural history it could not yet rival the resources of Philadelphia, but it had a respectable botanical garden and a highly competent and conscientious professor of botany and natural history. Its library was perhaps the best in the nation, and its medical school, hitherto of no great consequence, could now boast an excellent faculty and first-rate facilities. Above all, its alumni had established a tradition of generous giving in support of whatever was needed to make Harvard the best. Surrounding Harvard like flying buttresses were supporting institutions and societies with varying degrees of interest in the sciences. The combined facilities, abilities, and prestige of these institutions, added to the growing preeminence of Harvard as a seat of science and learning, would enable Boston to equal and in some respects surpass Philadelphia as a scientific center by midcentury.

4

Science Along the Hudson

OF the major cities in the United States in the Jeffersonian era, New York experienced the greatest difficulty in establishing lasting scientific institutions. The city had never been famous for devotion to science and learning, as New Yorkers themselves admitted. The explanation, De Witt Clinton suggested, lay in a variety of factors: the practical motives of the original settlers; the debilitating effects of a royal colonial government that tended to alienate the governing classes from the governed; the cultural and linguistic diversity of the population, leading to political and religious factionalism; the absence of provisions for elementary education; the low condition of the professions of law, medicine, and divinity; the disastrous effects of the Revolution; the schisms within the medical community; and the dominance of the commercial spirit.[1]

Yet New York had many enterprising and talented devotees of science in the early years of independence. Among others there were Samuel L. Mitchill, editor of the wide-ranging and highly successful *Medical Repository,* physician, professor, poet, gentleman farmer, chemist, naturalist, ethnologist, state legislator, congressman, and founder or cofounder of most of the city's scientific institutions; David Hosack, initiator of the Elgin Botanic Garden and coeditor with John W. Francis of the *Medical and Philosophical Register* (1810–1814); Archibald Bruce, mineralogist par excellence and editor of the *American Mineralogical Journal* (1810–1814), the first specialized scientific publication in the United States apart from medical journals; John Griscom, the Quaker scientist who maintained a remarkably successful series of public lectures on chemistry and natural philosophy for more than two decades; John Scudder, proprietor of a substantial natural history museum from 1811 on; Robert R. Livingston, politician, diplomat, and president and moving spirit in the Society for the Promotion of Useful Arts; and De Witt Clinton, mayor, state legislator then governor, father of the Erie Canal, adept in geology and natural

91

history, and promoter of science and technology in his native city. Despite the efforts of these and other gifted and energetic men, the city's progress in forming a stable institutional structure for the pursuit of science was halting and erratic.

One of the most important factors producing a scattering of energies in the scientific community in the state of New York was the transfer of the capital from New York City to Albany in 1798. The effects of this can best be seen in the history of the Society for the Promotion of Useful Arts, which began as a New York City institution known as the Society for the Promotion of Agriculture, Arts, and Manufactures and ended as a section of the Albany Institute. The society was organized in 1791 "with a view of collecting from different parts of the State the different modes of Agriculture that are in practice; to suggest such improvements as may be found to be beneficial . . . and to introduce, as far as circumstances may render proper, an emulation for the establishment of useful Arts and Manufactures in those parts of the State, where they can be beneficially carried on."

Incorporated by the state legislature in March 1793, the society included among its charter members many of New York's leading citizens — Robert R. Livingston, John S. Hobart, Samuel Bard, Samuel L. Mitchill, Governor George Clinton, John P. Delancey, the artist William Dunlap, Philip Van Cortlandt, Jeremiah Van Rensselaer, Simeon and Moses De Witt, and other worthies of similar note. Livingston was named president, Hobart vice-president, and Mitchill and Samuel Jones secretaries. Meeting in the senate chamber of the city hall in December 1792, the members heard a report on methods of cultivating summer barley, examined a model of a cast iron ploughshare, and heard Mitchill communicate memoirs on the culture of daisy-grass, the rejuvenation of impoverished soils, the preservation of hedges, and a method of catching porpoises and manufacturing their skins into leather.[2]

As the foregoing list of topics indicates, the purposes of the society were highly practical. Agriculture in particular was to be the focus of attention in accordance with the preoccupations of gentlemen farmers like Livingston, Ezra L'Hommedieu, Van Cortlandt, Van Rensselaer, and Mitchill himself, who owned a farm on Long Island. Science was to be admitted only to the extent that it served to advance the practical arts. In Mitchill's mind, however, there was no doubt that chemistry, geology, mineralogy, and other sciences were essential to the progress of agriculture and industry. Robert Livingston was of the same opinion, and the society was soon engaged in projects reflecting the convictions of its president and secretary. In 1796 Mitchill was commissioned to make a geological survey of New York City and its vicinity. For five or six weeks he ranged the country from Long Island to the Catskill Mountains, clambering over the rocks along the coast, studying the rocky palisades along the west bank of the Hudson ("climbing aloft, or walking below, or examining them with a goody spying glass, as he sailed slowly by them"), carefully tracing the termination of the granitic formations and the beginning of the schist near New Cornwall, noting the strike and dip of these and the limestone formations in the highlands, collecting fossils in the latter, and ascending the sandstone ranges of the "Blue Mountains" behind Kingston. The result was the first geological survey of a region of the United States by an American. It was published in part in the society's *Transactions* and more fully in Mitchill's *Medical Repository* (see Chap. 9). In his report to the legislature of this survey Mitchill acknowledged that he had failed to find the coal deposits he was looking for, but he consoled himself and the legislators with the thought that he had discovered "some new and important facts . . . which throw great light upon the history and theory of the earth."[3]

In 1793 the society was forced by the terms of its charter to move to Albany when the state capital was transferred to that city. Provided with meeting quarters in the new capitol, the society reorganized itself in 1804 as the Society for the Promotion of Useful Arts and continued to devote itself primarily to practical concerns, including the awarding of premiums for wool cloth and other domestic manufactures with funds supplied by the state legislature, to whose sessions the meetings of the society were tied. The scientifically inclined members of the society continued to press the claims of chemistry, geology, and mineralogy, however, and in 1810 the society issued a circular announcing its intention to collect and preserve specimens of earths, fossils, and metallic ores from all parts of the state. In 1813 the society established a Committee on Chemistry, charged with promoting the study of mineralogy, metallurgy, dyeing, tanning, brewing, baking, malting, glass making, pottery, "and all the Arts connected with the science of chemistry." The committee published a circular urging the cities and villages along the Hudson River to appropriate funds for a geological survey of eastern New York.

By this time the society, though still concerned more with practical utility than with "recondite science," was placing less emphasis on agriculture than formerly and more on "the perfection of those arts depending on chemical processes." By 1819, however, the legislature had transferred responsibility for awarding premiums to the State Board of Agriculture, and the society was in decline. It was saved from extinction only when the newly formed Albany Lyceum of Natural History agreed in 1824 to merge with the Society for the Promotion of Useful Arts to form the Albany Institute. At the end of its evolution the society had lost its connection with scientific circles in New York City and had become part of a nascent rival scientific center in the Albany-Troy region. Besides the institute the new center included the Albany Academy, organized in 1813 by Stephen Van Rensselaer and others; the Troy Lyceum of Natural History, founded and energized by the naturalist and geologist Amos Eaton; and the Rensselaer School, forerunner of Rensselaer Polytechnic Institute and headed by Eaton under the patronage of Stephen Van Rensselaer.[4] The institutional development of science in New York State had become polarized at opposite ends of the Hudson River.

COLUMBIA COLLEGE, THE ELGIN BOTANIC GARDEN, AND THE COLLEGE OF PHYSICIANS AND SURGEONS

The tangled history of the relations among Columbia College, the Elgin Botanic Garden, and the College of Physicians and Surgeons illustrates another set of factors influencing the growth of scientific institutions in New York City: individual initiative, factionalism in the medical community, and the vagaries of legislative support for science and learning. Columbia, like Harvard, Yale, and the University of Pennsylvania, had a history reaching back into the colonial period. It had a medical school founded in 1767 and a small but able medical faculty. With the appointment of William S. Johnson as president in 1787, the college shook off the effects of the British occupation during the revolutionary war and began to move ahead. Financial support from the state legislature made possible the completion of the college building, enlargement of the library, establishment of new professorships, and purchase of scientific apparatus. The reopening of the New York Hospital in 1791 made clinical instruction available to medical students.

Visiting the campus in 1787, Manasseh Cutler found a single three-story stone building occupying a pleasant site close to the Hudson River not far from the city limits. In front of the building was a spacious square enclosed with a high fence. The college library on the second floor was in rather sorry condition owing to the pillaging of British troops. Over the library was the philosophy chamber where John Kemp lectured on mathematics and natural philosophy with the aid of a new and elegant philosophical apparatus consisting of a small reflecting telescope, a compound microscope, a camera obscura "on a new plan," a complete set of "mechanical powers," a set of celestial globes, and a machine for generating static electricity. There were about 150 students in the college at this time.[5]

In 1792, in response to a petition from the Society for the Promo-

Columbia College's only building before 1820 and portraits of two of its professors of natural history: David Hosack (by John Wesley Jarvis) (left) and Samuel Latham Mitchill (by James Sharples) (right). Courtesy of the New-York Historical Society

tion of Agriculture, Arts, and Manufactures, the legislature granted funds to establish a professorship of natural history, chemistry, agriculture, and "the other Arts depending thereon." Samuel L. Mitchill was appointed to the post, with instructions to teach "the Philosophical Doctrines of Chemistry and Natural History" under the following heads: "1. Geology, or the natural and chemical History of the Earth; 2. Meteorology, or the natural and chemical History of the Atmosphere; 3. Hydrology, or the natural and chemical History of Waters; 4. Mineralogy, or the natural and chemical History of Fossil Substances; 5. Botany, or the natural and chemical History of Plants; 6. Zoology, or the natural and chemical History of Animals." The course was to be coordinated with the offerings of the professor of mathematics and mechanical philosophy so as to form "a complete set of doctrines and facts in the department of experimental Physics."[6] Probably Mitchill and John Kemp, professor of mathematics and natural philosophy, worked well together, for Kemp, like Mitchill, stressed the utilitarian value of science. "In his lectures," James Renwick reported, "he prided himself on giving every subject a practical bearing, and thus the principles and history of canal navigation formed a favourite theme."[7] One of Kemp's first students was De Witt Clinton, father-to-be of the Erie Canal.

As for Mitchill, he described his lectures as being "not only the classification and arrangement of natural bodies, but also . . . a great variety of facts which form the basis of Medicine, Agriculture, and other useful arts, as well as of manufactures." The course was required for medical students but could be attended by undergraduates and even by persons outside the college. It was illustrated, said Mitchill, by a handsome apparatus and a considerable collection of fossils. In the summer Mitchill taught botany to medical students, introducing them to the Linnaean classification and nomenclature and to plant anatomy, physiology, and economy, particularly "as connected with GARDENING and FARMING." The chief need for improvement of botanical instruction, Mitchill noted, was a botanical garden; this was conveyed to the state legislature by the Society for the Promotion of Agriculture, Arts, and Manufactures.[8]

The botanical garden project was taken up with even greater enthusiasm by young David Hosack, who replaced Mitchill in the newly established professorship of natural history, chemistry, and agriculture in 1795. A native New Yorker, Hosack had studied medicine in Philadelphia, Edinburgh, and London. In London he had received private instruction in botany from William Curtis, editor of the *Botanical Magazine*. He had also attended the public lectures of James E. Smith, president of the Linnean Society of London and owner of Linnaeus's herbarium and library. Taking pains to acquire a chemical apparatus, an extensive library, a collection of minerals, and a herbarium that included some Linnaean specimens, Hosack returned to the United States in 1794, resolved to make himself the James E. Smith of American botany. In the summer of 1795 he taught his first class, a group of sixteen medical students and a few others, among them his friend and former classmate De Witt Clinton.

Hosack was determined that Columbia should have a botanical garden. When the college authorities pleaded lack of funds, he turned to the legislature, enlisting the help of Clinton (then a state senator), but without success. Undaunted, Hosack decided to establish his own botanical garden, hoping for eventual public support. On September 1, 1801, he bought twenty acres of land in that part of Manhattan Island now occupied by Rockefeller Center (then more than three miles outside the city), paying $4,807.36 and a yearly quitrent of sixteen bushels of grain for the site. Within a few years, Hosack had invested thousands of dollars in buildings, fences, personnel, clearing, and cultivation and had made his Elgin Botanic Garden an important scientific establishment. In 1806 the *American Medical and Philosophical Register* reported, the garden had "three large and well constructed houses, exhibiting a front of one hundred and eighty feet."

> The greater part of the ground is brought in a state of the highest cultivation, and divided into various compartments, calculated for the instruction of the student of botany and medicine, and made subservient to agriculture and the arts. The whole establishment is surrounded by a belt of forest trees and shrubs, and these again are enclosed by a stone wall.[9]

Seeds and plants for the garden were obtained from a great variety of sources. Martin Vahl of Copenhagen was a regular correspondent and contributor. In Paris the chief gardener of the Jardin des Plantes, André Thouin, who had been supplying Thomas Jefferson with seeds for many years, now added Hosack to his list of botanical beneficiaries, supplying him with "the rarest and most valuable plants of the continent," which, Hosack noted, "have always been received in such a state of preservation, as scarcely in a single instance to have frustrated the liberal intentions of the donor." Mitchill contributed living plants from the New York vicinity and dried specimens of plants he had collected around Washington, D.C., while serving in Congress; Hosack's assistants and pupils, including John Eddy, Caspar W. Eddy, John E. Le Conte, and James Inderwick, added their contributions. Le Conte's brother Lewis supplied native plants from Georgia. From 1809 to 1811, Frederick Pursh, an excellent botanist who had formerly been employed at the royal gardens in Dresden and William Hamilton's garden in Philadelphia, was placed in charge of the Elgin Botanic Garden by Hosack and proceeded to increase considerably the number of native American plants growing there.[10]

Foreign visitors were favorably impressed with Hosack's garden, herbarium, and botanical library. François Michaux found the garden site well adapted for its purpose, "especially for plants that require a peculiar aspect or situation." Alire Raffeneau-Delile, a French botanist living with the Hosacks, sent a copy of his host's catalogue of the garden to a friend in Paris in 1806, remarking that he and Michaux often met each other in the "beautiful library" of Dr. Hosack, "who shares his books and col-

*Engraving of Hosack's Elgin Botanic Garden (by Leney),
which played an important part in the development of botany
in New York. Courtesy of the New-York Historical Society*

lections with the same liberality as do Messrs. Desfontaines and de Jussieu
in Paris." The garden, he added, contained a great variety of plants,
especially trees, from distant parts of the United States, besides many exotic plants such as the Asiatic honey locust, a Chinese flowering apple,
and a gold-dust tree.[11]

Hosack gave a series of lectures every summer, discoursing on the
structure and functions of plants; comparisons of plants and animals; and
the classification, anatomy, physiology, and chemical analysis of plants
and their nutrition, reproduction, and cultivation. He closed with a history
of botany and botanical classification and instructions for forming a herbarium. A student who attended some of Hosack's medical lectures described him as "but little above the ordinary size, a little inclined to corpulency, quick in motion, dark complexion, a piercing black eye, black
hair and very heavy black eyebrows capable of assuming a withering frown
. . . a loud, clear, sonorous voice with very distinct articulation. . . . His
delivery was emphatic and at times truly eloquent. . . . He always seemed
thoroughly to understand his subject and appeared familiar with the whole
range of medical science. . . ."[12]

John W. Francis, a pupil of Hosack's and later coeditor with him
of the *American Medical and Philosophical Register,* recalled with especial

pleasure the strawberry festival with which Hosack terminated his botanical course. Linnaeus, Hosack told his students, had cured his gout and lengthened his life by eating strawberries. "The disciples of the illustrious Swede must have a foretaste of them, if they cost me a dollar apiece."[13]

By 1806 Hosack was well on his way to satisfying his wish of becoming the James E. Smith of America.

> Surrounded by his large and costly library, his house was the resort of the learned and enlightened from every part of the world. No traveller from abroad rested satisfied without a personal interview with him; and, at his evening soirées, the literati, the philosopher, and the statesman, the skillful in natural science, and the explorer of new regions; the archaeologist and the theologue met together, participators in the recreation of familiar intercourse.[14]

The Elgin Botanic Garden was a great success and, though privately owned, an important facility for the teaching of students at Columbia. Unfortunately, however, Hosack's income as physician and professor was not sufficient to sustain the enterprise. After trying unsuccessfully to arrange a state loan, Hosack suggested that the state buy the property as an adjunct to medical education in New York City, hoping that it would be assigned to the medical school of Columbia and funded by the state. In March 1810 the legislature authorized the purchase, and in January of the following year the garden passed into the custody of the Regents of the State of New York. But the regents did not assign it to Columbia for reasons connected with the factionalism that had developed within the medical community in the city.

The medical school of Columbia College had developed very slowly, partly because state law did not require a medical degree for a license to practice medicine and partly because of quarrels among the faculty. Enrollments were small, the building was dilapidated, and the unimpressive library was housed at the New York Hospital. Faculty morale was low. In 1807 Nicholas Romayne and several other physicians, disappointed by the school's slow growth, secured legislative approval and financial support for a rival medical school to be known as the College of Physicians and Surgeons. Romayne was named president; Mitchill, vice-president and professor of chemistry; Edward Miller (coeditor with Mitchill of the *Medical Repository*), professor of the practice of physic and lecturer on surgery and midwifery; Archibald Bruce, registrar and professor of mineralogy; Benjamin De Witt, professor of the institutes of medicine and lecturer on chemistry; and John A. Smith, adjunct lecturer on anatomy. Hosack was one of the promoters of the new venture, but he withdrew when he failed to get the professorship he had hoped for. For six years there was factional struggle within the new medical school and between it and the Columbia medical faculty. Finally in 1813 the legislature merged the two medical faculties to form a single College of Physicians and Surgeons independent of Columbia. Not until the 1860s did it become a part of, by then, Columbia University. Deprived of its medical school, Columbia could not hope to rival Harvard and the University of Pennsylvania as a center of scientific studies.

Once formed, the College of Physicians and Surgeons rapidly developed into an institution of scientific importance worthy of comparison with the medical schools in Philadelphia and Boston. Enrollments rose steadily, reaching nearly 200 in 1816–1817. In September 1817 a three-story building in an adjoining lot was added to the facilities. Jacques Milbert, a visiting Frenchman, was surprised to find "so adequate a building, accommodating a chemical laboratory, a library, a mineralogy room, an anatomical amphitheater and a large hall used by various professors lecturing to their students."[15] To these appurtenances and the botanical garden were added the facilities of the New York Hospital and the Lying-In Hospital, where the professors (most of them connected with one or both) were free to bring their students. The New York Hospital, chartered in 1771, stood at the square where Broadway, Barley, Church, and Catharine streets intersected. In 1811 its library was increased substantially by the purchase of Hosack's botanical books.

A circular issued in 1817 described the courses offered at the college. Besides the lectures on purely medical subjects by Hosack, Valentine Mott, Wright Post, James Stringham, John W. Francis, and William Hammersley, there were lectures on natural history by Mitchill and on chemistry by William J. MacNeven, an Irish refugee educated in Vienna. MacNeven introduced his students to John Dalton's atomic theory and gave them practice in the chemical analysis of minerals in his well-equipped laboratory (see Chap. 7). Hosack no longer lectured on botany, but Mitchill traversed the entire range of natural history in his lectures, displaying (according to Francis) "a wholesome natural theology, blended somewhat after the manner of Paley . . . and an abundance of patriotism, associated with every rich specimen of mineral wealth."[16]

To illustrate his lectures on botany and zoology, Mitchill took advantage of the newly acquired Elgin Botanic Garden and John Scudder's museum of natural history. The museum, launched by Scudder in 1811 while he was still a student at the College of Physicians and Surgeons, proved highly useful for scientific instruction. In the pages of the *Medical Repository* Mitchill praised Scudder's animal preparations, his white bear of Greenland, Hindustan tigress; collection of American quadrupeds; water pelicans, gannets, oyster crackers, loons, and owls; and collections of insects, zoophytes, worms, snakes, shells, and minerals. In the Botanic Garden, Mitchill demonstrated the component parts of flowers and ex-

View of City Hall Park (1825) showing Scudder's American Museum in the Old Almshouse (rear left) and portrait of John Scudder (artist unknown). Courtesy of the Museum of the City of New York (City Hall Park) and the New-York Historical Society (Scudder)

pounded the principles of the Linnaean system of classification, while two of Hosack's botanical protégés, Caspar W. Eddy and James Inderwick, indicated to the students the characters peculiar to different species and genera.[17]

Through the teaching of Mitchill, MacNeven, and (in earlier days) Hosack the College of Physicians and Surgeons became an important training ground for a considerable number of able chemists, mineralogists, botanists, and zoologists, including John Torrey and the Beck brothers— Lewis, Theodric, and John. The Elgin Botanic Garden, unfortunately, fell into ruin through lack of funds to maintain it. In 1814 the legislature transferred it to Columbia in response to the college's request for financial support, but the garden continued to decline. By the time Columbia acquired the College of Physicians and Surgeons a half-century later, the garden had disappeared, but the site itself was eventually to become one of the university's most valuable assets as New York City expanded northward.[18]

FURTHER EFFORTS AT ORGANIZATION

In New York as elsewhere, the patriotic fervor generated by the War of 1812 stimulated new undertakings in many fields of culture, including science, but once again the scattering of energies blunted the effect of civic effort. In 1816 the New-York Historical Society, organized in 1804 and dedicated to both civil and natural history, commenced a cabinet of natural history, established lectureships, and appointed committees to gather information and specimens in their various departments of science. Mitchill was made chairman for zoology and geology; Hosack, for botany and vegetable physiology; Griscom, for chemistry and natural philosophy; and George Gibbs, for mineralogy. Mitchill began laying plans for a zoological museum, and Gibbs conceived the idea of displaying minerals from every state in the Union in specially designed cases, "one case being devoted to each state, after the manner adopted in the national collection at the École des Mines at Paris." As the beginning, John G. Bogert contributed his ten-year collection of minerals and shells. Samuel Bard offered his collection of Scottish plants, and Hosack, who had given his herbarium to the New York Hospital when it bought his botanical library, managed to retrieve the herbarium for the Historical Society. This collection contained a considerable number of specimens from Linnaeus's own collections, given to Hosack by James E. Smith during Hosack's studies in London. Hosack also presented a bust of Smith to the society, to be placed over the Linnaean specimens in the natural history room.[19]

Meanwhile, the long-delayed project of forming a general scientific and literary society comparable to those in Philadelphia and Boston finally took shape. In March 1814 the state legislature incorporated the Literary and Philosophical Society of New York. De Witt Clinton was president, Hosack one of three vice-presidents, and Mitchill one of two corresponding secretaries. John W. Francis and John Pintard were curators; Archibald Bruce, William J. MacNeven, Wright Post, and Robert Adrain were among the counselors. The full roster of charter members included many of New York's leading figures. The jurist James Kent was listed, as were Robert Fulton, John S. Hobart, Rufus King, Cadwallader Colden, Samuel Bard, Gouverneur Morris, Simeon and Benjamin De Witt, and many others. The society was to meet twice a month to hear communications in four fields of inquiry: belles-lettres, civil history, antiquities, moral and political sciences; medicine, chemistry, natural philosophy, and natural history; mathematics, astronomy, navigation, and geography; and husbandry, manufactures, and the useful arts. Curators were appointed to take care of natural history specimens and models of instruments and machines.

In practice, however, the Literary and Philosophical Society was dominated by devotees of natural history, the tone being set by Clinton's "Introductory Discourse" with its voluminous footnotes on the geographical, geological, paleontological, botanical, zoological, and ethnological literature of North America. Only one full volume of *Transactions* was published during the society's decade and a half of existence. Its contents were contributed almost entirely by medical men — Mitchill on earthquakes and the fishes of New York, Hosack on fishes and contagious diseases, Francis on morbid anatomy, MacNeven on the chemical analysis of mineral waters, and Hugh Williamson on comets. The meetings, held at first in Mayor De Witt Clinton's office, were largely taken up with communications of a similar kind.[26]

Despite the attention paid to natural history in the New-York Historical Society and the Literary and Philosophical Society, Mitchill and his fellow enthusiasts felt the need for an organization devoted exclusively to their favorite subject. In January and February 1817 several of them held a series of meetings at the College of Physicians and Surgeons to discuss "measures for instituting a *Cabinet of Natural History.*" The outcome was the formation of the Lyceum of Natural History of New York, named after Aristotle's illustrious research organization. "Disciples of the 'mighty Stagirite',," Mitchill explained, "they determined, after his example, to be Peripatetics, and to explore and expound the arcana of nature as they walked." Mitchill himself was president. Professors, students, and graduates of the College of Physicians and Surgeons were the mainstay of the society, but their efforts were ably supported by naturalists like Frederick C. Schaeffer; botanist John E. Le Conte; mineralogists James Pierce and Amos Eaton; and Constantine Rafinesque, a French naturalist befriended by Mitchill after being shipwrecked off New London in 1815 and losing the entire collection of books and specimens he was bringing to the United States after ten years of research in Sicily.

Field trips were organized, and the collections of the Lyceum began to accumulate rapidly. As early as August 1817 the *Medical Repository* reported that the Lyceum had acquired a "fossil mastodon," a right whale, a swordfish, new kinds of carp and pike, and a mountain sheep from the Rocky Mountains. It had also begun a catalogue of plants growing within 100 miles of New York City and sent traveling committees to the Fishkill and Catskill Mountains to study the natural history of these regions. As a result, the Lyceum's herbaria were "replete with undescribed plants." "The members are chiefly in the vigour of life, and ardent in the pursuit of natural science," the report concluded.[21]

Fortunately for the Lyceum, New York City had by this time begun to provide accommodations for its proliferating learned societies and their collections. When the Government House, which had housed the New-York Historical Society, the American Academy of the Fine Arts, and the Literary and Philosophical Society, was torn down in 1815, the literati of the city, led by John Pintard and De Witt Clinton, persuaded the city council to grant the use of the Old Alms House, a three-story brick building on the north side of City Hall Park, rent-free for ten years to a group of organizations to be known collectively as the New York Institution of Learned and Scientific Establishments. In 1816 the paupers were moved from the Alms House to Bellevue, and the new tenants moved in. The Literary and Philosophical Society was among the first to occupy the new quarters. Its spacious chamber was soon decorated with gypsum busts of Franklin and Caspar Wistar; a painting of Mitchill by Samuel Jarvis; and prints of Isaac Newton, Erasmus Darwin, Joseph Warren, and other worthies.

The Historical Society had a suite of rooms, including one for its books and manuscripts and two others for cabinets of mineralogy, botany, and zoology. The Lyceum of Natural History had only one room at first, but it subsequently expanded to four when John Griscom failed to renew his lease on the rooms where he gave his lectures on chemistry and natural philosophy. The west end of the building was occupied by Scudder's Museum, which opened in its new quarters on July 2, 1817, with a brilliant display of exhibits. "It was the opinion of several gentlemen present [at] the first exhibition," wrote the editors of the *American Monthly Magazine,* "that neither London or Paris, which they had visited, possessed specimens in such high state of preservation; and that as he already excelled in the preparation, he would soon exceed in the number of his subjects, and the extent of his Museum, any similar establishment." The Academy of the Fine Arts was the only organization without any natural science orientation in the New York Institution, but Clinton managed to introduce a technological motif by discoursing at length on Fulton's steamboat in his address welcoming the academy to its new quarters.[22]

With the founding of the New York Institution of Scientific and Learned Establishments, New Yorkers began to feel that their city was

throwing off its reputation for neglect of science, art, and letters. The poet
Fitz-Greene Halleck took up his pen:

> To bless the hour the Corporation took it
> Into their heads to give the rich in brains
> The worn-out mansion of the poor in pocket,
> Once "the Old Almshouse," now a school of wisdom
> Sacred to Scudder's shells and Dr. Griscom. ["Fanny" 1821]

In a soberer vein, David Hosack, acknowledging that New York City's
literary and scientific reputation had hitherto not been commensurate with
the advantages the city's central situation in the United States and preeminence in foreign commerce seemed to warrant, went on to assure the
students of the College of Physicians and Surgeons that the founding of
new scientific and learned societies and their accommodation in the New
York Institution had "justly elevated the city of New-York to that rank
which in other respects she has long enjoyed."[23]

There had indeed been an intellectual renaissance in New York in the
second decade of the nineteenth century, as the proliferation of societies
and journals showed. Archibald Bruce's *American Mineralogical Journal* and the *American Medical and Philosophical Register* of David Hosack
and John W. Francis expired in 1814, but H. Biglow's *American Monthly
Magazine and Critical Review,* founded in the spring of 1817, provided
a temporary outlet for the proceedings and papers of New York's learned
societies. For two years Biglow's magazine carried fairly detailed accounts
of the activities of the organizations housed in the New York Institution.
Papers by Mitchill and Rafinesque were especially numerous, and
Rafinesque edited a section of the magazine entitled "Museum of Natural
Science," in which readers were kept abreast of researches by Rafinesque
and his fellow naturalists.

Despite these encouraging developments, the institutional base of
science in New York City was still rather shaky as the decade ended.
Biglow's magazine collapsed in 1819; the Historical Society turned increasingly to civil history, donating its natural history collections to the Lyceum
of Natural History; and the Literary and Philosophical Society lost its
original impetus with the formation of the Lyceum. Only the Lyceum,
known today as the New York Academy of Sciences, survived into the
twentieth century. The members were enthusiastic in their researches and
even managed to start their own journal in 1823. Unable to attract donations of the kind that supported the Academy of Natural Sciences of
Philadelphia and the Boston Society of Natural History, the Lyceum never
acquired a building of its own to house its collections, but the members
were fully as active in research and publication as those of the Boston
and Philadelphia societies.[24]

CONCLUSION

In the final analysis it was the full array of institutions along the Hudson River that proved significant for the future development of American science. In New York City, Columbia began to offer better instruction in the physical sciences with the appointment in 1813 of the gifted Robert Adrain as professor of mathematics and astronomy and of James Renwick as instructor in natural and experimental philosophy and chemistry. Natural history flourished at the Lyceum, and the College of Physicians and Surgeons continued to give good training in chemistry and mineralogy as well as in medical subjects. Farther up the Hudson River the United States Military Academy at West Point, reorganized along the lines of the École Polytechnique in 1817, began to give solid instruction in engineering and the related sciences. By the 1830s the Rensselaer School was doing likewise and turning out able geologists and mineralogists as well. At the Albany Academy, Joseph Henry began the researches that were to lead him to the discovery of electromagnetic induction in 1831, the Albany Institute brought together the cultivators of both physical and natural sciences, and Amos Eaton undertook a geological survey of the Erie Canal route that was to have important consequences for the development of American geology (see Chap. 9). All in all, the prospects for American science in New York State were much better as the Jeffersonian era drew to a close than they had been at its inception.

5

Outposts of Science
in the South and West

IN America as in Europe science developed in the cities, but these were few and far between in the southern and western regions of the new nation. In the South, Charleston was the largest city in 1780, and it was not without some pretensions as a center of scientific activity. There Mark Catesby, the English naturalist whose *Natural History of Carolina, Florida, and the Bahama Islands* established a vogue for handsomely illustrated natural history folios, had lived and worked from 1722 to 1725. There too, Alexander Garden, physician-naturalist and correspondent of Linnaeus, Gronovius, and other European botanists, had carried on his researches in natural history from 1753 to 1783. Garden was but one of an able group of Scottish physicians—Lionel Chalmers, John Moultrie, John Lining, and others—who settled in Charleston in the mideighteenth century and pursued a variety of scientific interests ranging from natural history to studies of yellow fever, meteorology, and human metabolism.[1]

In 1737 St. Philip's Hospital was founded, in 1748 the Charles-Town Library Society, which served as a meeting place for physicians, merchants, and planters interested in literature and science. The society developed rather ambitious plans, including establishment of a public library, endowment of an academy "to encourage and institute youth in the several branches of liberal education," sponsorship of lectures on experimental natural philosophy, and formation of a museum of natural history.

In 1773 the museum project took shape with the announcement in the *South Carolina Gazette* that a committee consisting of Charles C. Pinckney, Thomas Heyward, Jr., Alexander Baron, and Peter Fayssoux stood ready to receive on behalf of the society "all the natural Productions, either Animal, Vegetable, or Mineral, that can be had in their several Bounds, with Accounts of the various Soils, Rivers, Waters, Springs, &c. and the most remarkable appearances of the different Parts of the Country."[2] In 1774 the society agreed to pay David Rittenhouse £350 to construct an orrery similar to the one he had built for the College of Philadelphia and offered to pay his expenses in coming to Charleston to install it.

In that same year David Ramsay, a young Pennsylvania physician recently graduated from the School of Medicine in Philadelphia, arrived in Charleston bearing a highly flattering recommendation from his idol and teacher Benjamin Rush, who declared him to be "far superior to any person we ever graduated at our college." Tall, handsome, and athletic, Ramsay had a reputation for universal knowledge and a prodigious capacity for work. Within a year he had been admitted to the prestigious Charles-Town Library Society. It was he who delivered the flamboyant "Oration on the Advantages of American Independence" before the assembled notables of Charleston on July 4, 1778, predicting that the politicians and warriors of Britain's rebellious colonies would, when independence had been achieved, "engage in the enlargement of publick happiness, by cultivating the arts of peace, and promoting useful knowledge, with an ardor equal to that which first roused them to bleed in the cause of liberty and their country."

But Ramsay's prophecies and the society's projects were soon shattered by hard realities. Six months before Ramsay ascended the orator's rostrum, a fire had destroyed the society's library and natural history collections. In May 1780 British troops entered Charleston after bombarding it by sea and land. Ramsay and nearly forty patriots were arrested and sent to St. Augustine, Fla., for detention. Not until December 1782 did the British troops depart, taking 4,000 Loyalists and more than 5,000 slaves with them and leaving the remaining inhabitants to rebuild their war-torn city and its institutions as best they could.

The work of reconstructing and adding to the city's literary and scientific institutions went slowly. The Library Society revived soon after the war and began replacing its collections of books and natural history specimens, but it did not develop into a scientific society. The College of Charleston, with an associated high school–level academy, was chartered by the state legislature, but twenty-five years later Ramsay conceded to Jedidiah Morse that it was "only a tolerably decent grammar school, . . . without funds and almost without pupils."[3] South Carolina College began operations in Columbia in 1805, thanks to the efforts of Governor John Drayton and others, but scientific subjects did not gain a firm hold there until after the arrival of Thomas Cooper in 1820. South Carolina was without a medical school until the Medical College of South Carolina was established in 1824.

In the absence of scientific societies, a medical school, and a vigorous college, the work of promoting scientific studies fell primarily on a handful of physicians and planter-politicians, aided from time to time by visiting naturalists attracted to South Carolina by its rich vegetation. Among these visitors none was more important than André Michaux, a botanical emissary of the French government. He arrived in Charleston in 1786 with his fifteen-year-old son François and established a garden and tree nursery about ten miles from the city, from which he shipped thousands of plants and seeds to France for the royal gardens and forests. He soon became acquainted with the Carolina planters (especially Pinckney, after whom he named the genus *Pinckneya*) and urged them to cultivate useful Old World plants. He continued his botanical explorations in Carolina and as far west as Kentucky until 1793, when he returned to France, taking his son with him and leaving his botanical establishment in the care of a neighbor.[4] Both Michaux and his son, who returned to America in 1801 and traveled widely in search of new forest trees, were to make important contributions to American botany (see Chap. 10).

Among the South Carolina planters there were three who showed a marked interest in science for its own sake during the Jeffersonian period: Thomas Walter and John Drayton in the early years of independence and Stephen Elliott somewhat later. Arriving in South Carolina from England shortly before the Revolution, Walter began work on a *Flora* of the region. He established a botanical garden on his plantation on the Santee River and planted more than 600 species of plants collected within a radius of fifty miles. He was greatly aided in this enterprise by a visiting Scottish nurseryman, John Fraser, who ranged the countryside as far as the foothills of the Appalachians, engaging in a friendly botanical rivalry with Michaux. Fraser added several hundred plants to Walter's garden before returning to England in 1788, carrying Walter's herbarium and the manuscript of his *Flora Caroliniana* with him. Published in London in 1788, Walter's *Flora* was the first such regional work by an American.[5]

Drayton was the first planter who attempted to follow in Walter's footsteps. With loving care he put together his own "Carolinian Florist," based primarily on Walter's work but drawing also on the publications of Benjamin S. Barton, William Bartram, André Michaux, and others. An early version of the manuscript, illustrated with watercolors, was presented to the Charleston Library Society. Drayton continued to work on the project in the midst of his duties as governor of South Carolina and in 1807 presented a second manuscript to the library of the College of South Carolina, where it lay unpublished until 1943.[6] Much of the information contained in it, however, was included in Drayton's *View of South Carolina* (1802) and in David Ramsay's *History of South Carolina* (1809). It must also have proved useful to Stephen Elliott when he began work on his *Sketch of the Botany of South Carolina and Georgia* (1816–1824) (see Chap. 10).

For most South Carolina planters, however, interest in science was confined to its bearing on agriculture. This interest led a group to gather in Charleston's Assembly Hall in 1785 to organize the South Carolina Society for Improving Agriculture and other Rural Concerns, later known as the Agricultural Society of South Carolina. The organizers included Thomas Heyward, Jr., Charles C. Pinckney, Thomas Pinckney, John and Edward Rutledge, William Drayton, and other leading planters. They levied stiff membership fees and agreed to meet in Charleston every third Monday in August, November, February, and May for dinner, business, and an occasional oration. Thomas Jefferson was elected an honorary member, and he was soon sending grape seeds, acorns of the cork oak, and cuttings from olive trees from France and urging experiments in viniculture and olive production. "Cotton and olives have lately engaged the attention of our farmers and planters," David Ramsay reported to Benjamin Rush, "the latter on the recommendation of Mr. Jefferson."[7]

For a while it seemed as if the Agricultural Society might play a role in South Carolina science similar to that played in New York by the Society for the Promotion of Useful Arts. Ramsay reported in 1809 that the society had conducted a successful lottery and bought forty-two acres of land in the vicinity of Charleston "in which agricultural experiments are occasionally made," but, according to the historian, little was done with this land on Charleston Neck except to build a large wooden building where the society could meet. Its members, including bankers, brokers, merchants, and professional men as well as planters, met several times a year to vote premiums for improved techniques of producing cotton, rice, cottonseed oil, olive oil, and the like. Apparently there were more ambitious activities as well. The "Address" to the public in 1821 announced that the society, "having surmounted the derangement of their finances, occasioned by an abortive attempt to establish an extensive vineyard in the interior of the state, and having disposed of their pattern farm (the occupation of which had been also prejudicial to their finances), and fixed their meetings at the house on the Old Race Ground, . . . have determined to apply their resources, in future, to other more feasible . . . objects of their association."[8]

In these agricultural endeavors there seems to have been little of the scientific spirit that infused the activities of the Society for the Promotion of Useful Arts, the Albany County Agricultural Society, and similar organizations in New York. The Carolinians knew of the activities of these societies, however. In their report for 1819 the curators quoted at length a pamphlet entitled "An Examination into the Expediency of Establishing a Board of Agriculture in the State of New York," substituting "South Carolina" for "New York" wherever that name occurred. "The sun of science," read one passage, "illumines our paths. The wonderful advances in chemistry, mineralogy, geology, and botany, which mark the present period with lustre and glory, cannot fail of leading to investigations and experiments in the art of agriculture." Then followed figures on New York's expenditures in support of education, medical science, botany, and natural history. South Carolina, the curators seemed to be saying, had fallen behind its northern sisters in providing a scientific base for agriculture.[9]

Meanwhile, the Charleston physicians had been doing their part to promote the development of science in the city. On Christmas Eve 1789 eleven, including Alexander Baron, Peter Fayssoux, and David Ramsay, organized the Medical Society of South Carolina. The doctors were primarily concerned with medical practice and problems, but they also showed an interest in meteorology and natural history in keeping with the tradition established by John Lining, Lionel Chalmers, and Alexander Garden in colonial times. Inspired by the example of the Chemical Society of Philadelphia, they invited the citizens of South Carolina to send in mineral specimens to be forwarded to Philadelphia for analysis. Benjamin Simons enjoyed a short-lived popularity with public lectures on chemistry. Both he and Ramsay urged without success the establishment of medical lectures for the young men of Charleston, who, said Simons, "amuse themselves in taverns and places of much less respectability."[10]

The most ambitious project of the Medical Society was the establishment of a botanical garden on land donated in 1805 by Richard Savage's widow for that purpose. The donation was timely, for by this time the garden and tree nursery established by André Michaux and his son in 1786 had been sold and was in decline. A Botanic Society was formed by public subscription, and a circular was sent to other states describing the plans and purposes of the institution. In 1808 Ramsay reported that the garden was flourishing and was "enriched with a considerable number of plants, both indigenous and exotics, arranged according to the Linnean system," and that "additions are constantly making to it by the citizens and from foreign countries." To the experienced eye of John Lyon, formerly in charge of William Hamilton's garden in Philadelphia, the garden seemed less than impressive. Visiting it with the Ramsays and Stephen Elliott in February 1809, Lyon noted in his journal that it was "a poor and insignificant ill conducted thing."[11] Financial difficulties were soon encountered, and all efforts to resolve them failed, including a lottery and the sale of the original garden plot in favor of a more extensive one with better soil. The subscribers lost interest, the society dissolved, and by 1819 the site was "reduced to an equality with the surrounding farms." Michaux's old garden also had reverted to a state of nature.[12]

With the failure of the projects initiated by the Medical Society and the Agricultural Society, the prospects for science in South Carolina looked bleak, but its advocates did not give up easily. Charles D. Simons formed a Philosophical Society in Charleston in 1809 with some promise of success, but it collapsed when he moved to Columbia to assume the chair of natural philosophy at South Carolina College and was drowned shortly thereafter. Undaunted by this setback, John Linnaeus Edward W. Shecut, author of an abortive *Flora Carolinaaensis: Or, a Historical, Medical, and Economical Display of the Vegetable Kingdom* (1806), took up the task of organizing the literati of Charleston into an effective society. Shecut was an enterprising physician who had studied with Ramsay and later under Benjamin Rush in Philadelphia, though he never finished his degree and hence was not a member of the Medical Society. On May 20, 1813, he and several physicians met to consider his proposal for an Antiquarian Society of Charleston, dedicated, as he wrote later, "*primarily* to the collection, arrangement and preservation of specimens in natural history; and of things rare, antique, curious and useful; and *secondarily,* the promotion and encouragement of the arts, sciences and literature generally." In the following month they were joined by David Ramsay, Stephen Elliott, John Drayton, Benjamin Elliott, and others, and the decision was made to call the new organization the Literary and Philosophical Society of South Carolina. It was incorporated under that name in 1814.[13]

Of the new members the most impressive and able was Stephen Elliott. The son of a Charleston merchant, he graduated from Yale in 1791 and settled on a plantation near Beaufort, S.C., whence he served in the state legislature in 1796–1800 and 1808–1812. During his respite from politics after the Jeffersonian-Republican victory in 1800 he began collecting plants and minerals and was soon in touch with naturalists all along the eastern seaboard. In May 1803 John Lyon recorded in his journal that he had visited Elliott at his plantation "Silk Hope" and had spent the time "very agreeably in botanical researches round the neighborhood." In 1808 Elliott's long and fruitful correspondence with the Pennsylvania botanist Henry Muhlenberg began. Four years later he accepted the presidency of the newly founded Bank of Carolina, retired from the state Senate, and moved to Charleston. He quickly became active in the Charleston Library Society and other civic organizations and made his house a stopping place for itinerant naturalists, meanwhile building up his herbarium and mineral collection and working on the *Sketch of the Botany of South Carolina and Georgia.* In 1814 he was elected the first president of the Literary and Philosophical Society.[14]

With the founding of the Literary and Philosophical Society, Charleston at last had a scientific organization worthy of the name, though less vigorous than its sister societies in Philadelphia, Boston, and New York. The members were divided into nine classes — mathematics and mechanical

Three stalwarts of South Carolina science: Governor John Drayton (by Charles de St. Memin) (upper left), Stephen Elliott (artist uncertain) (above), and David Ramsay (by Rembrandt Peale) (left). Courtesy of the Gibbes Art Gallery, Charleston, S.C. (Drayton and Ramsay) and of the South Caroliniana Library, University of South Carolina (Elliott)

philosophy; chemistry, electricity, and mineralogy; zoology and botany; anatomy, surgery, physiology, and medicine; agriculture and rural economy; commerce, manufactures, and internal navigation; history, geography, topography and antiquities; belles lettres, ancient and modern languages, and public and private education; and fine arts. Unfortunately, the records of the society have not survived. There is not even a list of founding members. The only publication of any substance was Elliott's inaugural address, in which he reviewed the state of the arts and sciences and urged his fellow Carolinians to assemble the native productions of the United States in a natural history museum and to join with other Americans in establishing the "literary character" of the United States before the nations of the world.[15]

The first of these recommendations received prompt and enthusiastic support. In 1815 the Charleston Library Society voted to donate its museum collections to the new society, and in the same year Felix L. L'Herminier, a French physician-naturalist who had spent many years in Guadeloupe, arrived in Charleston with an excellent collection of natural history specimens he wished to sell. With aid from the state legislature, the city council, and private subscriptions the society was able to make the purchase and to install L'Herminier as curator of the collection. Thomas Sumpter, American minister to Brazil, presented the museum with an extensive collection of South American minerals, birds, and insects; other donations were soon forthcoming from Joel R. Poinsett, Henry Middleton, Stephen Elliott, James MacBride, and other members of the society. John Shecut waxed enthusiastic about the society and its prospects:

> If allowances are made for the . . . constitutional apathy, that has always . . . prevailed among the natives of warm climates, the Society may be said to be in a flourishing condition. It consists of 138 members, many of whom are of the first standing in society, and of acknowledged literary and scientific talents. Its museum is rich in extensive collections of minerals, fossils and shells. The departments of zoology, particularly those of ornithology, herpetology, ichthyology, and entomology, are extensively filled. It is also rich in coins, medals and castings. The specimens of art are also very considerable; and the whole are arranged in the most appropriate order by their late superintendant.[16]

As Shecut's reference to the "late superintendant" indicates, L'Herminier had resigned as curator of the museum and returned to France, discouraged in his efforts to establish a chemical laboratory and drugstore in Charleston and unable to subsist on his salary at the museum. Apparently the museum continued to flourish. According to *Statistics of South Carolina* (1826) the museum was situated on Chalmers Street nearly fronting the city square and was "well stored with curious objects in natural history, Indian antiquities, foreign and native works of art, &c."

In Shecut's account of Charleston — its successes, failures, and prospects for the future — one senses the ambivalent feelings with which the devotees of science and literature regarded the progress of those departments of human endeavor in South Carolina during the first forty years of the state's existence. With considerable pride Shecut described the growth of commerce and population in Charleston and the evolution of its institutions. There were now about 27,000 people in the parishes of St. Philip and St. Michael, including 13,834 whites, 1,300 free Negroes, and

12,000 slaves. The Charleston Library Society had 280 members, 13,000 volumes, a funded debt of $10,000, and a yearly income of $3,000. The Literary and Philosophical Society was flourishing under the leadership of Stephen Elliott. But these signs of cultural progress could not, Shecut insisted, conceal the "gross and shameful neglect of the arts, sciences and literature" that had characterized the development of South Carolina during much of its history and the relative feebleness of its scientific institutions and achievements when compared with those of northern states. "Until the Carolinians are aroused to the formation of a permanent national character; and until the utility and vital importance of the arts, sciences, and literature, form a predominant feature of that character, these things must and will remain, the reproach of South-Carolina."[17]

Thus a beginning had been made by 1820, but the future was still uncertain when Thomas Cooper, shaking the dust of Philadelphia from his feet, arrived in Columbia to become professor of chemistry, mineralogy, and geology, and soon after, president of South Carolina College. In response to his initiatives and those of the Charleston physicians, the Medical College of South Carolina was established in Charleston in 1824, and the prospects for a broad range of scientific activity were correspondingly increased. Charleston would not rank with Philadelphia, New York, or Boston in the future development of American science, but it would continue to be the leading scientific center in the South until the Civil War.

CINCINNATI AND LEXINGTON

In the region west of the Allegheny Mountains, Cincinnati, Ohio, and Lexington, Ky., took the lead in founding scientific institutions. In Cincinnati these owed their early progress primarily to Daniel Drake, a native of New Jersey brought up on a Kentucky farm and apprenticed at the age of fifteen to a Cincinnati physician, William Goforth. From Goforth young Drake acquired not only sound medical training but a taste for natural history as well. Much interested in the "antiquities" of the western country, Goforth dug up a large collection of fossil bones at Big Bone Lick in Kentucky in 1803. Unfortunately, he entrusted them to a British traveler, Thomas Ashe, who took them to England and sold them in Liverpool, giving neither credit nor cash to the man who had collected them. But Drake learned much from his preceptor's experience and eventually made substantial contributions of his own to the study of western "antiquities."

Cincinnati was not a very promising site for developing scientific institutions when Drake arrived in December 1800. Its approximately 750 inhabitants were divided between the "upper town," perched on a former bank of the Ohio River, and the "lower town," situated on another old

A view of Cincinnati, Ohio, as it appeared when Daniel Drake arrived there in 1800 to begin his medical apprenticeship. Courtesy of the Ohio Historical Society

bank of the river. The inhabitants of the city, according to Drake, were "emigrants from every civilized country." They had arrived as individuals or families "without a more specific common object, than the pursuit of fortune and happiness." In such a population, Drake reflected, there could be little devotion to literary and scientific projects.[18]

Despite these unpromising circumstances and his own youth and inexperience, Drake was soon busy helping to create scientific and cultural agencies in his adopted city. The inspiration for these efforts may well have come from his acquaintance with Philadelphia institutions, gained during his studies at the medical school of the University of Pennsylvania in 1805–1806. Within a few years of his return to Cincinnati in 1806 Drake became involved in forming a debating society, a theatrical group, the Cincinnati Lyceum, a circulating library, a medical society, and an association for literary and scientific improvement called the School of Literature and the Arts. The scientific side of this last organization is indicated by the titles of some of the papers presented by members during its first year (1813): "On the Earthquakes of 1811, 1812, 1813," "On Light," "On Carbon," "On Air," "On Agriculture," "On Caloric," "On Gravitation," "On the Mechanical Powers," "On the Geology of Cincinnati and Its Vicinity," "On Hydrogen." Drake himself was the author of several, for he had begun the researches into the antiquities, natural history, geology, mineralogy, meteorology, diseases, economic life, and general history of Cincinnati and vicinity that were to be presented to the public in his *Notices of Cincinnati* (1808) and *Natural and Statistical View, or Picture of Cin-*

cinnati and the Miami Country (1815). In his meteorological researches Drake was aided greatly by Jared Mansfield, surveyor-general of the United States for the Northwest Territory, who had established a temporary observatory at his mansion, Ludlow Station, at the east end of Cincinnati. There Drake found congenial company in Mansfield and his secretary Joseph G. Totten, a West Point graduate with wide-ranging scientific interests. There, too, Drake met Harriet Sisson, Mansfield's niece, whom he married in 1807. By the summer of 1816 he was solidly established as Cincinnati's leading physician and promoter of the sciences. When Thomas Nuttall arrived in Cincinnati that summer on a western tour, Drake entertained him and introduced him to the botanists of the region.[19]

A second winter of study in Philadelphia in 1815–1816 whetted Drake's appetite for building scientific institutions even further. Welcomed by the literati of Philadelphia as the author of the *Picture of Cincinnati,* Drake attended the meetings of the American Philosophical Society and the famous Saturday evening gatherings at the home of Caspar Wistar. Returning to Cincinnati with medical degree in hand, Drake threw himself into more new projects for making Cincinnati the Philadelphia of the West. In the next few years he played a leading role in the founding of Cincinnati College, the Medical College of Ohio, the Commercial Hospital and Lunatic Asylum, and the Western Museum Society.

The museum project was suggested by William Steele, a Cincinnati merchant; but Drake, with his energy, enthusiasm, and knowledge of Peale's Museum in Philadelphia, shaped the Western Museum's development. Through the efforts of Drake; Steele; Elijah Slack, president of Cincinnati College; James Findlay, an attorney; and Jesse Embry, a land agent, subscriptions amounting to $45,000 were secured at $50 a share. An "Address to the People of the Western Country"(1818) invited the public to contribute specimens of minerals, fossils, plants, animals, and Indian relics. Drake donated his own sizable collection of minerals, mammoth bones, and antiquities, and exchanges were solicited from individuals, museums, and societies throughout the United States. Robert Best, a young instrument maker, was appointed to care for the collections and "philosophical" apparatus, and John J. Audubon was brought from Louisville to assist Best in stuffing, mounting, and displaying the specimens. Audubon, whose business ventures in Henderson and Louisville had failed, was only too glad to accept this job. "So industrious were Mr. Best and I," Audubon wrote in his journal, "that in about six months we had augmented, arranged and finished all we could do for the museum."[20] In the opinion of one of Audubon's biographers, Donald C. Peattie, it may have been his work at the Western Museum and Drake's public praise of his crayon portraits of American birds — "[he] has, in his port folio, a large number that are not figured in Mr. [Alexander] Wilson's work, and many

Daniel Drake in later life (engraving by Alexander H. Ritchie) (from his book Pioneer Life in Kentucky*) and John James Audubon (self-portrait) when he was employed in Drake's Western Museum. Courtesy of the Ohio Historical Society*

which do not seem to have been recognized by any naturalist" — that suggested to Audubon the idea of portraying the entire bird fauna of America.

Like Charles W. Peale before him, Drake had an exalted conception of a museum's function. It should enlighten the public through its library, lectures, and exhibits; delight the naturalist, the antiquary, and the mechanician; bring together the productions of nature in America and compare them with those of other continents; throw light on the history of the earth and its inhabitants; foster the economic development of the country; stimulate invention by the exhibition of model machines; encourage the development of science and the fine and practical arts; and provide a working laboratory and demonstration hall for the professors of nearby institutions of higher learning.[21]

By December 1819 the Western Museum seems to have begun operation, at least on a limited scale. On the tenth of that month it was announced that two courses of popular lectures, one on mineralogy and geology by Daniel Drake and the other on the elements of natural philosophy by Elijah Slack, would be given "in the Western Museum at the College Edifice." Eight days later Drake opened his series with an introductory lecture on the utility and pleasures of the study of mineralogy and geology. One auditor, a journeyman shoemaker named Francis Mason, reported that Drake illustrated his lectures on the geology of the western country with "many fossils that had been dug up in the neighbourhood." Drake's audience was small, said Mason, owing to the entrance fee.[22] In due course there were other lectures on natural history, astronomy, chemistry, medical jurisprudence, aerostation, and ornithology. Experiments with "laughing gas" (nitrous oxide) were a favorite feature of the chemical lectures.

The museum's collections were expanded. By 1823 it claimed to possess 100 mammoth and Arctic elephant bones; 50 megalonyx bones; 33 quadrupeds; 500 birds; 200 fishes; 5,000 invertebrates; 1,000 fossils; 3,500 minerals; 325 botanical specimens; 3,125 medals, coins, and tokens; 150 Egyptian and 215 Indian artifacts; 112 microscopic designs; several views of American scenery and buildings; about 500 specimens of the fine arts; an elegant organ; and the head of a South Sea Island chieftain "preserved and beautifully tatooed by the Cannibals of New Zealand." For twenty-five cents, the museum advertised, the visitor could inspect "one of the largest and best-arranged scientific collections in America."[23]

Despite these promising beginnings, the Western Museum was soon in trouble. The severe business depression that gripped the entire nation beginning in 1819 made it impossible to collect the subscribed funds. In 1823 the institution passed into the hands of Joseph Dorfeuille, a wandering French naturalist and polymath in whom the talents of a genuine naturalist were blended with those of a master showman. In his hands the museum soon ceased to be primarily a scientific institution. Moved to a new location near the public landing, it became a commercial success known more for its "Chamber of Horrors" than its geological and entomological specimens. Like Peale's "Great School of Nature," the Western Museum succumbed to the profit motive.

Meanwhile, the Medical College of Ohio was experiencing difficulties of a different kind. Organized by Daniel Drake, Coleman Rogers, and Elijah Slack in the summer of 1818 after Drake's return from a winter of lecturing at the medical school of Transylvania University in Lexington, the college was chartered by the Ohio legislature in December of that year. By November 1820 twenty-four students had registered for instruction, which was to be given in a large room above the drugstore run by Drake's father. Drake was president and professor of the institutes and practice of medicine; Benjamin Bohrer, professor of materia medica and pharmacy; Jesse Smith of New Hampshire, professor of anatomy and surgery; and

Slack, registrar, treasurer, and professor of chemistry. Robert Best was appointed to assist Slack "on account of that gentleman's engagements as President, and Professor of Mathematics, Mechanical Philosophy, Chemistry and Mineralogy, in the Cincinnati College; Lecturer on Natural Science in the Museum; and one of the Pastors of a religious society in Kentucky."[24] To provide the professors and students with hospital instruction, the Ohio legislature appropriated $10,000 in depreciated bank paper for the Commercial Hospital and Lunatic Asylum of Ohio, although it was several years before it was constructed.

But just as Drake's dream of a medical school and hospital associated with the Cincinnati College seemed about to be realized, internecine quarrels and personal rivalries among the medical faculty brought Drake's connection with the Medical College of Ohio to an abrupt end. At a faculty meeting on the evening of March 7, 1822, Jesse Smith rose trembling and offered a resolution "that Daniel Drake, M.D. be dismissed from the Medical College of Ohio." The motion carried without opposition, and Drake was deposed as president and professor. "I could not do more than tender them a vote of thanks, nor less than withdraw," wrote Drake, "and performing both, the Doctor [Jesse Smith] politely lit me downstairs."[25] Efforts were made by outraged Cincinnatians to reinstate Drake, but he had had enough. Disheartened by the turn of events, he decided to accept a longstanding invitation to rejoin the faculty of the medical school at Transylvania, where he had made a strong impression in the winter of 1817–1818. Drake was to return to Cincinnati in 1827 to continue his work of institution building, but for the moment he saw a brighter future in that rival emporium of the West, Lexington, Ky. The medical institutions Drake had founded were eventually to fulfill his hopes, however. In 1896 the Medical College of Ohio became the medical department (now the College of Medicine) of the University of Cincinnati, the lineal descendant of Cincinnati College. The hospital founded by Drake continues to this day as the Cincinnati General Hospital.

By contrast to Cincinnati, Lexington had the reputation of giving generous support to the arts and sciences. In the presteamboat era Lexington was an entrepôt of trade between East and West, linked by road with Cincinnati on the Ohio, with Virginia and the Carolinas, and with Tennessee and the trails leading southward to Natchez and New Orleans. It was also a center of manufactures, producing cordage, bagging, and sail from Kentucky hemp, and cotton and woolen goods as well. By 1815 it had over 6,000 inhabitants, and *Niles' Register* predicted its bright future as "the greatest inland city in the western world."[26] Proud of their city, its merchants and manufacturers gave liberally in support of its cultural institutions.

The most important of these was Transylvania University, founded

*Transylvania University 1823 (by Matthew Harris Jouett),
with its president, Horace Holley (lithograph from a painting
by Gilbert Stuart) (left), and its leading medical professor,
Benjamin W. Dudley (by Jouett). Courtesy of the Transylvania University Library*

in 1780 by a merger of two private academies representing the rival interests of Presbyterians and Episcopalians. The university's development was hampered by sectarian strife among its religious supporters until in 1818 the Kentucky legislature reorganized the Board of Trustees and placed the institution under state control. To the consternation of both religious factions the board invited Horace Holley, a Boston Unitarian minister with a reputation for eloquence and learning, to become president. Arriving in Lexington in May 1818 to look at the situation, Holley was shown around the town by no less a personage than Henry Clay. He found the streets "broad, straight, paved, and clean," with rows of trees on either side. The houses were brick, many of them quite impressive. At the Athenaeum there was a paucity of books but an abundant supply of newspapers and periodicals. As for the university: "Everything is to be done, and so much the better, as nothing is to be reformed. Almost the whole is proposed to me to arrange," Holley wrote to his wife.[27]

Apparently Holley liked what he saw. In November 1818 he was inducted as president, and Transylvania entered on the most brilliant period of its existence. A new college building three stories high, with thirty rooms, chapel, lecture halls, and an ornamental cupola, was completed in 1818. Enrollments, which had formerly been less than 100 students, soon reached an average of 350 pupils annually.

The showpiece of the reorganized university was the medical school, which had begun to attract an able faculty under the leadership of Benjamin W. Dudley. A native Lexingtonian of talents similar to those of Daniel Drake, Dudley had taken a medical degree in Philadelphia and made his way to Paris and London for further studies. Returning to Lexington in 1814, he had joined the medical faculty at Transylvania and begun to build its faculty. Drake was invited to spend the academic year 1817–1818 in Lexington, and in 1819 Samuel Brown, who had taught medicine at Transylvania from 1799 to 1806, was persuaded to leave his plantation in Alabama and rejoin the medical faculty. Brown was a man of broad scientific interests, a graduate of Dickinson College and the University of Aberdeen, and a member of the American Philosophical Society. Described by a younger contemporary as "a man of fine appearance and manners, an accomplished scholar, and gifted with a natural eloquence and humor that made him one of the most fascinating lecturers of his day," Brown seems to have impressed all who knew him by the range and variety of his interests. François Michaux visited him in Lexington in 1802 and accorded him "the first rank of physicians settled in that part of the country." According to Michaux, Brown received the latest scientific journals from London, collected fossils and other natural history specimens, analyzed mineral waters, and was responsible for the introduction of inoculation for smallpox in Kentucky.[28]

In 1820 Charles Caldwell was recruited from Philadelphia. Soon afterward the citizens of Lexington subscribed more than $3,000 to provide salaries for Caldwell and Brown, and an additional $13,000 was secured from the Kentucky legislature, the city of Lexington, and private donors to send Caldwell to Europe to buy books, apparatus, and medical specimens. In due course Caldwell returned with a handsome chemical apparatus, an anatomical figure modeled in wax, an obstetrical machine, and a library of several thousand "choice and rare" books. Arriving from Cincinnati in 1823, Daniel Drake was highly pleased with these acquisitions, the "large and commodious" lecture rooms, and the "pervading spirit of emulation, both among the students . . . and professors."[29]

In the faculty of arts and sciences the leading scientific figure was Constantine Rafinesque, a person so controversial both in his own day and since and so much involved in subsequent chapters that it will be well to introduce him fully here. Born in 1783 in Constantinople of French and German parents, Rafinesque grew up in Marseilles and Leghorn, displaying from childhood a voracious appetite for books, especially travel literature, and a passionate interest in natural history. On a three-year visit to the United States in 1802–1805 he formed a friendship with a Philadelphia merchant, John D. Clifford, who shared his interest in natural history. Clifford later transferred his business operations to Lexington, became a member of the Board of Trustees of Transylvania, and began assembling a museum of natural history collections and Indian relics.

Visiting Philadelphia in 1817, Clifford again met his friend Rafinesque, who had returned to America after ten years in Sicily and begun a new career as a wandering naturalist and sometime businessman with the friendly assistance of Samuel L. Mitchill and other New York literati. Rafinesque was now a well-known naturalist in both Europe and America, a member of the Academy of Natural Sciences of Philadelphia, the Lyceum of Natural History of New York, and various European scientific societies. Invited by Clifford to visit the western country, Rafinesque set off on foot in the spring of 1818, passing through York, Chambersburg, Bedford, and Greensburg on his way to Pittsburgh. From there he descended the Ohio on a flatboat, collecting specimens of plants, fishes, shells, and fossils as he went. An idea of how he appeared to the inhabitants of Kentucky and Ohio may be gained from Audubon's description after Rafinesque's visit in Henderson, Ky. Somewhat bedraggled from his peregrinations, Rafinesque was dressed in a "long loose coat of yellow nankeen, much the worse for the many rubs it had got in its time, [which was] stained all over with the juice of plants [and] hung loosely about him like a sack."

> A waistcoat of the same, with enormous pockets, and buttoned up
> to the chin, reached below over a pair of tight pantaloons, the lower

parts of which were buttoned down to the ancles. His beard was . . .
long . . . and his lank black hair hung loosely over his shoulders. His
forehead was so broad and prominent that any tyro in phrenology
would instantly have pronounced it the residence of a mind of strong
powers. . . . I observed some degree of impatience in his request to
be allowed at once to see what I had.[30]

Arriving in Lexington after exploring the Ohio to its junction with
the Mississippi, Rafinesque spent his time studying and drawing specimens
in Clifford's museum. "He [Clifford] wanted to increase it," Rafinesque
wrote in his *Travels,* "and he induced me to come and settle with him in
Lexington, promising to procure me a Professorship in the University and
to travel every year with me in the vacations to increase his museum and
my collections."[31] Pleased at the prospect of exploring the western country in company with a congenial and wealthy friend, Rafinesque agreed
to Clifford's proposal and returned to Philadelphia to wind up his business
ventures and arrange for shipment of his books and collections to Lexington. Meanwhile, true to his word, Clifford persuaded the Trustees of
Transylvania to offer Rafinesque a professorship of botany and natural
history "without any salary from the institution except the privilege of
boarding in commons free of expense – any remuneration he may receive
for his lectures to be at such reasonable rates as his students and himself
may voluntarily agree to."[32]

Unfortunately for Rafinesque and Transylvania, Clifford died suddenly in 1820, only a few months after Rafinesque's return to Lexington and
the beginning of their joint explorations. Clifford's museum was placed
in the custody of the university, but Rafinesque was left without a close
friend and colleague in his projects. He stayed on at Transylvania for
another six years, however, lecturing on botany, natural history, materia
medica; medical zoology and mineralogy; the natural and moral history
of mankind, agriculture and gardening, and various other topics. He gave
private courses in French, Spanish, and Italian; served as librarian and
secretary to the faculty at Transylvania and as secretary to a literary and
philosophical society known as the Kentucky Institute; organized a
botanical garden; collected specimens and explored Indian mounds for
miles around, sending memoirs and specimens to scientific societies, journals, and museums in Europe and the eastern United States. He published
a flood of scientific, literary, and philosophical pieces in short-lived western
journals and newspapers and enjoyed the hospitality of President Holley
and his family, cutting a fine figure with his slender figure, dark silky hair,
long dark eyelashes, small delicate hands, and refined European manner.[33]

As a teacher Rafinesque was known for his total absorption in his subject matter, his careless dress, and his vulnerability to practical jokes. One

This portrait of Constantine Samuel Rafinesque, the only one of certain authenticity, appeared as the frontispiece of his book Analyse de la Nature. *Signed by Falopi, probably a Sicilian illustrator of natural history specimens, it depicts Rafinesque at the age of 27. Courtesy of the American Philosophical Society*

of his students remembered him as "a man of peculiar habits and very eccentric, but . . . one of the most interesting men I have ever known." Among the general public, however, Rafinesque's passion for discovering new species of shells, plants, fossils, and fishes seemed ludicrous.

> The art of distinction is entirely unknown to them [Rafinesque complained]. . . . Thus our Cat-fishes, eels, shads, sturgeons, &c. are for them mere fish to fill their stomach! and moreover they are all of European breed, and were carried here by Noah's flood direct from the Thames, the Seine and the Rhine. I let them rail to their heart's content, and I laugh at them.[34]

These jibes did not endear Rafinesque to the citizens of Lexington or smooth his path to academic preferment. Transylvania did confer the Master of Arts degree on him in 1822, but he was never able to obtain a secure berth in the medical school, where student fees might have rendered his financial situation less precarious. His project for establishing a botanical garden by public subscription started out well, but the business depression of the early 1820s and his departure for Washington to try to patent his idea of using stock certificates as a circulating medium left the garden without adequate funds or supervision. In March 1826 the Transylvania Botanic-Garden Company was dissolved and its assets sold.[35]

Returning to Lexington in 1825, Rafinesque found that President Holley, impatient at his long absence from his duties, had removed his collections from the college rooms, deprived him of his post as librarian, and withdrawn his boarding privileges. Discouraged by these setbacks, Rafinesque stayed through the 1825–1826 term, then dispatched his books and collections to Philadelphia and headed east in the spring of 1826, traveling by way of Ohio and Niagara Falls to acquaint himself with more regions of the United States.[36]

Although Rafinesque did not know it, Transylvania's heyday as the leading cultural institution of the western country was nearing its end. In 1827 Daniel Drake, dean of the medical school, resigned to resume his career in Cincinnati; and Horace Holley, tired of defending himself from the charges of atheism and infidelity leveled against him by his Presbyterian critics, resigned the presidency only to fall victim to yellow fever on his trip east by way of New Orleans. Lexington's days as the "Athens of the West" were over.

CONCLUSION

Transplanting scientific institutions into the regions west of the Appalachians was a slow and difficult enterprise, demanding enormous energy, optimism, and perseverance. Just as the science of the eastern sea-board stood in provincial relationship to the mature institutions of Europe, so that of the western country imitated and depended on eastern models, especially the Philadelphia pattern. In both regions physicians played leading roles in organizing scientific activity.

In the South agricultural modes of life militated against rapid develop-ment of scientific institutions. Charleston had made promising beginnings in colonial times, but little was built on these in the early national period. The contrast between David Ramsay's soaring hopes when he arrived in Charleston fresh from his medical education in Philadelphia and the rather sad condition of South Carolina science and medicine when he met an un-timely death at the hands of a deranged patient in 1815 was symptomatic of a broader cultural contrast between the urbanizing North and the plantation-dominated South. Science was predominantly an urban enter-prise, and the South's heavy dependence on a few cash crops left little room for the development of cities.

6

From Newton to Laplace
in American Astronomy

THE foregoing chapters have described the scientific, social, and intellectual context of American science in Jefferson's time and the difficulties encountered in building
an institutional framework within which scientific researches could be pursued successfully. In turning now to actual researches and the publications
that resulted, we must not forget the circumstances and attitudes that
shaped the development of American science in its formative years and
at the same time conditioned and limited scientists in their efforts to add
to the body of scientific knowledge. Intellectual curiosity was their primary
motivation, but national pride and a desire to promote the economic
development of their country also played a part. For nearly all, however,
science was a part-time activity, something to be studied in hours that could
be spared from teaching, preaching, surveying, medical practice, or
politics. Books and scientific instruments were hard to come by, and the
tariff placed by Congress on the importation of these aids made research
even more difficult.

Public patronage for science was virtually nonexistent. Above all,
Americans lacked the cultural confidence that springs from a long history
of national achievement in science and letters. We should not be surprised,
therefore, if progress in the various sciences was spotty, if promising beginnings sometimes failed to produce decisive results, if patriotism sometimes
colored scientific judgment, if visiting foreign scientists unimpeded by practical duties often stole the march on American naturalists in describing
American flora and fauna, if American science was for the most part more
descriptive than theoretical. The thrust of our story lies less in brilliant
scientific discoveries and innovations than in the achievements of individuals working under difficult circumstances and the slow but steady
progress of American science toward full partnership in the scientific enter-

prise of the Western world. American contributions to science must be understood and judged with reference to that great enterprise.

In no field of inquiry were British and European developments and the institutional setting of American scientific research more important in shaping American achievement than in astronomy and the related fields of mathematics and mathematical physics. Astronomy is both an observational and a theoretical science, and in both respects Americans of the Jeffersonian period found themselves at a disadvantage. In 1761 and again in 1769 they had cooperated with British astronomers in observing the transit of Venus across the sun and had won the praise of both British and European scientists for their observations.[1] In the postrevolutionary era, however, observational astronomy was revolutionized by William Herschel's giant reflecting telescope, by means of which he discovered a new planet (Uranus), two new satellites of Saturn, and the white polar caps of Mars and inaugurated the study of double stars, star clusters, and diffuse nebulous appearances in the heavens. Soon afterward, substantial improvements were made in refracting telescopes, and Britain, which had led the world in observational astronomy in the eighteenth century, began to feel the competition of European observatories.

The United States had no observatories in the full sense of that word. In the 1780s and 1790s and again in the second and third decades of the nineteenth century the American Philosophical Society laid plans for establishing an observatory in Philadelphia, but these projects were never completed. At Harvard College, feasibility studies aimed at creating an observatory were undertaken in 1805 and again ten or twelve years later, but the cost proved prohibitive and nothing was done until 1839. Yale did not acquire a modern telescope until 1828. In Washington, D.C., William Lambert and others urged Congress to authorize a national observatory, but without success. The United States Navy had to make do with an improvised observatory at the Depot of Charts and Instruments in the 1830s until Congress saw fit to authorize a naval observatory in 1842. About the same time, the United States Military Academy at West Point acquired its observatory.[2] In the Jeffersonian era, American astronomers were largely dependent on their own instruments.

On the theoretical side, the Americans were handicapped by lack of familiarity with the new developments in mathematics and mathematical physics that had been taking place on the continent of Europe, where mathematical geniuses like Pierre Simon Laplace and Joseph Lagrange (using the Leibnizian notation for the calculus instead of the much clumsier Newtonian notation that lingered on in Britain) were making rapid progress in both terrestrial and celestial mechanics. The spectacular achievements of these scientists in demonstrating the motions of the solar system by systematic application of Newton's law of universal gravitation, as in Laplace's *Mécanique céleste,* were reported in the United States, but the circumstances of life and learning in the new nation precluded all but a very few from gaining a thorough understanding. Professors of mathematics and natural philosophy in American colleges were dependent on British texts, but Britain was now half a century behind the Continent in these sciences. At the University of Pennsylvania, Robert Patterson edited American editions of James Ferguson's *Lectures on Mechanics* and his *Astronomy Explained upon Sir Isaac Newton's Principles and Made Easy to Those Who Have Not Studied Mathematics* for the use of his students. At Yale, Jeremiah Day produced a series of textbooks covering algebra, mensuration of surfaces and solids, plane trigonometry, navigation, and surveying that proved enormously popular in American colleges, partly because they were well adapted to the elementary level of student preparation and partly because they explained basic concepts clearly and in great detail. In astronomy and natural philosophy Day used William Enfield's *Institutes of Natural Philosophy,* a popular British text.[3]

At Harvard the mathematical textbooks used until about 1820 were Euclid's *Elements* and Samuel Webber's *Mathematics Compiled from the Best Authors . . . ,* a two-volume work by the Hollis Professor of Mathematics and Natural Philosophy that contained brief expositions of arithmetic, logarithms, algebra, geometry, plane trigonometry, mensuration of surfaces and solids, gauging heights and distances, surveying, navigation, conic sections, dialing, spherical geometry, and spherical trigonometry. According to a biographer of Webber's successor, the students never reached the end of Webber's text: "Pure mathematics was a very unwelcome study. Nine tenths of every class had broken down in quadratic equations; seven eighths did not get so far. In physics, the book used in recitations was Enfield's *Elements* [*Institutes*] *of Natural Philosophy,* an insufficient and unskilful compend originally, and already far behind the progress of discovery."[4]

It should be remembered that all students took a prescribed course of study and most had little mathematical training before coming to college. Admission requirements at Harvard did not include arithmetic until 1802, and it was not until 1816 that thorough training in this subject was required. According to President Willard's son, the mathematics course did not extend beyond algebra and conic sections. The professor of natural philosophy lectured to the junior and senior classes with experimental demonstrations, but attendance at these lectures was not compulsory and there was no examination. The same professor gave a short course of lectures on astronomy to the senior class when his other duties permitted.[5]

Not until late in the Jeffersonian period, and then chiefly at Harvard and West Point, were American students introduced to French texts in mathematics and mathematical physics. In 1818 John Farrar began the introduction of a series of French texts in his own translation for the use of Harvard students under the general title "Cambridge Course in Mathematics." First came Sylvestre Lacroix's *Algebra,* then Adrien Legendre's *Geometry* and Lacroix's *Trigonometry,* and finally in 1824 Étienne Bézout's *First Principles of Differential Calculus,* which gave Farrar's pupils their first acquaintance with the Leibnizian notation for the calculus. These works were somewhat behind the latest advances in mathematics (Bézout's work dated from before the French Revolution), but they were a marked improvement on the English treatises by Samuel Vince and Charles Hutton and American textbooks like those of Day and Webber.[6]

West Point was the only other educational institution to introduce French ideas and methods in mathematics and physics before 1820. Ferdinand Hassler, the gifted Swiss geodesist-engineer whom Jefferson chose to plan and initiate the United States Coast Survey, attempted to introduce some elements of Continental mathematics during his brief tenure at the academy during 1807–1810, but the cadets were poorly prepared in mathematics and the academy itself was in a chaotic condition. In 1817, however, it was reorganized by Sylvanus Thayer, who had come back from his tour of European military establishments with a profound respect for the educational methods of the *École Polytechnique* and a firm resolve to introduce them at West Point as quickly as possible. The professor of mathematics, Andrew Ellicott, and the professor of natural philosopy, Jared Mansfield, were elderly surveyor-astronomers brought up on British textbooks and manuals; but Claude Crozet, the new professor of engineering, was a graduate of the *École*. Although he had been hired to teach engineering, Crozet soon discovered that he must first teach his students mathematics, including a new branch known as descriptive geometry. Lacking a textbook, Crozet adopted a device he had become acquainted with at the *École* – blackboard and chalk – until he could bring out his own *Treatise on Descriptive Geometry,* published in 1821.

A "tall, somewhat heavily-built man, of dark complexion, black hair and eyebrows, deep-set eyes, remarkable for their keen and bright expression, a firm mouth and square chin, a rapid speech and strong French accent," Crozet made a lasting impression on his students. One of them, Benjamin Latrobe, oldest son of the Benjamin Latrobe who helped design the national Capitol, wrote of him:

I can, even after the lapse of between sixty and seventy years, fancy
that I see the man before me. He had been an engineer under Napoleon
at the battle of Wagram and elsewhere, and the anecdotes with which
he illustrated his teaching were far more interesting than the "Science
of War and Fortification" which was the name of the text-book at the
time. When he left the Academy he became chief engineer of the State
of Virginia, which is indebted to him for the system that made her
mountain roads the best, then, in America.[7]

Of the various professors of mathematics and natural philosophy in
the American colleges in this period only one, Robert Adrain, displayed
marked originality in the fields of study he professed, and it seems appropriate to give a brief account here of his life and work. Born in Car-
rickfergus, Ireland, on September 30, 1775, Adrain early showed unusual
mathematical ability, beginning his career as a teacher at the age of fifteen.
In 1798 he joined an Irish uprising against British authority. Wounded
severely and left for dead, he survived and managed to escape and make
his way to the United States with his wife and child. For the next ten years
he served successively as headmaster at academies in Princeton, York, and
Reading, contributing meanwhile to the *Mathematical Correspondent,* the
first journal of its kind in the United States, edited in New York City by
George Baron. Adrain's contributions showed originality and a wide familiarity with current work in mathematics. Acquainted with Laplace's
works, Adrain, in one of his papers, tackled the problem of the motion
of a ship on a rotating earth. In another he proposed a problem concerning the form taken by a homogeneous, flexible, nonelastic string rotating
about two fixed points with a constant angular velocity, the attraction of
gravity being disregarded — a problem that led to elliptic integrals. It was
not solved for another half-century but eventually became a favorite in
textbooks on elliptic functions. His best known contribution to the *Correspondent,* however, was his "View of Diophantine Algebra," the first
American article devoted to this branch of mathematics.

In 1808, after trying vainly to keep the *Correspondent* afloat by taking over the editorship himself, Adrain launched a new journal in Philadelphia entitled *The Analyst, or Mathematical Companion.* It attracted
contributions from several well-known American mathematicians, including Nathaniel Bowditch and Robert Patterson. In the second number
of the *Analyst* Patterson posed a problem that elicited Adrain's most
original contribution to mathematics, namely, how to correct the measurement of a polygon whose successive sides are given in length and direction but, when plotted, does not close up. According to the Harvard
mathematician Julian L. Coolidge, Bowditch proposed a limited solution
based on somewhat doubtful assumptions, but Adrain set out to find a general law governing errors in measurement. His solution to the problem was

THE ANALYST;

OR,

MATHEMATICAL MUSEUM.

CONTAINING

NEW ELUCIDATIONS, DISCOVERIES AND IMPROVEMENTS,

IN VARIOUS BRANCHES OF THE

MATHEMATICS,

WITH COLLECTIONS OF QUESTIONS

PROPOSED AND RESOLVED

BY INGENIOUS CORRESPONDENTS

VOL. I

PHILADELPHIA:

PUBLISHED BY WILLIAM P. FARRAND AND Co.
FRY AND KAMMERER, PRINTERS.

1808.

Title page of Robert Adrain's Analyst, *a journal that did much to encourage serious work in mathematics and mathematical physics in the United States. Courtesy of the American Philosophical Society*

the first demonstration of the exponential law of error, usually attributed to Carl F. Gauss who published it a year later. Adrain's proof of the law, based on the assumption that the simplest solution was the correct one, was the first of a series of attempts, none of which has proved entirely satisfactory. After stating his error formula, Adrain went on to deduce from it the process known as the method of least squares, which had been published two years earlier by Legendre, but without proof. Adrain owned a copy of Legendre's memoir and probably developed his law of error as a means of justifying Legendre's method. In any case, the method of least squares and the related law of error provided astronomers with a powerful tool for achieving reliable results from a multiplicity of observations.[8]

The *Analyst,* like its predecessor, expired after one volume, but Adrain's contributions to it won him an appointment as professor of mathematics at Queen's College (Rutgers) in 1809. Three years later he moved to Columbia University, where he taught mathematics and natural philosophy until 1826. He edited three editions of Charles Hutton's *Course in Mathematics* for the use of his students, adding footnotes of his own. To the third edition (1822) he appended an essay on descriptive geometry, a new branch of mathematics elaborated by Gaspard Monge in 1794 and introduced at West Point two decades later by Claude Crozet. Crozet's

Descriptive Geometry may have suggested to Adrain the idea of adding an essay on this subject to the third American edition of Hutton's work, but the exposition was Adrain's own.

Elected to the American Philosophical Society in 1812, Adrain published two papers in its *Transactions* for 1818, one on the figure of the earth, the other on its mean diameter, which showed his command of the ideas and methods of Laplace and Lagrange. He continued to publish mathematical papers and even launched a new journal in 1825, *The Mathematical Diary,* but his main contributions were accomplished. In Coolidge's opinion, they outshone those of any of Adrain's American contemporaries:

> Others may have given better demonstrations of the exponential law of error, Adrain gave the first. His methods in Diophantine analysis were not novel, but he did obtain some new theorems. He may have frittered away in problem solving a talent that might have produced noteworthy contributions to our science, but his problems were far ahead of those suggested by others. . . . What he might have accomplished under more favorable circumstances must . . . remain a subject of conjecture. What he did accomplish entitles him to the glory of a pioneer in the development of American mathematics.[9]

ASTRONOMY AT THE AMERICAN PHILOSOPHICAL SOCIETY: ANDREW ELLICOTT AND HIS ASSOCIATES

The two main avenues of publication for American astronomers were the *Transactions* of the American Philosophical Society and the *Memoirs* of the American Academy of Arts and Sciences. Having taken the lead in astronomy by arranging for observations of the transit of Venus in 1769 and publishing these and other American observations in the first volume of its *Transactions* in 1771, the Philadelphia society did its best to maintain its preeminence in this field, relying at first chiefly on the contributions of David Rittenhouse. During most of the 1780s Rittenhouse was busy with his duties as state treasurer and with canal and boundary surveys for the state of Pennsylvania, but he managed to observe Herschel's new planet shortly after its discovery in 1781, a transit of Mercury in 1782, and in 1784 a new comet for which he computed the orbit. To facilitate these observations, he introduced two important technical innovations: the use of a spider web as a cross hair and a collimating lens system to facilitate observations with his meridian telescope.[10]

In 1788, having built a three-story mansion at Seventh and Mulberry streets with facilities for his meteorological and astronomical instruments, Rittenhouse began more systematic observations, keeping daily records that have been preserved to the present, recording significant events in the solar system, and publishing his most important observations and technical innovations in the *Transactions*. These contributions were appreciated in Europe. The French astronomer Joseph Lalande, in his *Astronomie,* made use of Rittenhouse's observations of the annular solar eclipse of 1791, and Franz von Zach in his *Tabulae Motuum Solis* incorporated Rittenhouse's method of calculating the sun's meridian time. The most gratifying recognition, however, came in 1795, one year before Rittenhouse's death, when he learned he had been elected a Fellow of the Royal Society of London. It must have been a great satisfaction to Rittenhouse to know that his candidacy for membership in the society had been supported not only by the Astronomer Royal Nevil Maskelyne and the physicist Henry Cavendish but also by Jesse Ramsden, who like Rittenhouse was an instrument maker as well as a "philosopher."

After Rittenhouse's death in 1796 the leading contributor of astronomical papers to the *Transactions* was his friend and fellow astronomer-surveyor-instrument maker Andrew Ellicott. Since Ellicott knew and collaborated with most of the other contributors on astronomical subjects and like Rittenhouse depended on political connections for surveying appointments, the following account of his career will serve not only as an outline of his astronomical work but also as a guide to the contributions of other astronomers connected with the American Philosophical Society and to the political and economic hazards involved in combining astronomy with surveying.

The son of a Quaker clockmaker, Ellicott exhibited the same combination of mechanical and mathematical talents that had distinguished Rittenhouse. Sent to a Quaker school in Solesbury, Pa., at the age of fifteen, he had the good luck to study mathematics, astronomy, and physics with Robert Patterson, a Scotch-Irish immigrant who later became professor of mathematics and natural philosophy at the University of Pennsylvania and director of the United States Mint. Patterson and Ellicott formed a strong friendship, and Ellicott soon put his mathematical studies to use in almanac making and surveying. Appointed one of the Virginia commissioners for running the boundary between Virginia and Pennsylvania in 1784, he found himself in the company of men of high scientific caliber. His associates on the Virginia commission included James Madison, president of the College of William and Mary, and John Page, a scientific correspondent of Rittenhouse and Franklin and later governor of Virginia. The Pennsylvania commissioners were John Ewing, provost of the University of Pennsylvania; Thomas Hutchins, official geographer of the United States; John Lukens, surveyor-general of Pennsyvlania; and David Rittenhouse. When the survey was finally finished after six months of grueling labor, Ellicott visited Philadelphia, where he was welcomed warmly by Rittenhouse, Franklin, and his old teacher Robert Patterson. With their support he was elected a member of the American Philosophical Society in September 1786. Three years later he moved to Philadelphia with his family.[11]

The life of a surveyor was not ideal for a would-be astronomer, but Ellicott managed to combine the two occupations despite difficulties that would have discouraged a lesser man. In 1786 he joined in the survey of the northern boundary of Pennsylvania. In 1787 and 1788 he surveyed the Presque Isle triangle south of Lake Erie for the federal government. In 1791 President Washington asked him to survey the newly established District of Columbia, a task he accomplished in winter weather under extremely difficult circumstances with the aid of his brothers and the sixty-year-old Negro almanac maker Benjamin Banneker, who worked in the field observatory tent, tended the astronomical instruments, and made periodic observations of the stars.[12]

In 1799 the observations made on these surveys were published by Ellicott in the *Transactions*. Reporting observations made at the eastern and western ends of the Pennsylvania line by the boundary commissioners, Ellicott explained that they had relied mainly on observations of Jupiter's first satellite because of the rapidity of its motion. For this reason, wrote Ellicott, "we thought proper to put as much dependence upon it, as upon the others collectively, and that the mean of those results . . . should be deemed the astronomical distance between the eastern and western observatories." In the second part of this memoir Ellicott explained in detail how he corrected his observations to take account of the changing attraction by the sun of the earth's equatorial bulge during the interval between the solstices and the equinoxes and of the apparent motion of the stars caused by the aberration of light and nodding of the earth's axis. In a memoir immediately following he explained his method of calculating the eccentricity of the planetary orbits, part of which he believed to be new.

In 1796 President Washington appointed Ellicott to work with the Spanish commissioners in running the boundary between the United States and Spanish Florida along the thirty-first parallel of latitude. In 1798–1800, working under incredible hardships, Ellicott and his colleagues surveyed the boundary line all the way from the Mississippi River to the Atlantic coast. During the first part of the survey he enjoyed the assistance of a Natchez planter named William Dunbar, youngest son of Archibald Dunbar of Morayshire, who was serving as one of the Spanish commissioners. Impressed by Dunbar's mathematical talents and general scientific knowledge, Ellicott called him to the attention of the American Philosophical Society and its president, Thomas Jefferson. Dunbar in turn expressed his admiration for "the transcendent scientific talents of my very particular friend, Andrew Ellicott, Esq. . . . to whose condescension and

*Andrew Ellicott (left) as depicted by "a Spanish lady" at the
time he worked on the boundary survey of the 31st parallel in
collaboration with William Dunbar (artist unknown) of
Natchez, one of the Spanish commissioners, with entries from
Ellicott's astronomical notebook and journal. With the kind
permission and assistance of Silvio Bedini, National Museum
of American History (Ellicott); Mississippi Department of Archives and History (Dunbar); and the American Philosophical
Society (notebook and journal)*

communicative disposition, I am indebted for much pleasure, information and instruction."[13] The two men became fast friends and corresponded for many years. Dunbar left the survey at the end of August 1798, but Ellicott pushed on through forests, swamps, and flooded rivers, observing the topography, flora, and fauna and making continual astronomical observations "*first* [as he later wrote], because observations accurately made never become obsolete, and may at some future day be found essentially useful, and *secondly,* to determine by experiment, what reliance might be placed in observations made at temporary stations, without any of the conveniences annexed to permanent observatories."[14]

Fortunately for the success of the survey, Ellicott was not only a highly competent scientist and intrepid woodsman but also a skilled instrument maker who could repair the astronomical equipment whenever, as frequently happened, it was damaged by weather and rough transportation. His journal of the survey, published in 1803, was an important contribution to astronomy and to the knowledge of the climate, soil, topography, flora, and fauna of the region he had traversed, but the federal government showed little appreciation of his work. Unable to collect full payment for his labors and expenses and twice refused an audience by President Adams, Ellicott was forced to sell his library of technical books and his theodolite to meet his financial obligations. When he visited at the White House in 1801, Jefferson offered to make amends for the financial injury Ellicott had suffered by appointing him surveyor-general of the United States. Ellicott was grateful for the offer but was reluctant to move his family to the western country, far from friends and scientific colleagues.

He remained in Philadelphia, preparing the journal of the boundary survey, making astronomical observations with Robert Patterson in a small octagonal observatory, corresponding with Bishop James Madison on the latest scientific developments in Europe, and urging on Jefferson the importance of astronomy for the development of American commerce. The latest improvements in navigation, he told Jefferson, derived chiefly from developments in lunar theory "effected by observations made for that purpose in different parts of Europe and [by] the accurate mode now practised for graduating astronomical and other instruments." If petty dukes and princes in Europe could maintain astronomical observatories, asked Ellicott, why could not the Congress of the United States maintain one or more as an aid to navigation and commerce? These institutions would soon pay for themselves; but in a country where the legal and military professions were the main avenues to fame and honor, it was all but impossible to convince either Congress or the public of this fact. As for himself, Ellicott concluded, "tho I am convinced from my own experience that scientific pursuits in the United States are little regarded I cannot help directing my attention to mathematics and astronomy and endeavouring to render both useful to man. . . ."[15]

In October 1801 Ellicott was appointed secretary of the land office of Pennsylvania and moved to Lancaster to take up his duties. He continued, however, to keep in touch with Patterson, Madison, Jefferson, and other correspondents. At the end of December he sent Jefferson some further astronomical observations, remarking that the number would have been greater except that "the duties of my office require so great a proportion of my time, that I have none left for the pursuit of any branch of science, but what I borrow from those hours generally devoted to sleep." In any case, he declared, he was determined to keep the subject of astronomy alive "till succeeded by some American, whose fortune may put it in his power to be more useful, by enabling him to devote his whole time to the improvement of so important a branch of science."[16]

In May 1802 Ellicott sent Jefferson his method of calculating the rising and setting of heavenly bodies, apparently at Jefferson's request. He added that he had managed to place a strand of spider's web in the focus of his transit instrument and intended to equip his telescope with a diaphragm to facilitate observations of the eclipses of Jupiter's satellites, as recommended by Lalande. Best of all, the Pennsylvania legislature had authorized his use of the large reflecting telescope that the Pennsylvania proprietors had sent to Philadelphia at the time of the transit of Venus in 1769 and that had since lain idle. The telescope, Ellicott wrote, was "much the best instrument of the kind upon this continent, and is sent to London to be put in complete repair and [at?] public expense, – the duties of my office prevented me from doing it myself."[17]

Ellicott was extremely eager to possess this instrument, for he had begun a correspondence with Jean-Baptiste Delambre, the foremost positional astronomer in Europe and soon afterward permanent secretary for the class of mathematical sciences in the National Institute of France. In 1801 Ellicott had sent Delambre his observations made during the survey of the thirty-first parallel. Delambre responded promptly, complimenting Ellicott on his work and requesting further assistance in making observations on the satellites of Jupiter and the disappearance of Saturn's ring. Delighted at the prospect of assisting the great French astronomer, Ellicott devoted every moment he could spare to making new observations. On March 10 he sent Delambre a twelve-page letter, directed in care of Robert R. Livingston, the American minister in France, containing a variety of observations and giving the longitude and latitude of Lancaster. In explanation of the circumstances in which he worked he added:

> In this country I have not a single Astronomical correspondent, neither is it a science which has ever been patronised by either of the States, or by the general government. . . . The economy of public money is considered as the standard of merit, and supposed to include every thing necessary for the honour, dignity, or reputation, of a nation.

> . . . The President of the United States is both a lover of Science, and
> a man of science himself; but he has no power by our Constitution
> to aid any branch of philosophy, mechanics, or literature unless it be
> done at his own cost.[18]

To Benjamin Rush, who had persuaded him to publish an account
of his boundary survey and had agreed to edit the manuscript, Ellicott
reported in October 1802 that the work was nearly completed, adding:

> The appendix, . . . which contains 152 pages [and] includes all the as-
> tronomical observations . . . and mathematical operations, was
> printed 18 months ago, and a copy has been reviewed by the national
> institute of France, and highly complimented. An account of the re-
> view was forwarded to me by the celebrated de Lambre — one of the
> scientific secretaries of that learned body, with a request that I would
> second their exertions in completing the theory of the 4th satellite of
> Jupiter. On this subject we have been corresponding and comparing
> our observations for 8 months.[19]

Fortunately for Ellicott, he had begun the study of French during his southern expedition and was able by dint of continued study to read the letters
and astronomical works sent to him by Delambre and Lalande. In 1808
he wrote to Robert Patterson urging that Delambre be elected a member
of the American Philosophical Society. The society did so promptly. A
year later Ellicott sent Delambre Part One of the latest volume of the
society's *Transactions,* containing the observations Ellicott had made at
Lancaster from 1801 to 1804.

By this time the telescope, having been repaired in London, was returned to Lancaster, and Ellicott applied himself to his observations more
enthusiastically than ever, requesting Patterson to make corresponding
observations in Philadelphia whenever possible. He also kept in touch with
William Dunbar in Natchez and with the roving Spanish astronomer José
J. de Ferrer, an officer in the Spanish marine whose acquaintance Ellicott
had first made during the Spanish-American boundary survey. In 1801
William Dunbar, pleased with his election to the American Philosophical
Society, had procured a reflecting telescope, a reflecting circle, and other
astronomical instruments from London at considerable expense and had
asked Ellicott's advice about the construction of an observatory sufficient
to house his library, microscopes, electrical machine, and chemical apparatus as well as his astronomical equipment. Ellicott replied promptly
with helpful suggestions and urged him to send the society as many
memoirs as possible. The result was the publication of several memoirs
by Dunbar in the sixth volume of the *Transactions,* including his observations of the solar eclipse of 1806 along with those of Ellicott and Ferrer.

THE

JOURNAL

OF

ANDREW ELLICOTT,

LATE COMMISSIONER ON BEHALF OF THE UNITED STATES
DURING PART OF THE YEAR 1796, THE YEARS 1797,
1798, 1799, AND PART OF THE YEAR 1800:

FOR

DETERMINING THE BOUNDARY

BETWEEN THE

UNITED STATES

AND THE

POSSESSIONS OF HIS CATHOLIC MAJESTY
IN AMERICA,

CONTAINING

OCCASIONAL REMARKS

ON THE

SITUATION, SOIL, RIVERS, NATURAL PRODUCTIONS,
AND DISEASES OF THE DIFFERENT COUNTRIES

ON THE

OHIO, MISSISSIPPI, AND GULF OF MEXICO,

WITH

SIX MAPS

Comprehending the Ohio, the Mississippi from the mouth of the Ohio to the Gulf of
Mexico, the whole of West Florida, and part of East Florida.

TO WHICH IS ADDED

AN APPENDIX,

Containing all the Astronomical Observations made use of for determining the boundary,
with many others, made in different parts of the country for settling the geographical positions of some important points, with maps of the boundary
on a large scale; likewise a great number of Thermometrical
Observations made at different times, and places.

𝔓𝔥𝔦𝔩𝔞𝔡𝔢𝔩𝔭𝔥𝔦𝔞:

PRINTED BY BUDD & BARTRAM,

FOR THOMAS DOBSON, AT THE STONE HOUSE,
NO. 41, SOUTH SECOND STREET.

1803.

The most extensive of the eclipse memoirs was that of Ferrer. Finding himself in Albany shortly before the eclipse, he traveled fifteen miles south to Kinderhook, N.Y., to observe the event, "taking for that purpose an excellent chronometer of Arnold, N. 63; a circle of reflection; and an achromatic telescope, constructed by Troughton according to particular directions." In the *Transactions* Ferrer reported not only his own observations but also those of John Garnett in New Brunswick, Simeon De Witt in Albany, Ellicott in Lancaster, Ferdinand Hassler in western Philadelphia, Dunbar in Natchez, James Madison in Williamsburg, and Joseph McKeen at Bowdoin College. In a second memoir he added observations by Philip Bauza in Madrid, J. M. de la Cuesta on the Island of Leon, and three unnamed persons in Boston, Mass. These results, he added, had been communicated to Lalande in Paris, who had made his own computation of the conjunctions in mean time at the various locations, enabling Ferrer to present an accurate table of the latitude and longitude, west from Paris, of all the locations from which observations had been made. Not since the transit of Venus in 1769 had there been such extensive American participation in the observation of a celestial event.[20]

Ferrer was also in touch with Franz von Zach, director of the ducal observatory at Gotha in Germany and editor of the prestigious *Monatliche Correspondenz zur Beförderund der Erd-und Himmels-kunde.* In 1802 Zach had published in his journal an account of astronomical observations made by Ferrer during a journey from Pittsburgh to the Gulf of Mexico by way of the Ohio and Mississippi rivers; these observations included some made by Ellicott. In the following year the *Monatliche Correspondenz* carried a portrait and brief biography of David Rittenhouse, obtained from C. Daniel Ebeling, a German scholar in Hamburg interested in North American geography (see Chap. 8). The biography stressed Rittenhouse's role as an astronomer and president of the American Philosophical Society. A further communication from Ferrer giving American latitudes and longitudes appeared in Zach's journal in 1805. In subsequent issues Zach gave an extended account of various articles of astronomical and geographical interest published in the sixth volume of the *Transactions,* including several of Dunbar's memoirs and the longitude determinations of Ferrer and Ellicott based on observations of the solar eclipse of 1806. This volume had been sent to the Gotha observatory by the society, and Zach, supposing that it might be the only copy in Germany, decided to share its contents with his readers as an example of scientific activity in "that rapidly and powerfully burgeoning empire," the United States of America. He was especially pleased with the American observations of the 1806 eclipse, which could now be compared with those made in Europe.[21]

Ellicott's reputation as an astronomer had been growing steadily. At Jefferson's request he instructed Meriwether Lewis in the art of making astronomical observations. In 1804 Alexander von Humboldt, fresh from his travels in Latin America, visited Ellicott in Lancaster. Members of the state legislature and other government officials came to look through his telescope, which was installed in a building on the land office premises. Encouraged by these evidences of public esteem and by Delambre's assurances that his work was valued by the French astronomers, Ellicott must have felt that his dedication to science had finally won the recognition it deserved and that his future career as an astronomer would be easier.[22]

If so, he had not reckoned sufficiently with Pennsylvania politics. In 1810 a dissident faction of Jeffersonian Republicans won control of the state government. Governor Thomas McKean was defeated, and the new governor promptly dismissed Ellicott from the land office. To add insult to injury, the legislature passed an act depriving him of use of the state telescope, thus bringing to an end the series of observations he was carrying on in cooperation with the National Institute of France. In Philadelphia the members of the American Philosophical Society made plans for bringing Ellicott to that city and placing him in charge of an observatory to be erected in State House Square under the sponsorship of the society and the University of Pennsylvania, but the vindictive legislature soon thwarted that plan. Deprived of his means of livelihood, Ellicott accepted an invitation from Georgia to survey the boundary between that state and North Carolina. He spent the winter months sleeping on frozen ground and eating corn meal and pork. Once again he suffered the indignity and hardship of being refused his just pay, since his survey had vindicated the claims of North Carolina, not Georgia.

In July 1813 Ellicott's fortunes took a turn for the better. He was appointed professor of mathematics at West Point and continued in that position until his death in 1820. He would probably have preferred the professorship of astronomy and natural philosophy, but that had been awarded to another surveyor-astronomer, Jared Mansfield. Born in New Haven, Mansfield had been expelled from Yale in his senior year for "discreditable escapades," but he was subsequently awarded the M.A. and LL.D. degrees by his alma mater. His *Essays, Mathematical and Physical* (1801), described by Florian Cajori as "the first book of original mathematical research by an American," won him an appointment as captain of engineers in the U.S. Army. In the following year, Mansfield accepted the post of surveyor general of the United States, which Ellicott had declined two years earlier, and proceeded west to direct the survey of Ohio and the Northwest Territory. During his years in Cincinnati, 1805–1812, he made frequent astronomical observations and sent some of them to the Connecticut Academy of Arts and Sciences for publication. Arriving at West Point in 1814 after serving in the War of 1812, Mansfield looked about for pupils in astronomy and natural philosophy. Eventually five students were found who seemed to have adequate preparation for these studies. Since there were as yet no recitation rooms at the academy, Mansfield taught his pupils in his own living room, using Enfield's *Institutes of Natural Philosophy* as a textbook. Ellicott in turn did his best to teach the cadets the elements of mathematics from Charles Hutton's well-known text. Thus the two began the process of giving the West Point cadets the kind of scientific training that would enable them to carry on the tradition of scientific exploration inaugurated by Lewis and Clark with the help of Philadelphia scientists like Ellicott.[23]

Meanwhile, both professors kept up their contacts with the world of science. When the *Transactions* resumed publication in 1818 after a nine-year lapse, it carried contributions by Ellicott and Mansfield along with others by Patterson, Hassler, and Adrain. The Philadelphia society had not abandoned its claim to preeminence in the mathematical sciences, but it was meeting severe competition from its Boston rival, the American Academy of Arts and Sciences.

THE AMERICAN ACADEMY
AND THE WORK OF NATHANIEL BOWDITCH

Like its sister society in Philadelphia, the American Academy of Arts and Sciences began its career of publication with a strong showing in the field of astronomy. The first volume of its *Memoirs,* published in 1785, contained six astronomical papers reflecting widespread observational activity in the Cambridge-Boston area and in eastern New England generally. John Winthrop IV had established a tradition of excellence in astronomy at Harvard during the colonial period, and this was carried on by his successor in the Hollis professorship, Samuel Williams, and by the new president of Harvard, Joseph Willard, assisted by various Harvard tutors, students, and professors and by amateur astronomers like Manasseh Cutler, John Prince, Phillips Payson, and Elijah Paine. The immediate task of Willard and Williams was to establish the latitude and longitude of Cambridge accurately. Using a Sissons astronomical quadrant of 2½ feet radius, Williams determined the latitude by observations on the meridian altitude of the sun's upper limb and on the pole star and various stars near the equator. In making these observations, Williams reported:

> I found it much easier to note the bisection of a star by the wire of the telescope, than to determine exactly the point of contact between the limb of the sun and the wire. On this account I esteem the obser-

vations of the meridian altitude of the stars more accurate than those of the sun: And therefore fix upon the mean of all the *sidereal* observations as the true Latitude of Harvard-Hall at Cambridge, 42°23′28″, 46.[24]

Finding the longitude of Cambridge was a more difficult matter, requiring comparisons of observations of the eclipses of Jupiter's moons, solar and lunar eclipses, transits of Mercury and Venus, and occultations of stars by the moon with those made at the Greenwich and Paris observatories. In February 1781 Willard sent Nevil Maskelyne a computation of the longitude of Cambridge based on a comparison of Winthrop's observations of the solar eclipse of August 5, 1766, and the transits of Venus and Mercury in 1769 with corresponding observations made at Greenwich and Norriton, Pa. Willard added that he hoped the Royal Society would not reject his contribution "on account of the political light in which America is now viewed by Great Britain. . . . I think political disputes should not prevent communications in matters of mere science; nor can I see how any one can be injured by such an intercourse."[25] Published in the *Philosophical Transactions* of the Royal Society for 1781, Willard's memoir was the first American contribution to appear in that publication after the revolutionary war and the last such contribution until a letter by the entomologist Thomas Say appeared in 1819. In the *Memoirs* Willard recapitulated the argument of his letter to Maskelyne, adding further computations based on a transit of Mercury observed by Winthrop in November 1743 and on the solar eclipse of June 24, 1778, observed by himself, Cutler, and Prince at Beverly, Mass. Apparently Willard had some contact with the Paris Observatory as well as with Greenwich, for he acknowledged his indebtedness to Lalande for sending Continental data on the solar eclipse of 1778. Samuel Williams also was in correspondence with Lalande.

The celestial event most widely observed in New England in this decade was the solar eclipse of October 27, 1780. In keeping with the tradition established by Winthrop in connection with the transit of Venus in 1761, Harvard and the American Academy asked the Massachusetts government to provide a ship to convey Williams and his assistants to Penobscot Bay, Long Island, to observe the eclipse. Since the British were still in control there, it was necessary to arrange safe passage for the expedition. On October 9 Williams embarked in company with Stephen Sewall, professor of Oriental languages, James Winthrop, librarian of Harvard, Fortesque Vernon, and several students, taking with him the Sissons quadrant, a reflecting circle, and telescopes. There were problems with the British military authorities, who ordered them to have no communication with inhabitants of the island and to leave on the day after the eclipse, but they managed to make reasonably good observations. The most significant observation concerned a phenomenon, now known as "Baily's beads," which they saw at the end of their watch:

> Immediately after the last observation, the sun's limb became so small as to appear like a circular thread, rather like a very fine horn. Both ends lost their acuteness, and seemed to break off in the form of small drops or stars; some of which were round, and others of an oblong figure. They would separate to a small distance; some would appear to run together again, and others diminish until they wholly disappeared[26]

The same eclipse was observed by Willard, Cutler, and Prince at Beverly, tutors Caleb Gannett and John Mellen and Edward Wigglesworth at Harvard, Joseph Brown and his brother in Providence, R.I., Guillaume de Granchain at Newport, R.I., and John Clarke at St. John's Island in the Gulf of St. Lawrence. All these observations were published in the *Memoirs,* along with data on other solar and lunar eclipses and various transits of Mercury.

After 1800 the astronomical papers in the *Memoirs* were largely the work of one man, Nathaniel Bowditch of Salem, Mass. Like Rittenhouse, Bowditch was self-educated. Apprenticed to a ship chandler at the age of twelve, he devoted the early morning and late evening hours to studying whatever books he could lay his hands on; fortunately, the gentry of Salem were generous. At the home of John Prince he had access to the philosophical library that Prince, Willard, Cutler and others had formed by purchasing the books of the Irish chemist Richard Kirwan when these were sold at auction by an American privateer (see Chap. 3). Delighted with this literary treasure, Bowditch read through four folio volumes of Chambers' *Cyclopaedia* and copied by hand for his own library many mathematical papers in the *Philosophical Transactions* of the Royal Society, Emerson's *Mechanics,* Hamilton's treatise on conic sections, and extracts from works by Willem s'Gravesande and Benjamin Martin, Daniel Bernoulli, and other writers. He taught himself Latin and worked his way through a copy of Newton's *Principia* that he had obtained through William Bentley. He composed an almanac at the age of fifteen, assisted in a survey of Salem in 1794, and in the following decade traveled the seas in Salem ships, first as a clerk, then as supercargo, and finally as master and supercargo jointly.

On sea as on land Bowditch devoted every spare moment to study, teaching himself and his shipmates the elements of astronomy and navigation and acquiring a working knowledge of French, Spanish, Italian, and Portuguese. To assist shipmasters in navigating the seas, he brought out two corrected editions of a British treatise on navigation. Then in 1800 he published his own *New American Practical Navigator,* destined to become the most widely used nautical guide ever published; it is still in print. In 1804 he left the sea to become president of a marine insurance company in Salem. He remained in that post until 1823, when he moved to Boston to become actuary at the Massachusetts Hospital Life Insurance Company. Throughout all these years he continued to devote his leisure hours to his favorite studies.[27]

Like Joseph Willard, whom he greatly admired, Bowditch devoted much time and energy to improving the determination of the latitudes and longitudes of locations in the United States from which astronomical observations were being made. Two eclipses of the sun, one on June 16, 1806, the other on September 17, 1811, proved useful for this purpose. Bowditch observed these eclipses in the garden adjoining his house and made every effort to secure the notations of other observers to assist him in his computations. For the eclipse of 1806 he obtained observations made near the northern limit at Pawlet, Vt., by a Captain Potter. These were forwarded to Bowditch by Samuel Williams, who was now teaching at the University of Vermont. At the southern limit Bowditch obtained observations made at Tarpaulin Cove, Martha's Vineyard, and Falmouth, the last of these forwarded to him by President Webber of Harvard. By courtesy of Ferrer, Webber also sent him the observations made at Albany, Kinderhook, N.Y., Philadelphia, Lancaster, and Natchez. Using these data and Willard's determination of the longitude of Cambridge, Bowditch computed the longitudes of the various places from which the eclipse had been observed.

The eclipse of 1811 was widely observed in the United States, and Bowditch took advantage of the opportunity to improve the accuracy of his previous determinations of longitudes. He began by recomputing the longitude of Cambridge, taking into account not only the observations Willard had used but other more recent ones and the new values astronomers had arrived at for the difference of meridians between Greenwich and Cambridge and for the ratio of the polar to the equatorial diameter of the earth. Using four solar eclipses, a transit of Mercury in 1743, and the transit of Venus in 1769, all of which had been observed in both Europe and America, Bowditch calculated the difference between the Greenwich meridian and that at Harvard Hall at 4h.44′29″7, declaring his belief that "this longitude is more accurately ascertained than that of any other place in the United States." Interestingly enough, Bowditch's result differed by only one second from the value arrived at by Willard.[28]

In his memoir on the solar eclipse of 1811 Bowditch recorded his own observations at Salem and those made by Walter Folger, Jr., at Nantucket; Thomas Jefferson at Monticello; Seth Pease in the District of Columbia; George Blackburn at the College of William and Mary; John Garnett at New Brunswick; John Kemp at Columbia; Jeremiah Day at Yale; Samuel

Williams at Rutland, Vt.; James Dean at Burlington, Vt.; the Reverend Mr. Nicholls at Portland, Maine; and Parker Cleaveland at Bowdoin College. Combining these data with earlier observations of the transits of Mercury and Venus in 1769; a transit of Mercury in 1789; and the solar eclipses of 1778, 1780, 1791, 1805, and 1806, he computed a table of latitudes and longitudes for forty-one locations, ranging from Hudson's Bay and Quebec in the north to Washington, D.C. and Williamsburg in the south and reaching westward as far as Natchez.[29]

Not all Bowditch's computations were concerned with geographical locations, however. In the case of the meteor that was seen near Weston, Conn., on December 14, 1807, and the one that passed over Wilmington, Del., on November 21, 1819, Bowditch undertook to determine the height, direction, velocity, and magnitude of the meteors as a contribution to resolving the spirited controversy that had developed in Europe and America concerning the nature and origin of "stones which have fallen from the sky," as the French called them. A brief review of American opinions on this subject will help to place Bowditch's Weston memoir in proper perspective.

As early as 1780 Rittenhouse had expressed his views on the matter in a letter to John Page of Virginia concerning a meteor they had both seen in January of that year.

> The extraordinary Meteor you mention was likewise visible here, the air being serene and clear. I did not see it until the bright streak was become very crooked, it then bore S.70°W. nearly, from Philadelphia, and comparing this course with that observed by you, I find it must have fallen on or near the Ouashiota mountains mentioned in Lewis Evans' map, about 80 miles from Philadelphia and 365 from Williamsburg. And taking its altitude 7°, as observed by you, adding 2½ degrees for the depression of that place below your horizon, its entire apparent altitude above the spot where it fell was 9°; which, on a radius of 365 miles, will be 61 miles perpendicular height. The breadth of the luminous vapour was, I think, in some places, when I saw it, not less than a quarter of a degree; this at 480 miles distance must have been at least 20 miles. It was certainly a grand appearance near the place where it fell, if any human eye was there.[30]

Rittenhouse went on to speculate that meteors, or "shooting stars," were "bodies altogether foreign to the earth and its atmosphere, accidentally meeting with it as they are swiftly traversing the great void of space." That they were generated in the atmosphere seemed unlikely to him because of its thinness at the great heights at which some meteors had been seen. Their motions, he added, could not be controlled by gravity, since they descended in all directions and seldom in a line perpendicular to the

The largest surviving fragment of the meteor that passed over Weston, Conn., on December 14, 1807 (present weight 28 lb). Courtesy of Yale Peabody Museum of Natural History

horizon, and their velocities exceeded anything that the force of gravity could produce. Their luminous tails, he thought, might be generated electrically or otherwise by their passage through the earth's atmosphere.

In October 1805 Andrew Ellicott informed Thomas Jefferson that he had recently received a pamphlet from France containing documents collected by the National Institute establishing beyond doubt the fact that stones had actually fallen from the atmosphere at various places. In explanation of these sky stones Ellicott speculated that the materials of which they were composed were present in the atmosphere and could be "collected and brought into contact by some chemical operation with which we are yet unacquainted; — but in which the *meteor* from which the stones fell was the principal agent." But Jefferson was doubtful that stones could be formed in the atmosphere and fall to the earth. "Chemistry," he told Ellicott, "is too much in its infancy to satisfy us that the lapidific elements exist in the atmosphere and that the process can be completed there. I do not know that this would be against the laws of nature and therefore I do not say it is impossible; but as it is so much unlike any operation of nature we have ever seen it requires testimony proportionably strong."[31]

Further evidence on the subject fell from the heavens on the morning of December 14, 1807, when a meteor flashed over Weston, Conn., at 6:30 A.M., illuminating the northern sky, rousing the inhabitants with three loud explosions like cannon shots, and raining numerous fragments of rock on the earth with a heavy thudding. A few days later Benjamin Silliman and James L. Kingsley of Yale arrived on the scene and collected as many of the fragments as they could find for chemical analysis. Silliman's analysis, which established his reputation as a chemist in both the United States and Europe, will be discussed in Chapter 7.

On the astronomical side, the Silliman-Kingsley memoir drew on the treatise of a former president of Yale, Thomas Clap, entitled *Conjectures upon the Nature and Motion of Meteors,* published posthumously in 1781. From various accounts of "high meteors" and from the known laws of physics Clap reasoned that these bodies must be about half a mile in diameter and solid, at least in their external parts, since they were able to withstand the shock of tremendous explosions without disintegrating or losing their globular shape. The rumbling noise heard during their passage Clap attributed to friction with the earth's atmosphere: "for a cannon ball, of six inches in diameter, passing through the air, with 1-25th part of the velocity of the Meteor, will make a humming noise, which is generally heard two miles." The same friction, he argued, must generate an electrical charge capable of producing the explosions reported by observers. No known laws of nature, however, could raise such bodies to a height of 100 miles above the earth's surface and give them a projectile velocity twenty times that of a cannonball. They must, therefore, like comets and planets, move through space in orbits determined by the laws of projectile and centripetal force, "and as all the coelestial bodies are so remote that they can have no sensible influence upon them, when they are within 100 miles of the earth, it is evident that the earth must be the attractive central body, round which they revolve, as the secondary planets revolve round the primary, or rather as comets revolve round the sun in long ellipses, near to a parabola." Adopting Halley's method of computing the trajectories of comets, Clap concluded that there were probably three such "meteors" moving around the earth, "whose mean distances are about as great as the moon's, and, therefore, performing about 36 revolutions in a year; then one of them will appear in each country of 500 miles square, once in 27 years: And so often at least, they have been in fact seen in Old England and New."[32]

In 1810 Jeremiah Day, professor of mathematics and natural philosophy at Yale, published a paper on the origin of meteoric stones in the *Memoirs* of the Connecticut Academy of Arts and Sciences in which he supported and elaborated Clap's theory of the nature of meteors. From reports of the Weston meteor's apparent diameter he estimated that it must have been several hundred feet. It was possible, he conceded, that a mass of stone projected from a volcano on the moon might be drawn into the earth's gravitational field and describe a path similar to that observed in meteors, as some European theorists had suggested, but it was unlikely that thousands of such projectiles, each several hundred feet in diameter, were thrown off from the moon annually. Clap's theory, though as yet only a hypothesis, was consistent with observed phenomena. It involved some unverified assumptions about electricity, but these could be tested by experiment, Day concluded.[33]

With these varying opinions in the background, Bowditch took up the subject in his paper on the Weston meteor, published in the *Memoirs* of the American Academy in 1815. He had collected accounts of the meteor's passage with great care, journeying to Wenham, Mass., to interview a Mrs. Gardner, who had an excellent view of the meteor from her bedroom window. By comparing the data she supplied with those provided by Silliman and Kingsley and by William Page of Rutland, Vt., he was able to estimate its height, direction, velocity, and magnitude. He estimated its velocity at more than three miles per second and its weight, supposing the meteor to be of the same specific gravity as the fragments analyzed by Silliman, at more than six million tons. There could be no question, Bowditch declared, that the fragments that fell in Connecticut were but an inconsiderable fraction of the meteor, which must have continued on its course without falling to earth. The explanation of these phenomena he left an open question:

> The greatness of the mass of the Weston meteor does not accord either with the supposition of its having been formed in our atmosphere, or projected from a volcano of the earth or moon; and the striking uniformity of all the masses that have fallen at different places and times (which indicates a common origin) does not, if we reason from the analogy of the planetary system, altogether agree with the supposition that such bodies are satellites of the earth.[34]

On November 21, 1819, a brilliant meteor passed over Wilmington, Del., in a southwesterly direction at so great a height that it was seen in Baltimore and in Danvers, Mass., at the same time. Once again Bowditch collected newspaper and personal accounts of the event, determined the meteor's celestial latitude and longitude, and estimated its velocity and magnitude. In this case no fragments fell to the ground.[35]

Meteors were exciting phenomena, but the comets of 1807 and 1811 gave Bowditch the best opportunity to display his command of celestial mechanics. He first saw the comet of 1807 at 7:00 P.M., September 24, "near the foot of the constellation *Virgo*." On October 8 he began making systematic observations, measuring its distance from several of the fixed stars. In mid-December, when the comet had ceased to be visible to the naked eye, he was helped to continue his observations by John Prince, "who was so obliging as to fix several cross wires, at equal distances from each other, in the diaphragm of an excellent night-glass, having a vertical and horizontal motion in a stand; and by placing the wires nearly perpendicular to the horizon, we were enabled to estimate the difference of altitude and azimuth between the comet and any fixed star, near which it passed, more accurately than we otherwise could have done, though not with that degree of accuracy we could have wished."[36] With these data and others supplied by John Farrar in Cambridge and Walter Folger, Jr., at Nantucket, Bowditch calculated the elements of the comet's orbit, using the method described by Laplace in the first volume of his *Mécanique céleste.* He published them in the *Salem Gazette* for November 10, 1807, noting that these elements were different from those of any of the comets listed by Lalande in his *Astronome.*[37]

The comet of 1811 excited great interest at Harvard, where Farrar, John Davis, and others decided to republish John Winthrop IV's *Two Lectures on Comets,* delivered before the Harvard undergraduates in 1759 on the occasion of the return of Halley's comet, and Andrew Oliver's *An Essay on Comets,* originally published in 1772. They added a "Supplement" containing an extensive account of the observations of the 1811 comet in Europe and America and a vindication of Bowditch's conclusions against certain criticisms published in the *Richmond Inquirer* by a Mr. John Wood. Farrar's students were busy with the comet also, for the "Supplement" noted that one of them had presented to the Overseers of Harvard a calculation and graphical representation of the comet's elements.[38] Bowditch published his elements in the *Memoirs* of the American Academy in 1815, relying on his own observations and those of Farrar and Folger on Nantucket Island. In an accompanying memoir, Farrar called attention to some observations made by William Bond, "an ingenious mechanic of Dorchester, Massachusetts." This was the William C. Bond who twenty-four years later became director of the Harvard Observatory and soon afterward made it an institution of international importance. The son of a silversmith and clockmaker, Bond made use of astronomical observations in rating the chronometers he constructed for the use of ship captains. He discovered the comet of 1811 on April 21 less than a month after it was first sighted in Europe and six months before Bowditch began his observations. Not until he communicated his results to Farrar and Bowditch were they aware that a new star had entered the firmament of New England astronomy.

By 1820 New England could boast several persons interested and competent in celestial mechanics. Bowditch was the leading figure, but not the

Nathaniel Bowditch (by M. O. Woodbury), author of The New American Practical Navigator *and a translation with commentary of Laplace's* Mécanique céleste. *Courtesy of the Essex Institute, Salem, Mass.*

only one. The third and fourth volumes of the *Memoirs* contained not only Bowditch's numerous papers but also a computation of the elements of the comet of 1819 by Alexander M. Fisher of Yale and a memoir entitled "An Investigation of the Apparent Motion of the Earth, Viewed from the Moon, Arising from the Moon's Librations," by James Dean of the University of Vermont. To illustrate his argument, Dean described an experiment involving a pendulum suspended from two points. A heavy ball was subjected to oscillations in a vertical plane about its point of suspension at the same time that the suspension point oscillated in a perpendicular plane. Fascinated by the mathematical problems implicit in this experiment, Bowditch analyzed them by the methods of Laplace and demonstrated the various curves and line segments generated by the motion of such a pendulum. Years later the French physicist Jules Lissajous discovered these curves again in his work on vibrating strings. It is by his name, not Bowditch's, that they are known today.[39]

By this time Bowditch had achieved international recognition for his work in astronomy. He had been elected to membership in the Royal Society of London, the Royal Society of Edinburgh, and the Royal Irish Academy as well as the leading American scientific societies and had been offered professorships at Harvard and West Point. He declined these offers and a later one from Thomas Jefferson on behalf of the University of Virginia, preferring to remain a private scholar. In 1819 he resolved to acquaint himself with the contributions of German astronomers to the progress of astronomy. He purchased Gauss's *Theoria motus corporum coelestium,* Wilhelm Olbers's treatise on computing cometary orbits, and twenty-eight volumes of Zach's *Monatliche Correspondenz* and began the study of German in earnest. Reviewing the works of Gauss and Olbers in the *North American Review* in 1820, he stressed the growing importance of German science and scholarship and deprecated the general ignorance of the German language in both Britain and the United States.

Bowditch was dismayed not only by American ignorance of German science and literature but also by the lack of government support for science in the United States. In a second review of German publications for the *North American Review* he pointed out the "mortifying consideration" that there was not a single observatory in the New World from Canada to Cape Horn, "and while Great Britain alone can boast of more than *thirty* public and private observatories of considerable note, we have not in the whole United States *one* that deserves the name." As if this were not sufficiently discouraging to science, the federal government had imposed heavy tariff duties on the importation of scientific instruments and books. "This oppressive *tax* upon the literature and science of our country," wrote Bowditch, "seems not to comport with the nature of our government, which is founded upon the principle that the people are virtuous and intelligent; and it would therefore seem that it is upon the general diffusion of knowledge among the citizens that the safety of our excellent institutions of government must depend."[40]

In December 1822 Bowditch sent Zach, editor of *Correspondance Astronomique, Géographique, Hydrographique et Statistique,* copies of several of his astronomical papers and of the fifth edition of the *American Practical Navigator* in return for some books he had received. The memoirs on comets, Bowditch informed Zach, included a considerable number of unnecessary calculations, but these had been undertaken to see whether the cometary elements could be ascertained with nothing but a circle of reflection for purposes of observation. The results showed that they could, said Bowditch, for they agreed well with the determinations made in Europe.

Zach had been aware of Bowditch's work for some time. As early as 1808 he took note in the *Monatliche Correspondenz* of Bowditch's correction of an error in his *Solar Tables.* In 1808 he called the attention of the readers of this journal to Bowditch's method of reducing apparent lunar distances to true distances by taking account of parallax and refraction. Apparently Zach was well impressed with Bowditch's letter, for he published it in French translation with extensive notes in the *Correspondance Astronomique* for 1824, describing Bowditch as *"le premier, et jusqu'à-présent le seul grand géomètre en Amérique."* In the "Notes," Zach quoted at length Bowditch's remarks on the state of American science and the neglect of German science in Britain and America. He also included the table of latitudes and longitudes from Bowditch's *Practical Navigator* and the elements of the comets of 1807, 1811, and 1819 that Bowditch had published in the *Memoirs*. Promising his readers an early account of Bowditch's method of obtaining true lunar distances and his memoir on the motion of a pendulum suspended from two points, Zach closed by saying that he had requested a correspondence with Bowditch and had made arrangements for receiving American communications by way of Marseille and Leghorn, ports frequently visited by American ships.[41]

The climax of Bowditch's career as a mathematical astronomer was the publication of his translation with commentary of the first four volumes of Laplace's *Mécanique céleste,* a work designed, as Bowditch put it, "to reduce all the known phenomena of the system of the world to the law of gravity, by strict mathematical principles; and to complete the investigations of the motions of the planets, satellites, and comets, begun by Newton in his Principia." The translation was completed in 1814–1817, and the American Academy offered to pay for its publication, but Bowditch preferred to wait until he could afford to publish the work himself. The resulting delay gave Bowditch ample time to extend and revise his commentary, keeping it abreast of the latest developments in mathematical astronomy, including the publication of Laplace's fifth volume, published in 1827, the year of Laplace's death. Two years later Bowditch brought out the first volume of his translation and commentary, a massive volume of 746 pages. The second and third volumes appeared in 1832 and 1834. The fourth, nearly finished when Bowditch died in 1838, was published in 1839. The fifth volume was never translated, though Bowditch took some account of it in his commentaries.

What Laplace would have had to say concerning the work of his American translator if he had lived to read it will never be known, but the reception of the work by the European scientific community was all that Bowditch could have wished for. Legendre praised the work as "not merely a translation with a commentary" but rather "a new edition, augmented and improved, and such a one as might have come from the hands of the author himself . . . if he had been solicitously studious of being clear." Friedrich Bessel sounded the same theme in his letter to Bowditch from Königsberg in February 1836: "Through your labors . . . La Place's work is brought down to our own time. . . . You yourself enrich this science by your own additions. . . ." In England, Charles Babbage declared it "a proud circumstance for America that she has preceded her parent country in such an undertaking," and the Council of the Royal Astronomical Society praised Bowditch's edition as "unquestionably fitted to bring the

Mécanique céleste within the grasp of a number of students exceeding five times, at least, that of those who could master Laplace by themselves." This last tribute must have been particularly gratifying to Bowditch, for he had stated in the preface his purpose to make Laplace's work "accessible to persons who have been unable to prepare themselves for this study by a previous course of reading, in those modern publications, which contain the many important discoveries in analysis, made since the time of Newton."[42]

In this purpose Bowditch was completely successful, as later American astronomers like William Ferrel, Daniel Kirkwood, and Benjamin Peirce would testify from their own experience. Ferrel obtained a copy of Bowditch's translation and commentary in Liberty, Mo., in 1845 and went on to make a name for himself in geophysical fluid dynamics. Kirkwood studied the work while teaching school in Pottsville, Pa.; eventually he became director of the Kirkwood Observatory at Indiana University and left the phrase "Kirkwood's gaps" imprinted on the literature of planetary astronomy. Benjamin Peirce, Perkins Professor of Mathematics and Astronomy at Harvard from 1842 to 1880, cut his mathematical eyeteeth reading proof for his fellow townsman Nathaniel Bowditch on the Laplace translation. Later he dedicated his own treatise on analytical mechanics to "my master in science, Nathaniel Bowditch." Thus Bowditch not only raised himself to a position of international esteem by his own unaided efforts but also smoothed the way for the next generation of aspiring students of celestial mechanics.

CONCLUSION

In most respects American astronomy in the Jeffersonian era showed little advance on the modest achievements of the colonial period. Rittenhouse's continuing observations and publications won him election to the Royal Society of London the year before his death, but the colleges produced no astronomer worthy of comparison with Harvard's John Winthrop IV. Observational astronomy continued to be the part-time occupation of surveyors, professors, gentleman astronomers, instrument makers, and persons interested in navigation, all of whom operated with relatively simple equipment. The nearest approach to concerted observational activity like that which had attended the transit of Venus in 1769 was Ferrer's drawing together of observations of the solar eclipse in 1806. With the exception of Ellicott's collaboration with Delambre in observing the satellites of Jupiter, the efforts of American astronomers were devoted largely to establishing latitudes and longitudes and observing comets and eclipses. At the theoretical level they remained dependent on British texts and methods. Not until Nathaniel Bowditch began publishing his annotated translation of Laplace's *Mécanique céleste* in the year after Jefferson's death was a door opened through which American astronomers might enter into full participation in the science of celestial mechanics.

7

The Chemical Revolution
Comes to America

WHEN we turn from the observational science of astronomy to experimental sciences like physics and chemistry, we find, as in the case of astronomy, a mixed record of effort and achievement. In the colonial period Benjamin Franklin had won election to the Royal Society of London and a permanent place in the history of electrical science by his researches on the Leyden jar and his single-fluid theory of the nature of electricity. On his return to Philadelphia in 1785 he must have hoped that some of his fellow Americans would emulate his achievement in experimental physics. With the single exception of David Rittenhouse, however, he had no successor until Joseph Henry began his researches on electricity and magnetism in the 1820s. Chemistry rather than physics excited the interest of American scientists, hence chemical researches must be the main theme of this chapter. However, something must be said about the experimental work of Franklin's friend and fellow physicist David Rittenhouse.

AN EXPERIMENT IN PHYSICS

David Rittenhouse was an able experimenter, although he did not, like Franklin, pursue his experiments to a decisive result. His brief foray into problems involving the diffraction of light is a good example of his experimental ingenuity and his failure to make the most of his scientific opportunities. His friend Francis Hopkinson had posed the problem for him.

> Setting at my door one evening last summer, I took a silk handker-
> chief out of my pocket, and stretching a portion of it tight between
> my two hands, I held it up before my face and viewed, through the
> handkerchief, one of the street lamps which was about one hundred
> yards distant. . . . Agreeable to my expectation I observed the silk
> threads magnified to the size of very coarse wires; but was much sur-
> prised to find that, although I moved the handkerchief to the right
> and left before my eyes, the dark bars did not seem to move at all,
> but remained permanent before the eye.[1]

Baffled by this phenomenon, Hopkinson asked Rittenhouse to try the experiment himself and attempt an explanation.

On doing so, Rittenhouse saw that the four perpendicular and four
horizontal "dark bars" Hopkinson had seen were images formed by the
bending, or diffraction, of some of the light rays as they passed near the
threads of the handkerchief. Determined to investigate the matter more
precisely, Rittenhouse devised a square of parallel hairs, about fifty hairs
to the inch, and looked through them at an opening 1/10 of an inch wide
in the window shutter of a dark room. Encouraged by the results, he
substituted thicker hairs and increased the number per inch to 190.

> The three middle lines of light were now not so bright as they had
> been before, but the others were stronger and more distinct. . . . The
> middle line was still well defined and colourless, the next two were
> likewise pretty well defined, but something broader, having their in-
> ner edges tinged with blue and their outer edges with red. The others
> were more indistinct, and consisted each of the prismatic colours, in
> the same order, which by spreading more and more, seemed to touch
> each other at the fifth or sixth line, but those nearest the middle were
> separated from each other by very dark lines, much broader than the
> bright lines.[2]

Rittenhouse had constructed what is now known as a diffraction grat-
ing. Not content simply to observe the beam of light divided into "so many
distinct pencils," he set out to measure the angles they made with each
other by constructing a primitive spectroscope. Borrowing a small prismatic
telescope from Franklin, he fastened the frame of parallel hairs in front
of the object glass so as to cover its aperture entirely. Looking through
the telescope, he measured the space between the first "pencils" of light
and found the angular distance between their inner edges to be 13', 15"
and from the middle of one to the middle of the other 17', 45". The angles
subtended by the other lines were difficult to measure accurately, but Rittenhouse satisfied himself that "from the second line on one side to the
second on the other side, and so on, they were double, triple, quadruple,
&c. of the first angles."

It appears . . . that a very considerable portion of the beam of light passed between the hairs, without being at all bent out of its first course; that another smaller portion was bent at a medium about 7', 45" each way; the red rays a little more, and the blue rays a little less; another still smaller portion 15', 30"; another 23', 15", and so on. But that no light, or next to none, was bent in any angle less than 6', nor any light of any particular colour in any intermediate angle between those which arise from doubling, tripling, &c. of the angle in which it is bent in the first side lines.[3]

By hindsight it is easy to see that Rittenhouse, if he had continued to perfect his experimental technique, might have been able to measure the wavelengths of the different colors of the spectrum and to explain the dark lines and colored fringes as resulting from the interference of light waves with each other. But Rittenhouse was a disciple of Newton, and Newton had favored a corpuscular, or particle, theory of light as against a wave theory, observing that he could not understand how light waves could pass through narrow apertures without "an excessive bending either way" in the manner of water waves emerging from a narrow channel. Newton had observed the phenomenon of diffraction in the color fringes bordering shadows cast by hairs, but he preferred his own theory that light was subject to "fits" of easy transmission and easy reflection to a wave theory of light. Faithful to his scientific idol, Rittenhouse ended by quoting Newton instead of challenging him. New and interesting discoveries would be made by further experiments of the kind he had tried, Rittenhouse predicted, "but want of leisure obliges me to quit the subject for the present." He never returned to it. In England Henry Cavendish repeated Rittenhouse's experiments, but he also failed to draw important conclusions from them. It remained for Augustin Fresnel, Thomas Young, and Joseph Fraunhofer to push the study of diffraction to a decisive result in the opening decades of the nineteenth century.

One wonders what Rittenhouse might have achieved as an experimental physicist if he had had more time and the stimulation that scientific circles like those in London and Paris provided. Jefferson was well aware of the tendency of the new American nation to burden its scientific geniuses with practical duties. "Are those powers . . . intended for the erudition of the world . . . to be taken from their proper pursuit to do . . . common place drudgery . . . ?" he had asked in a letter to Rittenhouse in 1778. Rittenhouse was not the last American scientist to suffer this fate.[4]

THE CHEMICAL REVOLUTION
AND THE VOGUE OF CHEMISTRY IN AMERICA

Why was chemistry such a popular study in Jefferson's America? It was partly because of its presumed utility in the everyday concerns of life but also because this science was undergoing a major theoretical revolution in the closing decades of the eighteenth century and the opening ones of the nineteenth. To understand what all the excitement was about, we must first gain some idea of the nature of this revolution in chemical theory and see how it came to America.

By a strange quirk of history the beginnings of the revolution in chemistry coincided with the early stages of the American Revolution. On August 1, 1774, Joseph Priestley used a newly acquired burning glass twelve inches in diameter to focus the sun's rays on some red mercuric oxide (known to him as *mercurius calcinatus per se*) floating on top of a column of mercury in a tube inverted in a basin of mercury. As he had hoped, a gas was evolved from the red substance, pushing the mercury down the tube. This gas, he later discovered, was unusually favorable to both combustion and respiration. The problem was to determine its nature and give it a name. In this undertaking, Priestley's thinking was dominated by the phlogiston theory elaborated earlier in the century by Georg E. Stahl. Like hundreds of chemists before him, Stahl thought that burning bodies gave off some substance or principle of inflammability. To this hypothetical substance he gave the name *phlogiston,* from the Greek word meaning to burn. According to Stahl's theory, ordinary air supported combustion by absorbing phlogiston from the burning body. Since the air Priestley had obtained from calcined mercury was more conducive to combustion and respiration than common air, it must itself be largely devoid of phlogiston. On the basis of this reasoning Priestley decided to call it "dephlogisticated air," that is, common air that had been deprived of most of its phlogiston and hence could absorb large amounts of phlogiston from burning fuels, breathing animals, and calcining metals.

In October 1774 Priestley accompanied his patron Lord Shelburne to Paris. There he described his latest experiments to various French chemists, including Antoine Lavoisier, who had been trying for some time to discover why metals increased in weight when they were heated. If they gave off phlogiston on burning, as the phlogiston theory required, why did they not *lose* weight? Soon Lavoisier was repeating Priestley's experiments and testing the "eminently respirable air" given off by heated mercuric oxide. By the middle of 1777 he was convinced that Priestley's "dephlogisticated air" was in fact a gas that was taken on rather than given off in combustion and that formed a constituent part of all acids. In a memoir dated September 5, 1777, he called the new gas *oxygine* (acid former) and began building a new system of chemistry around it. Common air, he now could show, was a mixture of two gases, oxygen and azote (nitrogen). Water, as Henry Cavendish, James Watt, Lavoisier, and others soon discovered, was a compound of hydrogen and oxygen, capable of being separated into its constituent gases. Aristotle's four elements— air, earth, fire, and water—were not elementary at all. Phlogiston, the

protean substance chemists had invoked to explain everything from acid formation to combustion and respiration, was a figment of Stahl's imagination. The facts of chemistry could be explained without it. So said Lavoisier in his *Reflections on Phlogiston,* submitted to the Academy of Sciences in 1783.

To complete the chemical revolution Lavoisier joined with three converts to the new ideas — Louis B. Guyton de Morveau, Claude L. Berthollet, and Antoine F. de Fourcroy — in proposing a new chemical nomenclature based on the oxygen theory. Their *Méthode de nomenclature chimique* (1787) set forth the system of naming chemicals that is still used today. To cap the climax, in 1789 Lavoisier published his *Traité élémentaire de chimie,* the first modern textbook of chemistry.[5]

These events in the world of science did not go unnoticed by Thomas Jefferson, American minister to France in 1784-1789. Jefferson was cautious about innovations in science, especially in matters of nomenclature. He could take issue with the Count de Buffon on behalf of the general utility of chemistry, but the new chemical nomenclature was another thing. Chemistry was still "a mere embryon." The oxygen theory had yet to be proved true, and any attempt to reform established nomenclature on the basis of it was unlikely to succeed.[6]

In Jefferson's own country, however, the new chemistry made rapid progress. When the French traveler Brissot de Warville visited Harvard in 1788, he found Aaron Dexter using a textbook by Fourcroy and introducing his students to recent experiments of the French school. At Columbia, Samuel L. Mitchill proclaimed publicly that he taught the "antiphlogistic system" of chemistry, taking up the various topics in chemistry "very much after the manner in which they are arranged in the Table of Nomenclature agreed upon by the FRENCH ACADEMICIANS."[7]

In 1794 Mitchill published a little book entitled *Nomenclature of the New Chemistry,* designed, as he told his readers, "to explain and lay before you terms and phrases lately adopted and growing into use, and which indeed have already become so prevalent and fashionable among philo-

(Opposite page) Two English chemists who emigrated to Pennsylvania: Joseph Priestley (by Rembrandt Peale) (upper left) and Thomas Cooper (by Charles Willson Peale) (upper right) and their American colleagues James Woodhouse (lower left) and Robert Hare (both by Rembrandt Peale). Courtesy of the American Philosophical Society (Priestley); the College of Physicians of Philadelphia (Cooper); the Edgar Fahs Smith Collection, ACS Center for History of Chemistry, University of Pennsylvania (Woodhouse); and the Independence National Park Collection (Hare)

sophical people, that no accomplished chemist can remain ignorant of them. . . . These," he added, "form the Nomenclature of the new, and I believe I may almost venture to add, the true system of Chemistry." In setting forth the new nomenclature, Mitchill borrowed the scheme of presentation adopted by the German chemist Christoph Girtanner in his *Neue Chemische Nonmenclatur für die Deutsche Sprache,* published in Göttingen in 1791.[8] In the first column the new French names of chemical substances were given; in the second, Girtanner's German equivalents; in the third, the English equivalents; and in the fourth, synonyms and outdated names. The substances were divided into eight categories, each discussed briefly in the ensuing chapters.

Mitchill's preface concluded with remarks on the significance of the new chemistry for medicine. The discovery of the compound nature of water, he predicted, would have important consequences in the study of urine, perspiration, dropsy, and diabetes as well as for general medical theories. Mitchill ended his preface by informing his readers that the illustrious Joseph Priestley, chemist, theologian, and friend of American liberty, was considering emigrating to the United States, "as I learn by a letter from him not long since."[9]

Priestley was indeed coming to the United States, driven from England by the political and religious intolerance generated by the French Revolution. On July 14, 1791, the anniversary of Bastille Day, a mob had burned the Unitarian meeting house where Priestley preached and had attacked his house, demolishing his library and chemical apparatus and endangering Priestley himself, who fled to London. Bereft of his laboratory and ostracized by the fellows of the Royal Society, he decided to emigrate to Northumberland, Pa., where his sons and his friend Thomas Cooper had options to buy 700,000 acres of land and planned to establish a settlement for English liberals. In the spring of 1794, as Lavoisier was being consigned to the guillotine for his connection with the *ancien régime,* Priestley crossed the ocean to New York, arriving on June 4th. He was welcomed cordially by Governor George Clinton, Samuel L. Mitchill, and a variety of civic societies, including the Medical Society of the State of New York, the Tammany Society, the Democratic Society of the City of New York, and the Republican Natives of Great Britain and Ireland. The only jarring note was a vitriolic attack in the press by his fellow Englishman William Cobbett, the reactionary "Peter Porcupine" who kept up a steady drumfire against Priestley for several years.

In Philadelphia, Priestley was welcomed by the dignitaries of the city and the university and by the members of the American Philosophical Society, to whose ranks Priestley had been elected nine years earlier. They urged him to settle in Philadelphia, but Priestley was eager to reach Northumberland. The trip was not an easy one, he wrote John Vaughan.

> At Harrisburgh we hired a common waggon, and in it my wife and myself slept the two last nights. . . . The creeks were so swelled with the rains, that it was with great difficulty the waggon, loaded with stones, got over, and we and the baggage were ferried over in a canoe. Had we not been stopped by a countryman, . . . our driver would have driven in, and we should all have perished.[10]

Arriving at Northumberland, Priestley began to lay plans for a house, a laboratory, and a school. By the end of 1795 he had completed two papers designed to refute Lavoisier's chemistry, and in February 1796 he traveled to Philadelphia to attend the meeting of the American Philosophical Society at which they were read. In the same year he published the first part of his *Considerations on the Doctrine of Phlogiston,* a pamphlet soon translated into French and answered by the French minister to the United States, Pierre Adet. Adet had worked with Lavoisier and his colleagues on the new chemical nomenclature before coming to the United States. With his reply to Priestley the battle over the new French chemistry, already past its peak in Europe, was joined on American soil.

If Priestley had expected to win new recruits for phlogiston chemistry among American chemists, he soon realized that this was not to be. In the preface to his *Considerations* he candidly acknowledged that Lavoisier's system was already widely accepted in America:

> I hear of nothing else. It is taught, I believe, in all the schools on this continent, and the old system is entirely exploded. . . . So firmly established has this new theory been considered, that a *new nomenclature,* entirely founded upon it, has been invented, and is now almost in universal use, so that, whether we adopt the new system or not, we are under the necessity of learning the new language.[11]

The response of American chemists to Priestley's *Considerations* confirmed his estimate of the situation. John Maclean at Princeton began the attack with a pamphlet entitled *Two Lectures on Combustion,* published in Philadelphia in 1797. A Scotsman by birth, Maclean had studied chemistry not only at Glasgow, Edinburgh, and London but also in Paris, where he became a convert to the new system of chemistry then being promulgated. Like Priestley, a friend of republican liberty, he emigrated to the United States in 1795, arriving in New York less than a year after Priestley. On Benjamin Rush's advice he settled in Princeton but soon gave up the practice of medicine to accept an appointment at the College of New Jersey, teaching mathematics, natural philosophy, and natural history as well as chemistry. Silliman, who visited him a few years later, found him "a man of brilliant mind, with all the acumen of his native Scotland; . . . a sprinkling of wit gave variety to his conversation." Maclean was only twenty-six years old when he threw down the gauntlet to the aging Priestley in his *Two Lectures on Combustion.*[12]

The argument of Maclean's *Two Lectures* turned on the interpretation of certain experiments that Priestley regarded as a convincing refutation of the oxygen theory. The first group had to do with metals and what happened to them when they were heated. Lavoisier had argued that ordinary air was composed of nitrogen and oxygen in definite proportions and that metals were simple substances that took on oxygen from the air when heated gently, forming a metallic oxide and leaving the air from which the oxygen had been withdrawn with a high nitrogen content. When the oxide was subjected to intense heat, the oxygen was released and the metal restored to its elementary state. The released oxygen, when combined with the nitrogen from which it had been separated, reconstituted ordinary air. Priestley could not deny that he himself had produced oxygen ("dephlogisticated air") by heating the calx (oxide) of mercury with a burning lens, but he maintained that the behavior of other metals and other forms of mercury was not consistent with Lavoisier's theory. He asserted that turbith mineral ($HgSO_4 \cdot 2HgO$) could not be converted to mercury unless it was heated with the aid of some phlogiston-rich substance such as hydrogen or charcoal. He also argued that the nitrogen produced in experiments in heating metals was formed by the combination of phlogiston and dephlogisticated air (oxygen), that "finery cinder" (iron oxide) formed by heating iron did not contain oxygen, and that the hydrogen gas ("inflammable air") emitted when iron was dissolved in acid resulted from the decomposition of the iron, not from the decomposition of water in the acid, as Lavoisier had maintained.

Maclean's answer to these arguments was brief and to the point. He presented evidence from French experiments to show that turbith mineral could be converted to mercury by a simple increase in temperature; mercury was mercury however obtained. Nitrogen (which he, like Lavoisier, called *azote*) could not be formed by the union of oxygen with anything emitted from hot iron, and iron dissolved in sulphuric acid formed a substance indistinguishable from the black oxide formed by heating iron. He then went on to refute Priestley's objections to Lavoisier's contention that water was a compound of oxygen and hydrogen in fixed proportions. According to Priestley, the "finery-cinder" formed when water passed through a red-hot iron tube was a combination of iron and water, not of iron and oxygen. The hydrogen emitted in this experiment did not result from the decomposition of the water but rather from the union of the phlogiston in the iron with the water. To this argument Maclean replied that the French chemists could show that the weight of the water lost in the iron tube experiment was precisely the same as the joint weight of the

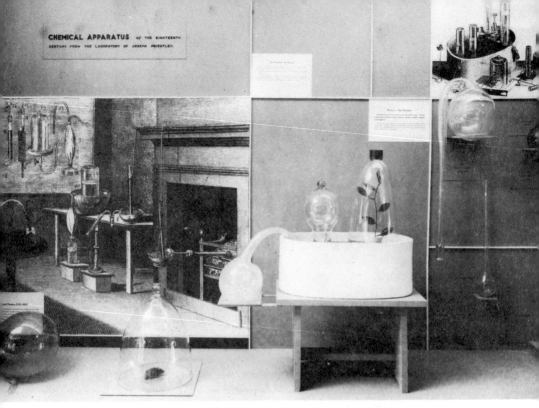

*Exhibit of Joseph Priestley's chemical apparatus, as displayed
at the National Museum of American History. Courtesy of the
Smithsonian Institution*

hydrogen formed and the weight added to the iron by the formation of
the oxide. The French chemists, Maclean pointed out, had conducted a
carefully controlled experiment lasting 185 hours in which 12 ounces, 4
gros, and 49.2270 grains of hydrogen and oxygen had been burned by
means of an electric spark, yielding 12 ounces, 4 gros, and 45 grains of
water. Priestley, on the contrary, could make no quantitative sense of his
own experimental results. As an example, Maclean, using Priestley's own
data and theoretical concepts, showed that Priestley's experiments on the
production of "carbonic acid gas" (carbon dioxide) from the combustion
of charcoal could not be made to yield coherent quantitative results whether
phlogiston and oxygen were both assumed to have weight or whether
weight was attributed to oxygen only.[13] Priestley's theory, Maclean informed his students, was "complicated, contradictory, and inadequate."
His experiments, insofar as they were reliable, served only to confirm the
theory of Lavoisier.

> You doubtless therefore will be inclined to prefer the antiphlogistic doctrine: Indeed you may adopt it with safety; for . . . it is dependent on nothing whose existence cannot actually be demonstrated; whose properties cannot be submitted to the most rigorous examination; and whose quantity cannot be determined by the tests of weight and measure.[14]

Maclean's spirited attack on Priestley's *Considerations* was soon followed by another written by James Woodhouse and published in the *Transactions* of the American Philosophical Society. Woodhouse was an enthusiastic chemical experimenter (see Chap. 2). Like Maclean, he was a young man just beginning his teaching career. He had benefited greatly from Priestley's arrival in America, visiting him in Northumberland and learning much from him. But as he repeated the experiments on which the great chemical controversy turned, he found himself driven to conclusions quite different from Priestley's. His experiments on turbith mineral showed it to be a sulphate of mercury from which he was able to produce oxygen and running mercury by ordinary heating, once the sulphuric acid had been removed. Further experiments in heating iron and copper showed that, "contrary to what had been said by Dr. Priestley, when a metal containing no foreign substance, is calcined in oxygenous gas, the pure air [oxygen] only is imbibed, no substance is emitted from the metal, and no azotic [nitrogenous] gas is formed." Again, when the calxes (oxides) of metals were heated in hydrogen, water was formed, indicating that the oxygen released from the calx had combined with the hydrogen to form water. Experiments on nitric acid and metals heated with charcoal were equally unfavorable to Priestley's views. In short, Woodhouse was compelled by the testimony of experiments to disagree with his revered mentor, though he assured his readers that Priestley's contributions to pneumatic chemistry would forever be considered "as making an aera in the science."[15]

Eager to answer his critics, Priestley found a convenient outlet for his views in Mitchill's *Medical Repository.* Mitchill in turn was only too glad to open the pages of his journal to so distinguished a chemist and so celebrated a controversy. More than forty contributions — articles, letters, reviews, and news items — relating to the phlogiston controversy appeared in the first six volumes of the *Repository,* twenty-nine written by Priestley, the rest by Mitchill, Woodhouse, and Maclean.[16] Although Mitchill had been among the first to introduce the new French chemistry into the United States, he attempted to play a mediating role between the contestants. In an open letter to Priestley published in 1798 he suggested that much of the controversy would be dissipated if the name phlogiston was adopted for the gas commonly called hydrogen. Postulating that all substances that burned with a flame contained hydrogen, Mitchill proposed using the word phlogiston to denote "the thing, in combustible bodies, which forms blaze or ignited vapour."

> The union of this with caloric [Lavoisier's term for the substance of heat], will make *phlogistous* or inflammable air, the air which burns with a blaze. The combination of phlogiston with oxygene, will constitute water or the oxyd of phlogiston, one of the products of inflammation, and like fixed air [carbon dioxide] and other compounds, formed during the same process, incombustible in common temperatures and circumstances.[17]

Thus Mitchill, although he recognized the role of oxygen in combustion, still felt the need to postulate a principle of inflammability in combustible bodies. Like Henry Cavendish and Richard Kirwan before him, he identified that principle with hydrogen. But Priestley would have none of the proposed compromise. "In my opinion," he replied, "there can be no compromise of the two systems. Metals are either necessarily simple or necessarily compound; and water is either resolvable into two kinds of air, or it is not."

The controversy continued to splutter in the pages of the *Medical Repository* until Priestley's death in 1804, with Priestley and Woodhouse as the main protagonists. Maclean withdrew after replying to some objections Woodhouse had raised to his arguments. Mitchill continued trying to find a middle ground, but even he became increasingly impatient with the "triune mystery" of Priestley's concept of phlogiston, declaring that, "according to it, carbon is phlogiston, and hydrogen is phlogiston, and azote is phlogiston; and yet there are not three phlogistons, but one phlogiston." Priestley was not amused by Mitchill's suggestion that in chemistry, if not in theology, he was a trinitarian. He explained that he had not meant to say that phlogiston was all of these chemical substances simultaneously but only that it was "a constitutent part of them all. . . . And where is the difficulty of conceiving that the same principle may be an ingredient in these different substances?" As his biographer Robert Schofield has noted, Priestley never grasped the central idea of the new chemistry. The knowledge of the elements that enter into natural substances he regarded as "but a small part of what is desirable to investigate with respect to them." The important thing was "the principles, and the mode of their combination; as how it is that they become hard or soft, elastic or non-elastic, solid or fluid, &c., &c."[18] For Lavoisier and his followers, on the contrary, chemistry was the quantitative study of the proportions of chemical elements combining to form compounds, assuming the principle of the conservation of matter. The task of the chemist was to resolve compounds into their constituent elements and then to recombine them.

The old chemistry, one of "principles" rather than material elements, was dying. With Priestley's death at Northumberland on February 6, 1804, it faded quietly away despite occasional attempts to revive it.[19]

THE UTILITARIAN APPEAL

Fully as important as theoretical novelty in generating American interest in the new chemistry was the general conviction of its potential utility. Jefferson was an ardent utilitarian in chemistry as in everything else, impatient of theory and insistent that chemists should apply themselves to improving such household arts as the making of bread, butter, cheese, vinegar, soap, beer, and cider. His learned friend Thomas Cooper held the same views and did his best to put them into practice. Educated in law and medicine at Oxford and London, driven from Paris as a Girondin heretic during the Reign of Terror, denied election to the Royal Society because of his religious and political beliefs, Cooper joined the exodus of political refugees from Britain shortly after the departure of his friend Priestley and came to America.

Appointed to a Pennsylvania judgeship after Jefferson's election in 1800, Cooper kept up his chemical interests through extensive reading and correspondence and close contact with Priestley, whose chemical apparatus he inherited when Priestley died. Appointed professor of chemistry at Dickinson College in Carlisle, Pa., in 1811, he proceeded to set forth his utilitarian view of science in an inaugural lecture that was as lengthy as it was erudite. Defining chemistry as the study of inorganic bodies "not considered as aggregates or masses, but as respects the composition and decomposition of the particles themselves that compose those masses," Cooper launched into an extended review of the history of chemistry from the earliest times to his own, explaining "by what means . . . that which was merely an Art, consisting of empirical processes, chiefly confined to the workshops of the manufacturers, put on the appearance and assumed the importance of a Science." One by one the chemical science and technology of the Jews, Egyptians, Romans, Greeks, medieval scholastics, and the modern age were passed in review with an impressive parade of authorities. Cooper ended with a consideration of the importance of modern chemistry in agriculture, medicine, mining, manufactures, baking, brewing, cookery, and printing. "Let any one examine the state of the arts and manufactures, fifty years ago, and compare it with the situation of the present day," he told his students, "and it will be found, that during these fifty years, more improvements have been made, originally suggested by chemical theories, and pursued under the guidance of chemical knowledge than in two thousand years preceding."[20] The moral was clear: men and nations must actively engage in the pursuit of scientific knowledge, which alone could render them respectable and respected.

While still at Carlisle, Cooper took over from John R. Coxe the editorship of *The Emporium of Arts and Sciences,* a Philadelphia journal, and turned it into a magazine of applied chemistry. Apologizing to his readers for the chemical cast of the articles on iron, steel, copper, brass, and tin manufactures, Cooper nevertheless defended his approach to these subjects by an appeal to British experience.

> The elements of natural philosophy and of chemistry, now form an indispensable branch of education among the manufacturers of England. They cannot get on without it. . . . The tradesmen of Great Britain at this day, can furnish more profound thinkers on philosophical [i.e., scientific] subjects, more acute and accurate experimenters, more real philosophers thrice told, than all Europe could furnish a century ago. I wish that were the case here [in America]; but it is not so.[21]

To illustrate the usefulness of chemistry, Cooper reported his analysis of nine samples of magnesian limestone collected in Chester County by Richard Peters, president of the Philadelphia Society for Promoting Agriculture, comparing his own methods of analysis with those used in Europe. In 1815 he published an exhaustive *Practical Treatise on Dyeing and Callicoe Printing,* described by the twentieth-century chemist Edgar F. Smith as "masterly." In 1818 his essay "On Bleaching" appeared in the *Transactions* of the American Philosophical Society.

Cooper was not the only advocate of industrial chemistry in the Philadelphia region. "If ever there was a time to bring the Arts and Manufactures to perfection in this country," declared James Cutbush in a series of articles on chemistry and its applications in the Philadelphia newspaper *Aurora,* "it is the present; for the season is approaching when, of necessity, which is the mother of invention, our internal resources . . . will advance a brilliant and glorious epoch in the annals of our Country. . . . Who is to establish the chain of manufactures—to convert the crude productions of Nature into useful articles, but you enlightened citizens, men of *science* and improvement, *artists* and manufacturers."[22] In a similar vein Thomas P. Smith stressed the importance of chemistry for mineralogy and mining in his annual oration before the Chemical Society of Philadelphia:

> Living as we do in a new, extensive, and unexplored country, separated by an immense ocean from all other civilized nations, we must feel ourselves deeply interested in a knowledge of its mineral productions. . . . Abounding as it does with the richest ores of the most valuable

metals, we should be committing a crime of the blackest dye, were we through *wilful ignorance* to trample under our feet these invaluable gifts of the CREATOR.[23]

In keeping with Smith's exhortation, the Chemical Society invited the public to send mineral specimens for chemical analysis free of charge. A similar notice was published by the American Mineralogical Society in New York, organized by Mitchill and his coeditors of the *Medical Repository* to promote "the investigation of the mineral and fossil bodies which compose the fabric of the Globe; and, more especially, for the natural and chemical history of the minerals and fossils of the United States."[24] Mitchill also championed the cause of chemistry in the Society for the Promotion of Useful Arts.

It appears, then, that chemistry was recommended to the American public as much on patriotic and utilitarian grounds as for its intrinsic scientific importance. Benjamin Silliman regarded it as "a national reproach" that most of the products manufactured with the aid of chemical knowledge were imported from Europe instead of being made at home. Such a situation was bound to hurt American interests, he declared. Not until chemists were trained in the United States and "induced to attempt the introduction and extension of the CHEMICAL ARTS" would the American republic attain that "pinnacle of national superiority, which Great Britain and France owe more to the *successful cultivation and application of natural science,* than the one does to the prowess of her armies, or the other to the triumphs of her marine."[25]

CHEMICAL TEACHING AND RESEARCH
IN THE AMERICAN COLLEGES

Until Benjamin Silliman established chemistry in the undergraduate curriculum at Yale, the training of American chemists was carried on largely by physician-chemists associated with medical schools. At Harvard, Aaron Dexter and John Gorham taught both medical students and undergraduates. Neither man did extensive laboratory research, although each contributed one of the few chemical papers published in the *Memoirs* of the American Academy of Arts and Sciences, and Gorham published chemical analyses of sugar, corn, and indiogene [indigotin] elsewhere.[26]

In New York City, too, chemical instruction was in the hands of physician-chemists, with the important exception of John Griscom, who did much to advance the cause of chemistry by his public lectures, private school, and teaching at Columbia and Queen's College (Rutgers). Mitchill taught chemistry at Columbia until 1801 and for a few months at the College of Physicians and Surgeons after it was organized in 1807. As we have seen, he had his own ideas about chemical nomenclature and theory. In his *Synopsis of Chemical Nomenclature and Arrangement* (1801) he substituted the term *phlogiston* for *hydrogen, anticrouon* for *caloric* [heat], and *septon* for *azote* [nitrogen] in Lavoisier's nomenclature and adopted Roger J. Boscovich's atomic conception in his chart entitled "Chemical Nomenclature and Arrangement."[27] The chart was divided into nine columns or categories, each expressing "some possible or actual condition of the sixty or more primitive forms of matter." Category V—"atoms rendered half sour, or quite sour, by oxygen"—reflected Mitchill's acceptance of Lavoisier's view that oxygen is the acidifying principle. Category IV—"denominated by the generic word *Septon*"—was especially dear to Mitchill's heart. In 1795 he had advanced the theory that all infectious diseases and several others as well were caused by the action of "gaseous oxyd of azote [oxidized nitrogen gas]" on the organs of the body. This theory, propagated enthusiastically by Mitchill in the *Medical Repository* and other publications and applied by him to problems of sanitation, tanning, soil fertilization, and the like attracted considerable attention in medical circles on both sides of the Atlantic in an age noted for general theories of disease; but Mitchill's innovations in chemical nomenclature proved as ephemeral as the theoretical notions underlying them. In other respects, however, his *Synopsis* was an excellent summary of the new French nomenclature and served to introduce American readers to atomic ideas in chemistry several years before John Dalton revived the atomic theory in his *New System of Chemical Philosophy* (1808–1810).

When Mitchill relinquished the chemical chair at the College of Physicians and Surgeons in 1807 in favor of the natural history chair, the way was opened for the appointment of a considerably abler chemist, William J. MacNeven, to the professorship of chemistry. Like the mathematician Robert Adrain, MacNeven was an Irish patriot. Born in County Galway, Ireland, in 1763, he received an excellent medical education at the University of Vienna through the liberality of his uncle, a physician at the court of Empress Maria Theresa. Returning to Ireland, MacNeven played a leading role in planning the Irish rebellion of 1798, was arrested for sedition, and spent four years in prison. After serving two years in the Irish Brigade of Napoleon's army in the hope of participating in a French invasion of Ireland, he emigrated to the United States, arriving in New York on July 4, 1803.

His talents were not long in being recognized. Columbia awarded him an honorary M.D. in 1806. He became an associate physician at the Alms House, gave a course of lectures on clinical medicine at the College of Physicians and Surgeons, and was appointed professor of midwifery at that institution. A year later he became professor of chemistry. He and

Mitchill became good friends. In 1815 they published together an analysis of the mineral springs at Schooley's Mountain in New Jersey. MacNeven also published one of the earliest American accounts of the preparation of potassium metal by the reduction of potash with iron filings in a gun barrel sealed off from the air.[28]

MacNeven was an outstanding teacher, well trained, fluent in several languages, adept at experiment, and fully abreast of the latest developments in chemistry. His *Exposition of the Atomic Theory of Chemistry,* published in New York in 1819 and republished in part in Thomas Thomson's *Annals of Philosophy* and in the *Journal de Physique,* succinctly set forth the theoretical speculations and experimental researches of John Dalton, William Higgins, William Wollaston, Thomas Thomson, Jöns J. Berzelius, Joseph L. Gay-Lussac, and Jacques Thenard by means of which the ideas of the Greek atomists had been resurrected and applied to the explanation of chemical phenomena in the opening decades of the nineteenth century. It was now well established, wrote MacNeven, that elementary substances always enter into chemical combination in determinate proportions. But this law of definite proportions must have a cause, which must be that the elementary particles of matter "are so constituted as to be exempt from decay or change, though they are capable of being variously compounded together and separated again, so as to give origin to the perpetual transitions of elementary into organized matter. . . . The productions of nature have not only succeeded one another in the same general order, but have been from the beginning invariably the same. . . . All this shews that the elements of bodies are permanent and unchangeable."[29]

To this treatise MacNeven appended an account of several chemical analyses performed by students at the College of Physicians and Surgeons under his direction. These were chiefly analyses of minerals, two of which were subsequently published in German translation under MacNeven's name in J. S. C. Schweigger's *Journal für Chemie und Physik.* Schweigger particularly liked the essay on a specimen of dolomite from the marble quarry near Kingsbridge on Manhattan Island, praising it for its "exactness and skilful development of the result."

In 1821 MacNeven published *A Tabular View of the Modern Nomenclature,* bringing his readers abreast of the discoveries and improvements in chemistry since the publication of Mitchill's *Synopsis* in 1801. Gone were Mitchill's phlogiston, anticrouon, and septon. Atomism was still the underlying conception, but it was the atomism of Dalton, not that of Boscovich. Oxygen was no longer the acidifying principle in view of Humphry Davy's discovery of the elementary nature and acidifying power of chlorine. The chemical knowledge of the day was presented in tabular form in a nomenclature little different from that which would be used today. In his prefatory paragraphs MacNeven explained the purpose of the chart and expounded with admirable lucidity the relationships between nomenclature, classification, and systematic theory.

> In physical science, there are three things to be distinguished. The series of facts which compose it, the ideas which recall those facts, and the words which express them. The words should recall the ideas. — The ideas should exhibit the facts: and as they are the words which preserve and transmit the ideas, it is impossible to improve the science without improving the language. However true the facts themselves, however true the ideas to which they give birth, they might, nevertheless, convey false impressions, had we not correct terms to express them.[30]

MacNeven conceded that certain aspects of the nomenclature and classification of the new French chemistry had been called into question by recent discoveries, especially by Davy's researches on chlorine. But he maintained that the general system of naming and classifying adopted by Lavoisier and his colleagues was still valid and highly useful. A system of classification based on a general theory was essential to scientific chemistry even though the classification of particular substances might be a topic of high controversy, MacNeven insisted. Davy's researches had shown the necessity of modifying Lavoisier's theory, "yet the present classification of both oxygen and chlorine (in consequence of their resemblance), is twice as probable as either could be alone. . . . By establishing a chain of connexion between the facts discovered, the most probable basis of the science is unfolded." Clearly, MacNeven had a firm grasp not only of the chemistry of his day but also of the principles underlying all scientific investigation.

Although both Boston and New York afforded competent chemical instruction, it was to Philadelphia that young Benjamin Silliman, newly appointed professor of chemistry and natural history at Yale, journeyed in 1802 to learn the elements of the subjects he had agreed to teach. Arriving in November, he lodged at Mrs. Smith's boardinghouse at the corner of Dock and Walnut streets and was soon attending Woodhouse's chemical lectures on the first floor of Surgeons' Hall. Woodhouse had organized the Chemical Society of Philadelphia, over which he still presided, and had helped to establish the credibility of the new oxygen chemistry against Priestley's objections. His *Young Chemist's Pocket Companion* (1797), containing instructions for assembling a portable laboratory and performing various experiments, was a highly useful book.[31] His research publications, other than those on the phlogiston-oxygen controversy, were mostly concerned with the chemical analysis of minerals. As a lecturer, Silliman found him unimpressive:

> Our Professor had not the gift of a lucid mind, nor of high reasoning powers, nor a fluent diction. . . . At the commencement of my first

course with him . . . he had just returned from London, where he had been with Davy and other eminent men. He brought with him a galvanic battery of Cruickshank's construction, . . . but as it contained only fifty pairs of plates, it produced little effect. Dr. Woodhouse attempted to exhibit the exciting effects of Davy's nitrous oxide, but failed . . . and the tubes were too narrow for comfortable respiration. He did not advert to these facts, but was inclined to treat the supposed discovery as an illusion.[32]

At his boardinghouse Silliman formed a warm friendship with another Philadelphia chemist, the "genial and kind-hearted" Robert Hare. The son of a well-known Philadelphia brewer, Hare had shown an early interest in chemistry and natural philosophy, attended Woodhouse's lectures, and joined the Chemical Society of Philadelphia. To this society he presented his first invention, the oxyhydrogen blowpipe, an account of which was published by the society in 1801. This device was the outcome of Hare's search for a means of supplying a steadier flow of air for the blowpipe than could be furnished by human breath or by a bellows. For this purpose he constructed a compartmentalized barrel in which gases could be stored under water pressure and released at will. It then occurred to him that if oxygen and hydrogen were released and ignited simultaneously at the orifice, a very intense flame capable of melting the most intractable substances could be produced. Lavoisier had achieved important results in fusing such substances with a pure oxygen flame, but Hare's oxyhydrogen flame was even hotter. With it he was able to fuse completely barite, alumina, and silex. Later, working with Silliman in the small laboratory Mrs. Smith permitted them to erect in a cellar kitchen, he added strontianite to the list of hitherto infusible substances melted by the blowpipe. Silliman was so impressed with Hare's invention that he later persuaded Yale to award Hare an honorary M.D. Meanwhile, Hare was elected to the American Philosophical Society, and his account of his blowpipe was published in Tilloch's *Philosophical Magazine* and in the *Annales de Chimie.*[33]

The early conjunction of the careers of Hare and Silliman was a fortunate one for both men and for the progress of American chemistry. Silliman gave Hare encouragement and moral support during the frustrating years preceding Hare's appointment as professor of chemistry at the University of Pennsylvania, and Hare provided Silliman with the technical inventions that made possible Silliman's researches on the fusion of metallic substances. Silliman established chemistry firmly in the undergraduate curriculum not only at Yale but also, through his many students, at colleges from New England to North Carolina and westward to Kentucky and Tennessee. Hare, in turn, trained many able chemists, including John W. Draper and Wolcott Gibbs, and made the medical

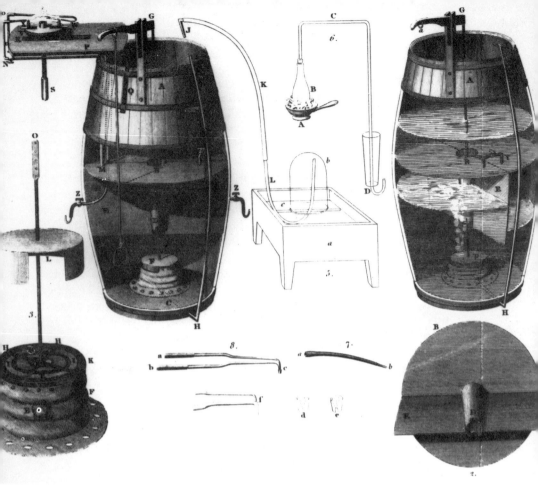

Hare's oxyhydrogen blowpipe. Gas is drawn from the glass jar
b inverted in the pneumatic trough a through the pipe LKJIH
by operating the bellows CF. When the bellows is compressed,
the gas is forced into one of the compartments into which the
cask is divided by the plate ER and stored there under hydro-
static pressure. Then another gas is drawn into the other com-
partment by turning the hood F of the bellows so as to direct
the gas into that compartment for storage under pressure.
Separate pipes M, N, O, and mno conduct the gases from the
two compartments to the platform P, where stopcocks (N and
n) are used to release the gases as desired at the point of
combustion. Hare's complete description of the blowpipe and
its operation may be found in Tilloch's Philosophical Magazine *14 (1802–1803), 238–45, 298–306, and plate VI at the*
back of the volume.

school at the University of Pennsylvania a leading center of chemical instruction throughout his distinguished tenure in the chemical chair.

Silliman was the first to establish himself academically. After two winters of study in Philadelphia, he persuaded Yale to send him to Britain to buy books and apparatus for his chemical lectures and to complete his scientific training. He took private instruction with Frederick Accum in London and arranged with him for the purchase of chemical books and apparatus, after which he spent the winter studying chemistry, mineralogy, and geology at the University of Edinburgh. In the chemical lectures of Thomas Hope, successor to the famous Joseph Black, Silliman found "a much higher standard of excellence than I had before seen."

Returning to New Haven in June 1806, Silliman resumed his chemical lectures, interspersing lectures on geology and mineralogy and giving a private course in mineralogy as well. At this point in his career an opportunity occurred that enabled Silliman to make his name known to scientists on both sides of the Atlantic. On December 14, 1807, a brilliant meteor passed over Weston, Conn., raining fiery fragments on the earth in a series of deafening explosions heard fifty miles away. When the news reached Silliman a few days later, he hurried to Weston with his colleague James Kingsley, professor of classical languages at Yale. Together they succeeded in collecting a considerable number of meteorite specimens, the largest of which weighed about six pounds.

The subject of "stones fallen from the sky" was a topic of lively debate among scientists everywhere at this time (see Chap. 6). Resolved to make the most of his opportunities, Silliman undertook a chemical analysis of the fragments, guiding himself by the memoirs of European chemists like Edward Howard, Louis Vauquelin, Martin Klaproth, and Antoine de Fourcroy, all of whom had published chemical analyses of meteorites within the past five years. The results of Silliman's analysis, published in the *Transactions* of the American Philosophical Society in 1809 and republished in the *Memoirs* of the Connecticut Academy of Arts and Sciences the following year, attracted widespread attention. The Silliman-Kingsley memoir was read aloud before the Royal Society of London and the Académie des Sciences in Paris, and notices of it appeared in several European journals. Silliman and Kingsley considered publishing a "Yalensian" book on the meteor, including a chemical and mineralogical account by Silliman, a historical essay by Kingsley, and a theoretical explanation based on former president Thomas Clap's theory that meteors were bodies orbiting the earth. The book was never published, but in 1814 Silliman inserted a section on meteors in the American edition of William Henry's *Epitome of Experimental Chemistry*.[34]

Apart from his analysis of the Weston meteor, Silliman's physical and chemical researches consisted largely of experiments on the fusion of refractory substances, using Hare's various inventions for producing intense heat. In a paper read before the Connecticut Academy of Arts and Sciences and published in their *Memoirs* in 1813, Silliman described his method of attaching Hare's oxyhydrogen blowpipe to the pneumatic cistern in his laboratory at Yale and his success in melting a number of "primitive earths" — glucine, zircon, lime, and magnesia — hitherto considered infusible. Other substances exposed to Hare's blowpipe were rock crystal, quartz, gunflint, chalcedony, oriental carnelian, red jasper, beryl, olivine, leucite, chrysoberyl, topaz, corundum, spinel ruby, and steatite. On the basis of these experiments Silliman concluded that "there is now, in all probability no body, except some combustible ones, which is exempt from the law of fusion by heat."[35] A decade later Silliman undertook still further researches on the fusion of carbon, graphite, anthracite, and diamonds with the aid of the oxyhydrogen blowpipe and two new heat-producing devices invented by Hare, the calorimotor and the deflagrator, described below. In his experiments on carbon, done with the deflagrator, Silliman noted the transfer of volatilized carbon from the positive electrode to the negative, an observation subsequently confirmed by European experimenters.[36]

However, it was as a teacher, not as a research scientist, that Silliman made his greatest contribution to the development of chemistry in the United States. Soon after his return to New Haven from his studies abroad, his laboratory in the basement of the Connecticut Lyceum was handsomely equipped with the apparatus he had purchased in England, and he had a young man named Foot as his laboratory assistant. In 1808 he brought out an American edition of William Henry's *Epitome of Chemistry* with extensive "Notes" of his own. Silliman continued publishing American editions of this work for class use until his own *Elements of Chemistry* was published in 1830–1831. His lectures, for which students paid a fee of twelve dollars and nonstudents sixteen dollars, were very popular, partly because of Silliman's commanding presence and knowledge of the world and partly because of the brilliant displays of pyrotechnics he achieved with the aid of Hare's oxyhydrogen blowpipe, calorimotor, and deflagrator.

> During the lecture hour there was no lull or intermission; all was . . . a constant appeal to the delighted senses. Here were broad irradiations of emerald phosphorescence, there the vivid spangles of burning iron, or the blinding effulgence of the compound blow-pipe, or the galvanic deflagrator. Strange sounds saluted the ear, from the singing hydrogen tube, the crackling decrepitation up to the loud explosions of mingled gases and detonating fulminates. As forms of matter once regarded simple were torn into their elements, or these again compounded in manifold ways, the very kaleidoscope of changes came

Pneumatic Cistern of Yale College

For description see Note 13.

Blow pipe tube

Sliding shelf inverted

Bent tube for transferring gases

Compound Blow-pipe for oxigen & hidrogen

Gas boxes inverted

Bellows

Drawn & engraved by A.Doolittle, from the Original constructed by Professor Silliman and invented by him & Mr Hare.

Published by W. Andrews Boston. Aug.t 1816.

Pneumatic cistern at Yale College (opposite page), incorporating Robert Hare's oxyhydrogen blowpipe and designed by Benjamin Silliman, whose portrait (by John Trumbull) is on the left. Portrait courtesy of the National Portrait Gallery, Smithsonian Institution; gift of Alice Silliman Hawkes

into view, of which the greatest was the transformation of the whole seeming phantasy into science, through the lucid rationale of the gifted lecturer.[37]

These displays, combined with Silliman's repeated references to the practical utility of chemistry in medicine, mining, agriculture, bleaching, tanning, cookery, and other practical arts and his constant acknowledgment of the beneficence of the Creator, made him one of the most popular lecturers of his day, whether in the classroom or on the public platform.

In 1816 Silliman began preparing to include lectures on the chemical analysis of minerals in both his chemical and mineralogical courses. In 1820 he was provided with spacious new quarters for his operations. At Silliman's suggestion a second story was added to the new commons building to accommodate a new chemical laboratory and the magnificent collection of minerals Silliman had procured from his friend George Gibbs of Newport, R.I. Yale's supremacy in chemistry, mineralogy, and geology was now firmly established, and Silliman had a plentiful supply of able student assistants to assist in his lectures and experiments. Through these pupils, Silliman's precepts and methods in chemistry, mineralogy, and geology were carried far and wide throughout the United States. Denison Olmsted went to North Carolina, George T. Bowen to the University of Nashville, Edward Hitchcock to Amherst College, Charles U. Shepard to Amherst and subsequently to the Medical College of South Carolina, Amos

Eaton to the Rensselaer Institute, Chester Dewey to Williams College, and Oliver Hubbard to Dartmouth College. Benjamin Silliman, Jr., went to the Medical School of the University of Louisville before returning to Yale to take his father's place and join with another Silliman student, James D. Dana, in editing Silliman's *American Journal of Science.* Through that journal and his students and public lectures, Silliman made a profound and lasting impression on the development of American science.[38]

Robert Hare's path to a distinguished career proved more difficult and roundabout than Silliman's. When Woodhouse died suddenly of a stroke in 1809, Hare enlisted Silliman, Robert Patterson, and others in support of his candidacy for the chemical chair, but Benjamin Rush secured the appointment of John R. Coxe. Deeply disappointed, Hare continued to assist his father at the brewery while keeping up his scientific contacts and studies as best he could. Silliman was his constant correspondent and confidant. Distressed at the efforts of Thomas Cooper, Joseph Cloud, and Edward Clarke of Cambridge University to deny Hare credit for invention of the oxyhydrogen blowpipe, Silliman published a letter in the *Connecticut Herald and Weekly Courier* vindicating Hare's claims. "It is proper that the public should know that Mr. Hare was the author of the invention, by means of which, in Europe, they are now performing the most brilliant and beautiful experiments; and that there are very few of these results hitherto obtained there, by the use of it . . . which were not, several years ago, anticipated here, either by Mr. Hare or by myself," Silliman wrote. Apparently the authorities at Harvard were of the same opinion, for in 1816 they awarded Hare a doctorate in medicine.

Finally, in 1818 the opportunity Hare had long desired materialized. Coxe resigned the chemical chair at the University of Pennsylvania medical school in favor of the professorship of materia medica, and Hare was elected to succeed him despite the opposition of those who insisted that only medical men should teach in the School of Medicine. Elated with his victory, Hare lost no time in demonstrating his talents as a student and teacher of chemistry. To his friend Silliman, who had just launched the *American Journal of Science and Arts,* he sent accounts of three new pieces of chemical apparatus he had devised. One of these was his sliding-rod eudiometer for measuring the volume of gases. This was essentially a cylinder and piston attached to the chamber containing the gas. The cylinder contained a graduated scale on which the experimenter could read the amount of gas admitted to or expelled from the cylinder. Hare described two forms of this device, one for use with nitric oxide or with liquids absorbing oxygen, the other for use with explosive mixtures.[39]

Hare's two other inventions were a continuation of his earlier search for ways of producing intense heat. The first of these, the calorimotor, was an electric battery designed to produce maximum heat flow. Guided by the idea that "the principle extricated from the Voltaic pile is a compound of caloric [heat] and electricity, both being original and collateral products of Galvanic action" and observing that the ratio of heat effects to electrical effects seemed to increase as the number of pairs in the pile decreased, Hare supported twenty copper and twenty zinc plates about nineteen inches square vertically in a frame, the different metals alternating at a distance of one half inch from each other. All the plates of the same kind of metal were soldered to a common slip, so that each set of homogeneous plates formed one continuous metallic surface. The copper and zinc surfaces were then connected with a wire and the whole apparatus was immersed in acid. The result was an intense ignition that consumed the wire. "The magnetic and heating effects of this instrument were surprising," wrote Edgar F. Smith in 1914, "and to this day no other form of voltaic apparatus appears to occasion the movement of so great a *volume* of heat with so low a projectile . . . force. By it, large rods of iron and platinum, when clamped between its jaws, are first fully ignited and then fused, with splendid phenomena, while at the same time its intensity is so low that hardly the least visible spark can be made to pass by it through poles of carbon."[40]

Hare's other heat-producing device was the deflagrator, described by him in the fourth volume of Silliman's *Journal* in 1822. Observing that the ignition produced by one or two galvanic pairs attained its highest intensity almost as soon as they were covered by acid, he decided to try the effect of immersing a large number of coils at once. Eighty concentric coils of copper and zinc were suspended by a beam and levers so that they could be immersed simultaneously in acid and withdrawn with equal dispatch. The sheets of copper and zinc were coiled so as not to leave a space wider than a quarter of an inch between them. Below each coil was placed a glass jar containing dilute acid. A small lead pipe was soldered to each pole of the apparatus, and a piece of charcoal about a quarter of an inch thick and an inch and a half long was inserted between these pipes. When the coils were suddenly lowered into the jars beneath them, the charcoal disappeared completely in the resulting ignition, and the pipe ends that had held the charcoal were destroyed.[41]

Silliman was greatly impressed with both the calorimotor and the deflagrator, examples of which he had received from Hare. The deflagrator, he declared, was "the finest present made to this department of knowledge since the discovery of the pile by Volta, and of the trough by Cruickshank."

> With your eighty coils . . . I obtained effects which . . . far surpassed the power of a battery of the common form of six hundred and twenty pairs of plates. . . . This . . . goes far to shake our previous theoretical opinions. . . . It is a great advantage of your *Deflagrator* that we can suspend the operation at any moment, with the same facility with which it was commenced.[42]

CALORIMOTOR.

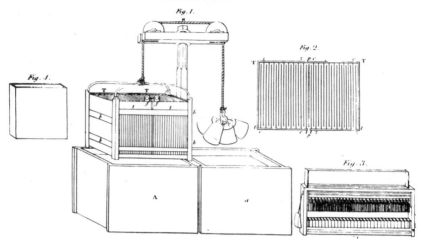

Robert Hare's calorimotor (above and left), an electric battery designed to produce maximum heat flow. Courtesy of the National Museum of American History, Smithsonian Institution

In 1839 the American Academy of Arts and Science belatedly made Hare the first recipient of the Rumford Medal for researches on heat, established by Count Rumford in 1796. European scientists also expressed appreciation of Hare's contributions in this field. Michael Faraday, after a long search for better methods of producing a strong flow of heat electrochemically, constructed a modified voltaic trough. "On examining . . . what had been done before," Faraday wrote in his memoir describing this apparatus, "I found that the new trough was in all essential respects the same as that invented and described by Robert Hare."[43]

Hare was a theorist as well as an inventor of new kinds of apparatus, but he was less successful in this line of endeavor than in the practical aspects of chemistry. Commenting on Silliman's discovery that the common galvanic battery and the deflagrator rendered each other ineffective when connected together, Hare speculated that the "voltaic fluid" was compounded of light, electricity, and caloric.

> The ordinary voltaic series employed in your experiments [he wrote to Silliman] may owe its efficacy more to electricity — and the deflagrator more to caloric. The peculiar potency of both may be arrested when they are joined, by the incompetency of either series to convey any other compound than that which it generates. The supply of caloric from the ordinary series may be too small, that of electricity too large; and vice versa.[44]

The idea that heat is a substance was an old one, supported by no less an authority than Lavoisier himself, and it enjoyed a considerable vogue in scientific circles well into the nineteenth century. Even those who were skeptical of Hare's theory of galvanism recognized that it had a certain plausibility. "The discordance of the ordinary pile with the Deflagrator," the Swedish chemist Berzelius wrote to Silliman, "appears inexplicable to me, except by the theory of Mr. Hare, which though ingenious, I find it difficult to admit, since the electromagnetic phenomena are in all their characters the same as the ordinary electricity."[45] Hare took issue with Berzelius, Davy, Liebig, Faraday, and other famous European chemists on a variety of theoretical topics. He was a mettlesome antagonist, but not one whose views eventually carried the day.

As a teacher, Hare was highly successful, especially in his experimental demonstrations. He spent freely from his own funds to acquire an extensive apparatus and made the most of his mechanical ingenuity in continually improving it. In the opinion of George Wood, a pupil and later a colleague of Hare, his chemical apparatus was "by the admission of all who had inspected it, unequalled in extent, variety, and splendor. . . . Individuals who have visited the schools of Germany, France, and Great Bri-

In this laboratory-lecture hall at the University of Pennsylvania Robert Hare displayed and used his chemical apparatus, several pieces of which were of his own invention. Wolcott Gibbs and John W. Draper were among his most distinguished pupils. Taken from Hare's
Compendium of Chemistry

tain agree in the statement, that they have no where met with a laboratory so amply furnished with all that is calculated to illustrate the science of chemistry as that of Dr. Hare."[46]

Hare's *Compendium of Chemistry,* first published in 1827, contained more than 200 copper plate engravings of this apparatus. It was a highly original textbook, ranging over the fields of physics, inorganic chemistry, animal chemistry, and plant chemistry. "In it" wrote Edgar F. Smith, "one observes the beginnings of gas analysis, the forerunners of apparatus employed by the brilliant Hofmann in the determination of the constitution of water, . . . the use of the mercury cathode in the electrolysis of metallic salt solutions, the first electric furnace ever constructed, in which calcium metal, phosphorus and calcium carbide were isolated or made." The *Compendium* also introduced the student to the leading theoretical issues in contemporary chemistry, issues on which Hare was engaged in controversy with Berzelius, Liebig, and others. Good students must have found the book highly stimulating. Wolcott Gibbs, Hare's ablest pupil and later professor of chemistry in the Lawrence Scientific School at Harvard, continued to recommend it to his students long after it had gone out of date. He referred the American chemist Ira Remsen to it when Remsen sent him an article on the nature of the chemical compounds known as double halides.

> Soon after the appearance of my article [wrote Remsen] I received a letter from Dr. Wolcott telling me that Robert Hare had expressed similar views in 1821. He sent me his copy of Hare's Chemistry and I was astonished to read the chapter that had been written fifty or sixty years before my article. The line of thought was practically identical with mine, and it was expressed beautifully. . . . Hare was both an investigator and scientific philosopher.[47]

Silliman made liberal use of the illustrations in Hare's *Compendium* in his own well-executed *Elements of Chemistry* (1830), and the two men continued fast friends to the end of their lives. Together they did much to raise the level of chemical teaching in the United States.

CONCLUSION

In chemistry even more than in astronomy the efforts of American scientists were devoted primarily to assimilating and communicating revolutionary developments in European science. The new chemical theories were more comprehensible to the general public than the abstruse formulations of mathematical astronomers and held out more promise of immediate practical application, hence the greater popularity of chemistry. At the theoretical level Priestley's emigration to the United States gave American chemists a sense of immediate involvement in the controversies generated by the new French chemistry. At the practical level there was more lip service than actual achievement in the application of chemistry to agriculture, medicine, mining, and industry; but some progress was made, and the enthusiasm generated by the prospect of practical applications did much to promote chemical study and teaching. In teaching and research physician-chemists led the way, but their example was soon emulated and in some respects surpassed by the work of Silliman and Hare. American chemistry could not compare with European chemistry in the Jeffersonian period or for several decades thereafter, but through the efforts of men like Woodhouse, Mitchill, Gorham, Griscom, MacNeven, Hare, and Silliman it had made a good beginning.

8

American Geography

WITH the victory over Great Britain and the formation of the United States of America, popular and scholarly interest in everything connected with the North American continent — its soil, climate, topography, minerals, plants, animals, and aborigines — received a powerful stimulus. Patriotism, economic interest, and scientific curiosity all dictated that the works of nature in the New World be described, catalogued, and placed at the service of the new nation. "With respect to the natural productions of this country," James Bowdoin told the members of the newly formed American Academy of Arts and Sciences, "they are perhaps as numerous as those of any other."

> But it doth not appear by any publications on the subject, that they have been examined to any great extent. . . . It is apprehended, however, that gentlemen of ingenuity and observation, have noticed and described many of them: and that their several descriptions and collections, brought into one stock, properly methodized and classed, would make a respectable figure; and encourage further examinations and researches, in order to our obtaining an extensive, and well digested body of *American* natural history.[1]

THE NATURAL AND CIVIL HISTORIES

As Bowdoin indicated, "gentlemen of ingenuity and observation" were taking up the task of describing the natural and civil history of their respective states, inspired by the example of Thomas Jefferson's *Notes on the State of Virginia.* In New England, Samuel Williams, formerly professor of mathematics and natural philosophy at Harvard College, published *The*

188

Natural and Civil History of Vermont (1794) with appropriate comments on the Count de Buffon's ideas concerning the origin and migration of species, and Jeremy Belknap devoted the third volume of his *History of New Hampshire* to "a Geographical Description of the State; With Sketches of Its Natural History, Productions, Improvements, and Present State of Society and Manners, Laws and Government." The materials for this volume, Belknap declared, had been collected "during a residence of twenty-two years in the eastern part of the State; from observations made in various places, and particularly in several journies to the northern and western parts; from original surveys of many townships and tracts of the country; from the conversation of many persons who have been employed in surveying, masting, hunting and scouting; as well as in husbandry, manufactures, merchandise, navigation and fishery."[2] Belknap had also circulated a letter of inquiry to the clergy and other "gentlemen of public character" throughout the state in search of information and had been granted fifty pounds in support of his enterprise by the New Hampshire legislature. For the identification and correct naming of the state's plants and animals he acknowledged his debt to Manasseh Cutler and William D. Peck.

The middle states produced no general accounts of their civil and natural histories, but the southern states followed the New England pattern. David Ramsay included a chapter on natural history in his *History of South Carolina*, and Governor John Drayton devoted nearly a hundred pages of his *View of South Carolina* to that subject, providing a botanical and zoological catalogue of the state and an account of the bones and teeth of an elephantlike creature dug up in Biggin Swamp during excavations for the Santee Canal.[3]

The available information for the western country was brought together in the third edition of Gilbert Imlay's *Topographical Description of the Western Territory of North America,* which included not only Imlay's account of the region, based on one year's residence in Kentucky, but also John Filson's *Discovery, Settlement, and Present State of Kentucky* (1784) and two works by the engineer-surveyor-cartographer Thomas Hutchins, "Geographer to the United States," entitled *A Topographical Description of Virginia, Pennsylvania, Maryland, and North Carolina* (1778) and *An Historical Narrative and Topographical Description of Louisiana and West-Florida* (1784) respectively.[4] An excellent surveyor and cartographer, Hutchins planned to undertake a general topographical description of the United States, but his assignment to the task of superintending the survey of the first seven ranges in Ohio under the Land Ordinance of 1785 precluded further literary efforts. He died in 1789, but not before authorizing others to use the information he had gathered on the western country.

THE GEOGRAPHIES OF MORSE AND EBELING

Meanwhile, the task of drawing together in a single volume the scattered materials on the topography, soil, climate, vital statistics, and civil and natural history of the United States had been undertaken by a person of quite different character, Jedidiah Morse, a young Congregational clergyman recently graduated from Yale, where President Ezra Stiles had recruited him for the Christian ministry. Impressed by the financial success of a small textbook, *Geography Made Easy,* he published in 1784 for the use of his students in a girls' school in New Haven, Morse decided to undertake a general work on American geography. Unlike Hutchins, Morse was not a field geographer and cartographer. He collected his materials by circular and correspondence and by amassing a private library of publications bearing on his subject, relying on professional mapmakers for the maps he needed. Jeremy Belknap, Samuel L. Mitchill, Thomas Hutchins, Noah Webster, Ebenezer Hazard of Philadelphia, and others helped him either by supplying information or by criticizing the first draft of his book.

In 1789 *The American Geography* issued from the press in Elizabethtown, N.J. In the preface Morse thanked the "men of Science" who had helped him, noting that he had frequently used their words as well as their ideas, "although the reader has not been particularly apprized of it." The fold-in map of the southern states, he explained, had been compiled by Joseph Purcell of Charleston, S.C.; that of the northern states by the engraver Amos Doolittle. The introductory section on astronomical geography was followed by an account of the discovery and settlement of America, with a summary of various opinions concerning the origins of its aborigines and native animals. Morse then described the geography, climate, and natural history of the United States before proceeding to a more detailed consideration of each state and territory. For his geographical account he relied chiefly on Thomas Hutchins and earlier geographers like Lewis Evans and Thomas Pownall and on Jonathan Carver's *Travels through the Interior Parts of North America in the Years 1766, 1767, and 1768* (1778). On the subject of American botany he confined himself to listing various cereals, fruits, vegetables, aromatic herbs, and medicinal plants, giving their common names only. His catalogue of animals, "collected principally from [Mark] Catesby, Jefferson, and Carver," was divided into beasts, birds, snakes, insects, and lizards. Following Buffon, the "beasts" were divided into those peculiar to the New World, those of the same genus but of different species from those in the Old World, and those indistinguishable from the Old World species. Several of these animals, including the mammoth (mastodon), opossum, buffalo, raccoon, and beaver, were described in separate paragraphs. One hundred and forty-

Three writers on American geography: Jedidiah Morse (by Nathaniel Hancock) (upper left), Jeremy Belknap (by Henry Sargent) (upper right), and Christoph Daniel Ebeling of Hamburg, Germany (engraving) (left). Courtesy of the Yale University Art Gallery (Morse), Massachusetts Historical Society (Belknap), and Houghton Library, Harvard University (Ebeling)

one birds were listed by their common names and a few described briefly. The snakes received a similar treatment. Morse's "insects," in keeping with Linnaeus's practice of lumping all invertebrate animals under "insects" and "worms," included snails, worms, spiders, and the sheep tick as well as flies, wasps, bees, mosquitoes, and the like. Morse's natural history closed with an account of the alligator and various Carolina lizards.[5]

The reception of *The American Geography* was highly encouraging, although rather critical in some quarters. Mitchill recommended an enlarged account of the American Indians and urged Morse to enumerate the fishes of the United States as well as the quadrupeds, birds, snakes, and "insects." He sent Morse a list of more than sixty species of fish found off the coast of Long Island, including the sepia, or ink-fish, and an article on the climate of New York.[6] Manasseh Cutler, who apparently had not been consulted before the *Geography* was published, found Morse's account of the plants and animals of the United States "very erroneous and defective" and offered his services in correcting the errors. To this end he enclosed several catalogues of trees and plants with their Latin names and a list of New England animals prepared in collaboration with Peck for Belknap's *History of New Hampshire*. The latter list, he acknowledged, was very imperfect, especially as to the insects.

> The best I could give you would be an exact copy from Dr. Belknap's history. . . . They have some insects at ye southward which are not found here—but it is not in my power to give you their Latin names— nor have I time at present to attempt an arrangement of them. . . . Perhaps it would not be best to attempt to be particular in this part of our natural history, of which so little is yet known.[7]

Belknap and Hazard were critical of Morse's maps, which were not the best, and nearly all the authorities admonished Morse for publishing without careful scrutiny information sent to him by his correspondents. Still, as Hazard conceded, the *Geography* was "the best publication of the kind that is extant."[8] Apparently publishers abroad agreed with this judgment, for they soon honored the work by issuing pirated editions in London, Dublin, and Edinburgh. Most encouraging of all, the German geographer Christoph D. Ebeling, professor of history and Greek at the Gymnasium of Hamburg and himself deeply engaged in a work on American geography, added his voice to the chorus of praise: As early as 1777, Ebeling wrote to Morse, "I had contemplated writing a Geography of America, but suspended my purpose because of the great and rapid changes then going on in your country. I am now glad of my delay; for I find I was misled by following English authors, and had imbibed from them prejudices which the perusal of your work has happily removed."[9]

Exhilarated by the success of his book, Morse immediately began work on a new edition that would correct the errors of the first and expand its coverage by adding a second volume on the geography of the Old World. In 1793 this new edition appeared under the title *The American Universal Geography: Or a View of the Present State of All the Empires, Kingdoms, States and Republics in the Known World, and of the United States of America in Particular.* The format of the American volume was much the same as in 1789, but there were improvements in the sections devoted to natural history. Francisco Clavigero's views on the peopling of America, set forth in his *History of Mexico,* were added to those drawn from William Robertson's *History of the Discovery and Settlement of America,* and their consonance with the biblical account of early human history was pointed out. Mitchill's narrative of his visit to the Seneca Indians and his account of the geology of the maritime regions of New York were quoted. Cutler's hand was apparent in the botanical section. American plants were divided into "Grain, Cultivated in the Eastern and Middle States," "Cultivated Grasses in the Eastern and Middle States," "Native Grasses in New England," "Wild Fruits in New England," "Nut Fruits," "Medicinal Plants in New England," "Flowering Trees and Shrubs in the United States," and "Forest Trees" and provided with Latin names. For a fuller account, "till a more perfect catalogue be furnished by Dr. Cutler and Dr. [Benjamin S.] Barton," Morse referred his readers to Catesby's *Natural History,* Jefferson's *Notes,* Cutler's account of New England botany in the *Memoirs* of the American Academy of Arts and Sciences, and two more recent publications: William Bartram's *Travels through North and South Carolina, Georgia, East and West Florida* . . . (1791) and Belknap's *History of New Hampshire.* Cutler also supplied Morse with a catalogue of American animals with their Linnaean names, and Joshua Fisher, a physician in Beverly, Mass., contributed some descriptions of various quadrupeds. The extensive catalogue of American birds was taken largely from Bartram's *Travels.* All were given Latin names, as were the "serpents" and "insects," but not the fishes, which Mitchill had offered to describe. The two main maps were the same as in the 1789 edition, but others were added, including a map of the District of Maine.

Morse had now established himself as the "father of American geography," challenged for that honor only by Ebeling, whose first volume of *Erdbeschreibung und Geschichte von Amerika: Die vereinten Staaten von Nordamerika,* devoted to the geography of New Hampshire and Massachusetts, appeared in the same year as Morse's *American Universal Geography.* But Ebeling, though a better scholar and geographer than Morse, was destined to offer little competition. His work was in German, which most Americans (including Morse) did not read, and it was never translated into English. Ebeling moved ahead slowly and cautiously, state by state, intending to add a general volume on the United States as a whole when he had completed the states and the western country. Moving southward from New England, he managed to finish his seventh volume, devoted to Virginia and dedicated to William Bentley, Samuel Miller, Samuel L. Mitchill, and Henry S. Tucker, one year before his death in 1817. For nearly forty years he had managed to keep abreast of the latest publications on every aspect of American geography and civil and natural history despite the obstacles interposed by war and revolution; but the task he had set himself was too great for a single lifetime, and the language in which he wrote insured that his volumes would have few readers in the United States. Nevertheless, through his correspondence with Americans like Morse, Belknap, Mitchill, William Bentley, James Freeman, Ezra Stiles, Noah Webster, Benjamin S. Barton, James Madison, David Ramsay, and others he supplied the learned world in the United States with a copious stream of information and publications from the continent of Europe and amassed a collection of Americana that was the admiration of all who saw it. "His library of American books, i.e., books on American affairs, is nearly as large as the Richmond Hill library," wrote Aaron Burr after a visit to Ebeling in 1809. "Geography is more particularly his department; and the extent and accuracy of his knowledge is truly astonishing."[10] Fortunately for Americans, Ebeling's collection of Americana was purchased at auction by Israel Thorndike, a Boston merchant, and given by him to the Harvard College Library, where it remains to this day.

As for Morse, he had chosen a different but equally difficult approach to the task of encompassing American geography in all its civil and natural ramifications. He would attempt a synoptic view of the whole, keeping abreast of the progress of knowledge in successive editions. He had few illusions about the magnitude of this undertaking. "The field before me is extensive, and I sometimes contemplate it with a misgiving heart," he wrote Ebeling in May 1794.

> I have but a slender constitution, a large and growing parish, many interruptions inseparable from my situation, and . . . an extensive correspondence. In such forbidding circumstances, to undertake the description of an unexplored, or but partially explored country, rising into importance with unexampled rapidity; and to attempt in successive editions of an Universal Geography, to keep pace with the progress of this age of discoveries, of changes, and of revolutions, are objects from which I shrink when I think of their magnitude.[11]

Morse's task was even larger than he realized. To carry it through on the model originally projected he would have to take account of the geographical, mineralogical, geological, paleontological, botanical, zoo-logical, and anthropological articles published in transactions of the American Philosophical Society, the American Academy of Arts and Sciences, and other learned societies; in the various medical journals, notably Mitchill's *Medical Repository;* and in the collections of various historical societies. He would have to pay attention to various local histories and guides such as Mitchill's *Picture of New York* (1807), James Mease's *Picture of Philadelphia* (1811), and Daniel Drake's *Natural and Statistical View; Or, Picture of Cincinnati and the Miami Country* (1815); Archibald Bruce's *American Mineralogical Journal* (1810–1814), the botanical works of Barton, Henry Muhlenberg, David Hosack, Jacob Bigelow, Thomas Nuttall, and Frederick Pursh, and the publications of foreign visitors or immigrants like William Maclure and the Count de Volney. After the Louisiana Purchase, which nearly doubled the territory of the United States, he would have to deal with the flood of information from the trans-Mississippi West that the scientific zeal of Jefferson, Morse's *bête noir* in politics, was making available. To what extent Morse succeeded in keeping abreast of new developments in geography and natural history will be seen later in this chapter. But first we must say something about the progress of scientific exploration in the western regions.

THE LEWIS AND CLARK EXPEDITION AND ITS SCIENTIFIC RESULTS

Long before the acquisition of the Louisiana Territory, Jefferson had been interested in western exploration for both national and scientific reasons. "While I was in Europe," he told William Dunbar, "I purchased everything I could lay my hands on which related to any part of America, and particularly had a pretty full collection of the English and Spanish authors on the subject of Louisiana."[12] In 1793, acting on behalf of the American Philosophical Society, Jefferson drafted instructions for an expedition up the Missouri River and thence to the Pacific Ocean by the French botanist and explorer André Michaux, who had volunteered to make such a trip. According to Jefferson's instructions, Michaux was to "take notice of the country, . . . it's general face, soil, rivers, mountains, its productions animal, vegetable, & mineral so far as they may be new to us & may also be useful or very curious; the latitude of places or materials for calculating it by such simple methods as your situation may admit you to practice, the names, numbers, & dwellings of the inhabitants, and such particularities as you can learn of their history, connection with each other, languages, manners, state of society & of the arts & commerce among them."[13] The expedition collapsed when it was discovered that Michaux had become an agent of the French government, but Jefferson continued to dream of western exploration.

In the summer of 1802, more than a year after his inauguration as president of the United States, Jefferson's interest in trans-Mississippi exploration was reawakened when he read Alexander Mackenzie's *Voyages from Montreal . . . through the Continent of North America, to the Frozen and Pacific Ocean.*

> Whatever course may be taken from the Atlantic [Mackenzie wrote], the Columbia [River] is the line of communication from the Pacific Ocean, pointed out by nature, as it is the only navigable river in the whole extent of Vancouver's minute survey of that coast: its banks also form . . . the most Northern situation fit for colonization, and suitable to the residence of a civilized people. By opening this intercourse between the Atlantic and Pacific Oceans, and forming regular establishments through the interior, and at both extremes, as well as along the coast and islands, the entire command of the fur trade of North America might be obtained, from latitude 48. North to the pole, except that portion of it which the Russians have in the Pacific. To this may be added the fishing in both seas, and the markets of the four quarters of the globe.[14]

The challenge implicit in Mackenzie's recommendations to the British government was not long in being taken up by Jefferson. Before the end of the year he had inquired of the Spanish and British ministers in Washington what their governments' reactions would be if the United States should send a party of travelers up the Missouri and across the mountains to the Pacific Ocean. "The President," the British minister informed his government, "has for some years past had it in view to set foot an expedition entirely of a scientific nature for exploring the Western Continent of America by the route of the Great River Missouri. . . . He supposes this to be the most natural and direct water-communication between the two Oceans, and he is ambitious in his character of a man of letters and science, of distinguishing his Presidency by a discovery, now the only one left to his enterprise—the Northern Communication having been so ably explored and ascertained by Sir Alexander Mackenzie's journeys."[15]

Such indeed was Jefferson's purpose, although he was careful to stress the commercial, military, and diplomatic benefits that would accrue from the expedition in his message to Congress requesting funds for the enterprise. To the Count de Lacépède, however, Jefferson described the expedition's aims as being "to enlarge our knoledge of the geography of our continent, by adding information of that interesting line of communication across it, and to give us a general view of it's population, natural

*Leaders of the famous Lewis and Clark expedition to the
Pacific Ocean by way of the Missouri River: Meriwether
Lewis (right) and William Clark (left) (both by Charles
Willson Peale). Courtesy of the Independence National His-
torical Park Collection*

history, productions, soil & climate. . . . It is not improbable," he added,
"that this voyage of discovery will procure us further information of the
Mammoth, & of the Megatherium . . . ," both of which animals Jefferson
thought must still be extant somewhere on the earth.[16]

Jefferson bent every effort to the attainment of these ends. His
secretary Meriwether Lewis was selected to lead the expedition not only
because of his knowledge of the western country but also on account of
his longstanding interest in botany, zoology, geography, and ethnology.
After receiving instruction from Jefferson himself, Lewis was sent to
Philadelphia and on to Lancaster, Pa., to receive further scientific training and suggestions from members of the American Philosophical Society, of which Jefferson was president. Barton was asked to brief Lewis concerning botany, zoology, and Indian ethnology and linguistics. Similar
requests went to Caspar Wistar and Benjamin Rush. Robert Patterson
and Andrew Ellicott were asked to give Lewis further training in making
astronomical observations and to advise him about the instruments he
should purchase for this purpose. Ellicott was especially helpful. "Mr.
Ellicot is extremely friendly and attentive . . . : he thinks it will be necessary

I should remain here ten or twelve days," Lewis wrote to Jefferson from Lancaster.[17]

Meanwhile, Jefferson was busy gathering information about the Louisiana Territory and planning still other expeditions into the region. In November 1803 he communicated to Congress "An Account of Louisiana" based on the results of a questionnaire drawn up by Jefferson and his secretary of the treasury, Albert Gallatin, and circulated among knowledgeable westerners like James Wilkinson, William Dunbar, and William C. Claiborne. In the next three years several expeditions were sent into the trans-Mississippi region under various auspices. Under orders from General Wilkinson, Zebulon Pike conducted two successive expeditions for the War Department, the first to the upper Mississippi Valley, the second to the headwaters of the Arkansas River. Jefferson himself took the initiative in organizing an expedition up the Red River under the leadership of Thomas Freeman and another to the Red River and its tributary, the Washita, under William Dunbar and George Hunter.[18] But these explorations were soon overshadowed by the far more dramatic and scientifically fruitful Lewis and Clark expedition.

On July 19, 1803, Lewis began recruiting his exploring party. From Washington he wrote to William Clark inviting him to join the expedition and outlining its objectives as set forth by Jefferson. The expedition would seek "an early friendly and intimate acquaintance" with the Indian tribes inhabiting the regions through which it passed and seek to open trade with them.

> The other objects of this mission are scientific, and of course not less interesting to the U. States than to the world generally, such as the ascertaining by celestial observation the geography of the country through which I shall pass; the names of the nations who inhabit it, the extent and limits of their several possessions, their relation with other tribes and nations; their languages, traditions, and monuments; their ordinary occupations in fishing, hunting, war, arts, and the implements for their food, clothing and domestic accommodation; the diseases prevalent among them and the remidies they use; the articles of commerce they may need, or furnish, and to what extent; the soil and face of the country; it's growth and vegetable productions, its animals; the mineral productions of every discription; and in short to collect the best possible information relative to whatever the country may afford as a tribute to general science.[19]

In Washington, Jefferson was making every effort to gather information about the projected route of the expedition. Gallatin, who turned out to be remarkably well informed about the western country, had the cartographer Nicholas King draw up a blank map on which could be entered the information derived from the voyages of Cook, Vancouver, and Mackenzie; from Aaron Arrowsmith's *A Map Exhibiting All the New Discoveries in the Interior Parts of North America . . . with Additions to 1802* (1802); and other sources. From William H. Harrison, governor of the Indiana Territory, Jefferson obtained extracts from the journal of Jean-Baptiste Truteau, agent of a St. Louis trading company, and a copy of John Evans's map of the region between St. Louis and the Mandan villages. Harrison also sent Clark a map that was probably a copy of one drawn by James Mackay during his travels on the Missouri in the summer of 1797. The Evans and Mackay maps gave accurate charts of the Missouri as far as its northernmost bend, but they disagreed as to the location of the Mandan villages.[20] A copy of British traveler David Thompson's 1798 map of the Mandan country was also procured.

On May 21, 1804, Lewis and his men began their ascent of the Missouri from the village of St. Charles "with a Barge of 18 oars, attended by two large perogues; all of which were deeply laden, and well manned," Amos Stoddard reported to the secretary of war from St. Louis. "I have heard from him about 60 miles on his route," Stoddard wrote, "and it appears, that he proceeds about 15 miles per day—a celerity seldom witnessed on the Missouri; and this is the more extraordinary as the time required to ascertain the courses of the river and to make other necessary observations, must considerably retard his progress. His men possess great resolution and they [are in the best] health and spirits."[21]

The party moved up the river without opposition and wintered at Fort Mandan. Before embarking for the headwaters of the Missouri in the spring of 1805, Lewis shipped the scientific information he had accumulated thus far to St. Louis. The materials included sixty-seven specimens of "earths, salts, and minerals" and sixty plant specimens carefully labeled to show when and where they had been collected. "These have been forwarded with a view of their being presented to the Philosophical society of Philadelphia, in order that they may under their direction be examined or analyzed," Lewis explained. He added:

> I have transmitted to the Secretary at War, every information relative to the geography of the country which we possess, together with a view of the Indian nations, containing information relative to them. . . . At this moment, every individual of the party are in good health, and excellent sperits; zealously attached to the enterprise, and anxious to proceed; not a whisper of discontent or murmur is to be heard among them; but all in unison, act with the most perfect harmoney. With such men I have everything to hope, and but little to fear.[22]

The accompanying invoice listed, besides the items mentioned by Lewis in his letter, the skins of male and female pronghorn antelopes with their

Route of the Lewis and Clark Expedition

1804-06

0	100	200	300	400

miles

Pacific Ocean

B R I T I S

drainage

CHINOOK

Fort Clatsop

TILLAMOOK

BLACKFOOT   Milk River

FLATHEAD   Marias R.

LEWIS AND C

Clark Fork   LEWIS

1806

Lewis and   LEWIS

Clark Pass   Great   R.

Falls   Judith R.

Clearwater   Lolo   Musselshell

1805   River   Pass

YAKIMA   Traveller's   Three Forks   Yellow

LEWIS AND CLARK   Rest

Columbia 1806   Bitterroot   Missouri R.   CLARK

Mt.   The Dalles   R.

Hood   NEZ   CLARK

PERCE   Salmon

R.   Bighorn

OREGON   Lemhi Pass

C O U N T R Y   SHOSHONI

Snake   Mou

drainage   River   divide

Great   M o

Salt   u

S P A N I S H   Lake

P O S S E S S I O N S

200

Reproduced from Donald Jackson's Thomas Jefferson and the Stony Mountains *by courtesy of the University of Illinois Press*

skeletons; the horns and ears of the black-tailed deer; the skins of a martin, a weasel, and three small squirrels; the skeleton of a "burrowing wolf"
[coyote] and of a white-tailed jackrabbit; the skin of a "brown, or yellow
bear"; the skin and skeleton of a badger; four mountain ram horns, thirteen red fox skins; several buffalo robes; and four live magpies, a live
prairie dog, and a live "hen of the Prairie" (sharp-tailed grouse). Lewis
promised to send a final message from the furthest navigable point on
the Missouri before crossing the Rocky Mountains, but nothing more was
heard from the expedition until it returned to St. Louis a year and a half
later in September 1806.

One can imagine the impatience with which Jefferson awaited the arrival of Lewis's shipment, the intense interest with which he received it,
and the care he took to preserve the specimens and arrange for their identification and description. On July 14 he sent Lewis's large map of the
Missouri country to the War Department to be reduced to a scale of twenty
miles to the inch for engraving. In August, having learned that a barrel,
four boxes, and a cage containing the prairie dog and a surviving magpie
had arrived in Washington, Jefferson sent instructions from Monticello
for the care and disposition of the skins, furs, horns, bones, seeds, and
animals contained in the shipment. "Be so good as to have particular care
taken of the squirrel & pie . . . that I may see them alive at my return,"
he wrote to his majordomo at the White House. "Should any accident
happen to the squirrel his skin & skeleton must be preserved."[23]

Returning to Washington in October, Jefferson sent Lewis's box of
minerals to the American Philosophical Society and the skins, skeletons,
horns, and the "burrowing squirrel" and magpie to Charles W. Peale for
his museum, keeping some of the Indian artifacts and wapiti horns for
display in the Indian Hall at Monticello.[24] Jefferson consigned the seeds
from the western country to William Hamilton, owner of Philadelphia's
most extensive and magnificent garden. Somewhat later, others were sent
to Bernard M'Mahon, a Philadelphia nurseryman with whom Jefferson
had long corresponded concerning horticultural matters. Jefferson relied
primarily on Benjamin S. Barton for the identification and description
of the plants and animals. Nothing was to be published until Lewis returned
to take charge of the report of the expedition, but Jefferson gave the public
a foretaste of the fruits of the expedition in a message to Congress on
February 19, 1806, supplementing his earlier "Account of Louisiana." The
message of 1806 communicated "Discoveries Made in Exploring the
Missouri, Red River and Washita, by Captains Lewis and Clark, Doctor
Sibley, and Mr. Dunbar; With a Statistical Account of the Countries Adjacent." The message began with an extract of Lewis's letter to Jefferson
from Fort Mandan dated April 17, 1805, and included "A Statistical View
of the Indian Nations Inhabiting the Territory of Louisiana and the Coun-

tries Adjacent to Its Northern and Western Boundaries," based on information gathered by Lewis and Clark.[25]

With the safe return of Lewis and his men to St. Louis in September 1806, bearing still new scientific trophies — "several skins of the Sea Otter, two skins of the native sheep of America, five skins and skeletons complete of the Bighorn or mountain ram, and a skin of the Mule deer beside the skins of several other quadrupeds and birds . . . a pretty extensive collection of plants, and . . . nine other [Indian] vocabularies" — Jefferson's hopes for the publication of a report that would constitute a memorable contribution to the annals of science were raised to a new pitch. On December 28 Lewis finally arrived in Washington accompanied by an Osage chief, the chief's wife and son, and an interpreter and his family. Clark had remained in Virginia to pursue his matrimonial intentions. Two days later Lewis and his party were received by the president, and Lewis resumed his former intimacy with Jefferson. What would one not give to have been an observer of the private conversations of the two friends in the ensuing days as Lewis relived the expedition in response to Jefferson's questions, Jefferson listening eagerly and undoubtedly wishing that he might have shared the great adventure.

Both men were concerned with publishing the narrative and scientific results of the expedition as soon as possible. When the necessary arrangements had been made to muster out the members of the expedition and reward them appropriately for their services to the nation, Lewis proceeded to Philadelphia to find a publisher and enlist the help of scientists in preparing his report. Arriving there in mid-April, he issued a prospectus of the forthcoming work through the publishing firm of Conrad and Son. At M'Mahon's tree nursery he had the pleasure of seeing seven different kinds of plants grown from seeds collected on the expedition. There he met the German botanist Frederick Pursh and made arrangements for him to describe the western plants, including the Fort Mandan collection, since Benjamin S. Barton had made little progress with the specimens entrusted to him.

At Peale's Museum, Lewis inspected the zoological and ethnological trophies of the expedition, sat for his portrait, and permitted Peale to take a life mask of his face for use in preparing a wax figure. In return, Peale agreed to make portraits of some of the animals collected by the expedition. For bird portraits Lewis engaged the services of Alexander Wilson, a Scottish immigrant who was then hard at work on his monumental *American Ornithology*. "It was the request and particular wish of Captain Lewis made to me in person that I make drawings of each of the feathered tribe as had been preserved, and were new," Wilson wrote in that work. For portraits of the Indian chiefs who had returned with the expedition, Lewis turned to Charles de St. Memin, a French refugee residing in Philadelphia. James Barralet, an Irish-born engraver, was given the task of preparing two illustrations of waterfalls. Finally, the Swiss mathematician and geodesist Ferdinand R. Hassler, then teaching at the U.S. Military Academy at West Point, was engaged to correct the longitude determinations made on the expedition, the shortcomings of Lewis's chronometer having rendered these suspect.

Late in July 1807 Lewis left Philadelphia to prepare himself for assuming his new duties as governor of the Louisiana Territory. Pleased with the progress of Lewis's negotiations for publishing the report of the expedition, Jefferson wrote to Lacépède describing some of the animals discovered by Lewis and Clark and announcing:

> Their description will be given in the work of Governor Lewis, the journal and geographical part of which may be soon expected from the press: but the parts relating to the plants & animals . . . will be delayed by the engravings. . . . The plants of which he brought seeds have been very successfully raised in the botanical garden of Mr. Hamilton of the Woodlands and by Mr. McMahon. . . . And, on the whole, . . . I can assure you that the addition to our knoledge, in every department, resulting from this tour . . . has entirely fulfilled my expectations in setting it on foot. . . . I will take care that the Institute as well as yourself shall recieve Govr. Lewis's work, as it appears.[26]

Alas for Jefferson's hopes and expectations! As governor of the Louisiana Territory, Lewis soon became embroiled in bitter controversies that unbalanced his mind and distracted his attention from the task of compiling a narrative of the expedition. Then, in October 1809 Jefferson received word that Lewis had died under mysterious circumstances at a roadside tavern in Tennessee, probably by his own hand. With a heavy heart he asked Clark to take charge of the projected publications and wrote to Peale, Barton, M'Mahon and others to inform them of the arrangements. Diffident of his literary abilities, Clark asked Nicholas Biddle, a Philadelphia businessman, diplomat, and literary figure, to write the narrative, with the understanding that Barton would prepare the natural history of the expedition.

After some hesitation Biddle accepted the assignment, but he soon found the task more difficult than he had anticipated. Clark supplied an excellent map of the western territory for the engraver, but endless difficulties impeded the work on the natural history and ethnography of the expedition. Barton, who was in poor health and had a reputation for promising more than he could deliver, seems to have done very little with the materials entrusted to him. Pursh worked diligently on the botanical specimens, but in April 1809, six months before Lewis's death, he had left Philadelphia for New York, taking with him copies of the drawings and descriptions he had made and a considerable number specimens as well. The Indian vocabularies Lewis had collected and sent to Jefferson were lost when the trunk carrying all Jefferson's laboriously collected lists to Monticello was stolen and the contents dumped into the James River. In 1812 the publisher of the expedition report went bankrupt, and Biddle had to make new arrangements for publication with the firm of Bradford and Inskeep.

Meanwhile, Biddle, having been elected to the state legislature in 1810, turned over the task of completing the narrative to Paul Allen, a fellow contributor to the *Port Folio* magazine. Jefferson, who had done all he could to bring about the publication of the combined narrative and scientific account of the expedition as rapidly as possible, wrote sadly to Baron von Humboldt in December 1813:

> You will find it inconcievable that Lewis's journey to the Pacific should not yet have appeared; nor is it in my power to tell you the reason. . . . I think, however, from what I have heard, that the mere journal will be out within a few weeks in 2 vols. 8vo. . . . The botanical & zoological discoveries of Lewis will probably experience greater delay, and become known to the world thro' other channels before that volume will be ready. The Atlas, I believe, waits on the leisure of the engraver.[27]

Jefferson's estimate of the situation was close to the mark. Early in 1814 the Biddle-Allen *History of the Expedition under the Command of Captains Lewis and Clark, to the Sources of the Missouri, Thence across the Rocky Mountains and down the River Columbia to the Pacific Ocean, Performed during the Years 1804-5-6* was published in Philadelphia. Besides the narrative of the expedition the work contained Jefferson's "Life of Captain Lewis"; a chapter on the customs and character of the Clatsop, Killamuck, Chinook, and Cathlamah Indians; a chapter containing a general description of the "beasts, birds, plants, etc." observed by the explorers; several brief appendices relating to the fur trade, Indian nations, geographical distances, and climate; and Clark's map of the country north and west of the junction of the Missouri and Mississippi rivers, showing the outbound and returning routes of the expedition. The atlas mentioned by Jefferson in his letter to Lacépède was never published, Bradford and Inskeep having failed financially in the year the *History* was issued.

Of the approximately 200 plant specimens collected by Lewis and Clark, 124 were identified and described scientifically in Pursh's *Flora Americae Septentrionalis,* which appeared in London in the same year the *History* was published. Pursh was well aware of the importance of the Lewis and Clark specimens. He created two new genera, *Lewisia* and

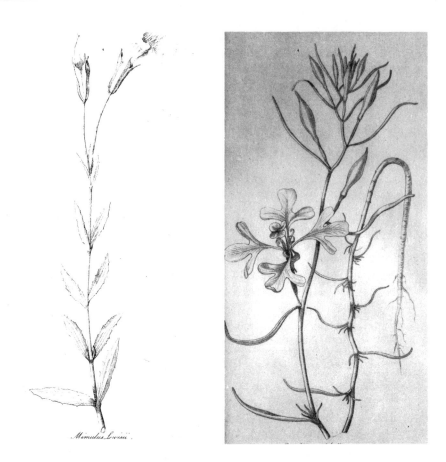

Mimulus Lewisii.

Plants named in tribute to Lewis and Clark in Frederick Pursh's Flora Americae Septentrionalis: *a specimen of the species* Mimulus lewisii *(left) and another of the genus* Clarkia. *Courtesy of the American Philosophical Society*

Clarkia, after the explorers, named three new species after Lewis — the wild flax (*Linum lewisii*), the monkey flower (*Mimulus lewisii*), and the syringa (*Philadelphus lewisii*) — and devoted 13 of the 27 illustrations in his *Flora* to specimens Lewis and Clark had collected. He also expressed regret that the plants collected between Fort Mandan and the Rockies and stored over the winter in a cache near the Great Falls of the Missouri River had been lost to science. "The loss of this first collection is the more to be regretted," wrote Pursh, "when I consider that the small collection communicated to me, consisting of about one hundred and fifty specimens, contained not above a dozen plants well known to me to be natives of North America, the rest being either entirely new or but little known."[28]

The zoological specimens brought or sent back by Lewis and Clark had the good fortune to be preserved in Peale's Museum, where naturalists could inspect them. Alexander Wilson named, described, and illustrated three of the birds—the western tanager (*Piranga ludoviciana*), Lewis's woodpecker (*Asyndesmus lewis*), and Clark's "crow" or nutcracker (*Nucifraga columbiana*)—in his *American Ornithology*. George Ord described and gave scientific names to the Columbian ground squirrel, the black-tailed prairie dog, the bushy-tailed wood rat, the grizzly bear, the mountain goat, the Oregon ruffed grouse, the Columbian sharp-tailed grouse, and the pronghorn antelope in an anonymous contribution to the second American edition of William Guthrie's *A New Geographical, Historical, and Commercial Grammar* (1815) and did likewise for the western gray squirrel and the eastern wood rat in subsequent memoirs. Constantine Rafinesque described and named the mountain beaver, the mule deer, the Oregon bobcat, and the prairie rattler. In 1823 Thomas Say added the short-tailed shrew, the coyote, the plains gray wolf, and the swift fox to the list of animals discovered for science by Lewis and Clark. The sage grouse was described by Charles L. Bonaparte in 1827, the mountain quail by David Douglas in 1837, the white-tailed jackrabbit by John Bachman in 1837, the poor-will by Audubon in 1839, and the piñon jay by Prince Maximilian of Wied-Neuwied in 1841.

The Lewis and Clark specimens in the museum also made an important contribution to such pioneering works as Wilson's *American Ornithology,* Richard Harlan's *Fauna Americana* (1825), John Godman's *American Natural History* (1826–1828), and J. and T. Doughty's *The Cabinet of Natural History and American Rural Sports* (1830–1832), all published in Philadelphia by naturalists who had access to Peale's Museum. Lewis and Clark had not discovered living mastodons or megalonyxes, as Jefferson and Wistar had hoped, but they had done the next best thing by finding the grizzly bear, the pronghorn, the mule deer, the prairie dog, and the magpie; their discoveries inspired a new generation of American naturalists—John Bradbury, Thomas Nuttall, Titian Peale, Thomas Say, John J. Audubon, and others—to follow in their footsteps.[29]

Western botany, zoology, and mineralogy could await further exploration by later investigators, but the ethnographic and linguistic researches of Lewis and Clark preserved for posterity information that would soon cease to be available as the tribes that provided it were decimated by the onward rush of western settlement. Peale's Museum was the first repository for the Indian artifacts collected on the expedition, if Jefferson's collection at Monticello is excepted. Some of the items on display were sent to Peale by Jefferson, others were presented by Lewis or Clark; still others were obtained for the museum from Jefferson's Bedford County estate, Poplar Forest, after Jefferson's death. Among Lewis's gifts were the peace

Three ornithological discoveries of the Lewis and Clark expedition, as delineated in Alexander Wilson's American Ornithology: *(upper left) the western tanager* (Piranga ludoviciana), *(upper right) Clark's nutcracker* (Nucifraga columbiana), *and (bottom) Lewis's woodpecker* (Asyndesmus lewis). *Courtesy of the American Philosophical Society*

pipes and ermine-skin mantle he had been given by Cameahwait, brother of Sacajawea, the young Indian woman who accompanied the expedition. Moved by Lewis's account of the reunion of Sacajawea and her brother and the consequent establishment of harmony between the Shoshonis and the exploring party, Peale created an exhibit displaying a wax figure of Lewis wearing the mantle of 140 ermine skins and holding the pipe of peace while he addressed words of peace and amity to the Indian chief. Other exhibits included a handsome painted buffalo robe, an Omaha tobacco pouch, Cree animal-skin garments painted with red and black symbolic figures, a Crow Indian cradle, a Chinook woman's skirt of cedarbark fiber, and an Indian hunting shirt made of buffalo skin and worn on the expedition by Clark.

Clark's personal collections were saved by him and placed in his own museum in St. Louis, which was founded in 1816 three years after his appointment as governor and superintendent of Indian affairs for the Missouri Territory. The museum occupied a long wing of his two-story brick house and served as an Indian council chamber as well as a museum. A visitor to St. Louis in 1816 recorded the scene there on a ceremonial occasion.

> The Gov. was of remarkably fair complexion and with gray locks and light blue eyes, hence the epithet White Chief. On the day of the solemn diplomatic session the Governor's large council chamber was adorned with a profuse and almost gorgeous display of ornamented and painted buffalo robes, numerous strings of wampum, every variety of work of porcupine quills, skins, horns, claws, and bird skins, numerous and large Calumets, arms of all sorts, saddles, bridles, spears, powder horns, plumes, red blankets and flags. . . . In the center of the hall was a large long table, at one end of which sat the Governor with a sword lying before him, and a large pipe in his hand. He wore the military hat and the regimentals of the army.[30]

Most of the Indian artifacts and zoological specimens collected by Lewis and Clark were lost to science when Clark's and Peale's museum collections were dispersed in the 1830s and 1840s.

The expedition's ethnographic and linguistic data had a different fate. When Jefferson learned of the death of Barton late in 1815, he wrote to the Abbé Correa da Serra in Philadelphia to enlist his help in recovering the Lewis and Clark materials that had been entrusted to Barton and others. The daily journals of the explorers, their descriptions of plants and animals, the Indian vocabularies they collected, the observations of latitude and longitude, Lewis's map "drawn on sheets of paper, not put together, but so marked that they could be joined together with the utmost accuracy"—all these, wrote Jefferson, were the property of the government, "the fruits of the expedition undertaken at such expense of money and risk of valuable lives. They contain exactly the whole of the information which it was our object to obtain for the benefit of our own country and of the world." Eventually, through Jefferson's urging and the willing cooperation of Correa da Serra, Biddle, Clark, and Mrs. Barton, many of these materials relating to the expedition were recovered and deposited with the American Philosophical Society, where they have remained ever since. In December 1817 Jefferson sent the society what remained of his own collection of Indian vocabularies after the theft of his trunk.

> I send with them [he wrote to Peter S. Du Ponceau] the fragments of my digest of them, which were gathered up on the banks of the river where they had been strewed by the plunderers of the trunk in which they were. These will merely show the arrangement I had given the vocabularies, according to their affinities & degrees of resemblance or dissimilitude. If you can recover Capt. Lewis's collection, they will make an important addition, for there was no part of his instruction which he executed more fully or carefully, never meeting with a single Indian of a new tribe, without making his vocabulary the 1st object.[31]

Apparently copies of some of the vocabularies had found their way to Europe before this time, possibly through Du Ponceau or Barton, for Part III of the third volume of *Mithridates: Oder Allgemeine Sprachenkunde,* published by Johann S. Vater in 1816, contained linguistic and other information derived from the observations of Lewis and Clark. Du Ponceau made good use of Lewis's vocabularies in his researches on comparative linguistics (see Chap. 14).

The daily journals of the expedition contained much valuable ethnographic information about the Indian tribes of western America, but it was more than a century before these were put to scientific use. Not until the first decade of the twentieth century were the journals published, and even then scholars were slow to utilize the ethnographic materials they contained. Writing in 1954, Verne F. Ray and Nancy O. Lurie praised the ethnographic observations of Lewis and Clark and urged their fellow ethnographers and ethnolinguists to make greater use of them. Lewis and Clark, they said, were "conscious of their role as social observers, collecting cultural data for scientific purposes." They identified the sources of their information and distinguished between first-hand information secured from tribe members and second-hand data derived from neighboring tribes. They treated the Indians as their intellectual equals and avoided explaining cultural differences by supposed racial differences. In the opinion of Ray and Lurie, the Lewis and Clark descriptions of the material culture of the various Indian tribes were "time and again equal or superior to accounts in modern ethnographies." The use of these data presents many problems to the modern ethnographer because of the difficulty in identifying place names, tribal names, cultural references, and the like; but ethnographers who are prepared to make the effort have been richly rewarded, as Ray's researches on the Lower Chinook Indians show. In the 1930s Ray found two usable informants from this tribe, long thought to be extinct, and used the extensive commentaries of Lewis and Clark as supplementary information in producing an ethnography "enriched by scores of observations of 1805–1806 dates."[32]

In the last analysis, however, any proper account of the scientific achievements of the Lewis and Clark expedition must stress its contribution to the geography of the North American continent. One need only glance at the Arrowsmith and Lewis map of that continent in the *New and Elegant Atlas* that accompanied the 1805 edition of Morse's *American Universal Geography* to appreciate the abysmal ignorance of western geography that prevailed in that year despite the efforts of Jefferson, Lewis, and Clark to obtain every scrap of information they could about the region the expedition was to travese. John L. Allen has described their mental picture of the region when the expedition set out.[33] They pictured "a fertile, well-watered area perfectly suited for occupancy by an agrarian people." The rivers of the area were thought to be long and navigable to their sources. These, it was believed, rose in a common pyramidal height of land or, possibly, on either side of a narrow mountain range. In either case, a short portage would effect an easy transition from the eastward to the westward flowing streams. Such a connection between the headwaters of the Missouri and those of the Columbia River would provide the long-sought water route to the Pacific.

These preconceptions had already undergone considerable modification by the time Clark drew the large map dispatched to the War Department from Fort Mandan in the spring of 1805, but the explorers were still hopeful of discovering a water passage to the Pacific interrupted by no more than a brief portage. The subsequent progress of the expedition destroyed that illusion and provided the basis for the new and more accurate picture of the western country depicted in Clark's great map for the *History of the Expedition*. This "cartographic masterpiece," as Allen calls it, was put together from the various maps Clark had drawn during the course of the expedition; but it also incorporated data from Pike's expedition into the upper Arkansas and Rio Grande basins and from the explorations of John Colter and George Drouillard in the Yellowstone region for the fur-trading company of Manuel Lisa. The result was a map that completely recast the geographical image of the "Stony" or "Shining" or "Rocky" Mountains, portraying them as a series of north-south ranges extending westward from the Black Hills; these were distinguished

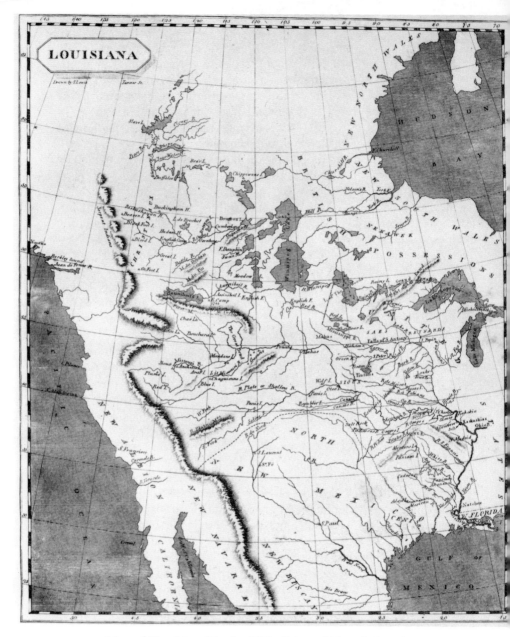

Map of "Louisiana" in the 1805 edition of Morse's American Universal Geography, *taken from Aaron Arrowsmith and Samuel Lewis,* A New and Elegant Atlas, Comprising All the New Discoveries to the Present Time . . . *(Philadelphia, 1804). Courtesy of the Harvard Map Collection, Harvard University*

from the coastal ranges nearer the Pacific Ocean. The great plain of the Columbia River was clearly indicated, the Continental Divide depicted for the first time, and the tributaries of the Missouri and Columbia rivers shown in something like their true complexity and inhospitableness to navigation in their upper reaches. The Missouri was shown to rise in the northwest, not the southwest as had been thought. The dream of a northwest passage to India was shattered forever.

To be sure, there were still errors and misconceptions. The longitudes, based on doubtful astronomical observations, were not reliable; the field maps from which Clark's general map was drawn were sketches rather than field surveys. The northern drainage of the Columbia River system was still problematical. The relationship of the Missouri and Columbia river systems was clarified, but the idea still persisted that the headwaters of the Yellowstone, Bighorn, Platte, Arkansas, Rio Grande, Colorado, Snake, and Willamette (which was called Multnomah and thought to be much larger than it actually is) descended from a common pyramidal height of land. But no future explorer of the Rocky Mountains would expect to stand like Balboa gazing in mute astonishment at the shining waters of the Pacific. Scientific knowledge had begun to supplant myth and rumor.[34]

The process of transforming traditional ideas of western geography was a slow one, however, as an examination of the sixth and seventh editions of Morse's *American Universal Geography* (1812 and 1819) will show. The 1805 edition had made some use of information gathered by William Dunbar and George Hunter in the Washita country and of Jefferson's "Official Account of Louisiana," submitted to Congress in 1803 and subsequently published in various newspapers; but the 1812 edition had little to say about the Lewis and Clark expedition and its discoveries, and that little was distinctly uncomplimentary. Under the heading *"Missouri,"* Morse portrayed Lewis and Clark as explorers whose science, judgment, and accuracy could not be relied on.

> The latitude and longitude of no one place is calculated; a connected chain of distances is not given; nor are we informed to what authority a great many facts, which the travellers did not witness, are reported: but, throughout the whole work, there is an attempt to swell the size, the length and the difficulties of the river, with the obvious view of increasing the importance of the discoveries made. . . .[35]

One wonders what account of the discoveries of the expedition Morse drew on in making these observations. His reference to a map showing the three branches of the Missouri as uniting at one place suggests that he relied on a spurious account published in Philadelphia and London in 1809 under the title *The Travels of Capts. Lewis & Clarke.* This book composed of extracts from the documents contained in Jefferson's message to Congress in February 1806, from Carver's *Travels Through the Interior Parts of North America*, and from Mackenzie's *Voyages from Montreal,* was prefaced by a map delineating the sources of the Missouri in the manner described by Morse. As a staunch Federalist, Morse had never been enthusiastic about the Louisiana Purchase, which his party thought would tilt the political balance between the commercial and industrial Northeast and the agricultural South and West heavily in favor of the latter. "This immense addition of territory to the United States," Morse wrote in the *Universal Geography,* "forms an important epoch in our history. What will be the ultimate effect upon the government, union, and happiness of our country, cannot be foreseen. . . . Time will be continually unfolding the consequences of this great event. All must contemplate them with solicitude for the honor and welfare of the nation."[36]

Quite apart from his political prejudices, by 1812 Morse seems to have abandoned his original intention of keeping his *Universal Geography* abreast of the latest researches in American geography, geology, and natural history. The *New and Elegant Atlas* that accompanied the sixth edition (1812) contained the same Arrowsmith and Lewis maps of North America and "Louisiana" as the 1805 edition had, exhibiting the "Stony Mountains" as a single range extending from the Arctic Circle southward to the headwaters of the Rio Bravo (Rio Grande), with gaps near the headwaters of the Missouri and the Platte. There was no trace of the coastal ranges. The sections on botany, zoology, and mineralogy were reprints of those in the fifth edition. The paragraphs devoted to "Louisiana" contained no reference to the Lewis and Clark expedition.

Some of these omissions in the 1812 edition might possibly have been excused on the ground that no official account of the expedition had yet been published, but no such excuse could be made for similar omissions in the 1819 seventh edition. By that time the Biddle-Allen *History of the Expedition* had appeared, and cartographers in Philadelphia, London, and Paris had begun to take account of the discoveries of Lewis and Clark and of Pike as well. Jefferson praised John Melish's map of 1816, based in considerable part on the discoveries and maps of these explorers. The American editions of William Guthrie's *New Geographical, Historical, and Commercial Grammar* and his *Universal Geography* profited from Clark's map of the western country and utilized the Philadelphia naturalists' knowledge of the Lewis and Clark specimens in describing the natural history of America. The 55-page account of American natural history in Guthrie's *Universal Geography* was replete with references to the discoveries of Lewis and Clark.[37]

Morse's rivals were making abundant use of the discoveries of Lewis and Clark, but he made no effort to do so. The *New and Elegant Atlas* that accompanied the seventh edition of his *Universal Geography* (1819) owed nothing to the explorers. The frontispiece map of North America displayed the Rocky Mountains as a single range broken only in Canada by the Unijigah, or Peace, River. No coastal range was shown. The region west of the Rocky Mountains was blank except for the Columbia River and its tributaries. The Lewis and Clark exploration of the Missouri was described in the same unenthusiastic manner as in 1812. A paragraph entitled "Passage from Missouri Territory across the Rocky Mountains to the Western Territory" drew on the Biddle account of the expedition, but without mentioning either the account or the explorers. In another passage, Lewis and Clark were described as having "sailed" down the Columbia River and its branches but as having failed to indicate clearly the longitude from which they embarked. Brief passages on the Indians and the natural history of the western country were derived chiefly from Biddle's *History of the Expedition*, but without acknowledgment. How far out of date Morse's "Natural Geography" section had become by 1819 is shown by the repetition of the notice in the 1805 edition that Cutler (who had long since given up his botanical projects) and Barton (who died in 1815) were collecting materials for works which "together will form a complete natural history of the American states."

If Jefferson ever read the account of American geography and natural history in Morse's 1819 edition, he must have been deeply discouraged and not a little disgusted. His efforts to promote the exploration of the western country had borne abundant fruit, but the fruit had been scattered far and wide by the failure to gather the harvest promptly. As he had predicted to Baron von Humboldt, the expedition's scientific discoveries became known only gradually through many channels. But they did become known and in so doing redounded to the credit of the explorers and of Jefferson himself. Even more important, Jefferson had established a tradition of government-sponsored exploration for scientific purposes that would generate a long series of expeditions linking the names of Dunbar, Hunter, Pike, Lewis, and Clark with those of Stephen H. Long, Henry R. Schoolcraft, Charles Wilkes, John C. Fremont, John W. Powell, and many others in an honor roll of American exploration. When John C. Calhoun, as secretary of war under President Monroe, sent instructions to Long for his exploration of the country between the Mississippi River and the Rocky Mountains, he modeled them on those Jefferson had given to Lewis, adding: "The instructions of Mr. Jefferson to Capt. Lewis, which are printed in his travels will afford you many valuable suggestions, of which as far as applicable you will avail yourself."[38] The exploits of Long and his successors, aided by civilian naturalists such as Lewis and Clark had done without, would add greatly to the growing body of scientific knowledge; but not until the Apollo 11 moon shot would the imagina-

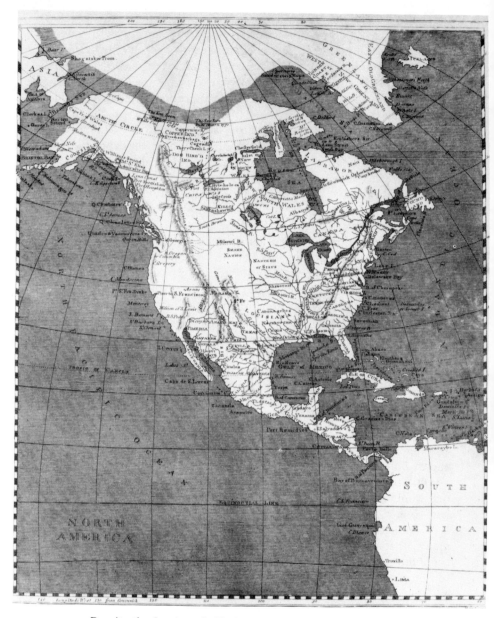

*Despite the Lewis and Clark expedition, this map from the
7th edition (1819) of Morse's* American Universal Geography
*shows little improvement on the 1805 edition. Courtesy of the
Harvard Map Collection, Harvard University*

tions of Americans be so deeply stirred as they were by the mission Jefferson sent out "to explore the Missouri river, & such principal stream of it, as, by it's course and communication with the waters of the Pacific ocean, whether the Columbia, Oregan, Colorado or any other river may offer the most direct & practicable water communication across this continent for the purposes of commerce."

CONCLUSION

Geography had not yet attained the status of a science in either Europe or America. Civil and natural history still rested comfortably side by side in state and regional histories, and treatises on "geography" blended cartography, history, geology, botany, zoology, ethnography, and political and economic description together as best they could. As knowledge accumulated concerning these various aspects of the North American continent, geographers like Jedidiah Morse and Christoph D. Ebeling attempted to incorporate the new information in their general accounts, but with diminishing success. With the Louisiana Purchase and the ensuing Lewis and Clark expedition, knowledge concerning the trans-Mississippi West expanded rapidly, but the information collected by these explorers was only gradually disseminated owing to delays and mishaps in publishing the results of the expedition. As planner and sustainer of that expedition and as author of the *Notes on the State of Virginia,* Thomas Jefferson played a leading role in promoting research into the geography, natural history, and ethnography of North America.

9

From the Theory of the
Earth to Earth Science

THE eighteenth was the century par excellence for theories of the earth. Most of these were highly speculative, unsubstantiated by careful empirical research. One group of writers, including William Whiston, William Woodward, and John Whitehurst, took as their model the biblical pattern of creation, deluge, and final conflagration and sought to show how these events resulted or would result inevitably from natural processes. Others took a more secular view of earth history. According to the Count de Buffon, the earth had originated as a mass of molten matter torn from the sun by a passing comet. As the mass gradually cooled, seas and mountains were formed; life appeared spontaneously and underwent gradual modifications, leaving huge fossil shells and bones as evidences of its former state.

On the American side of the Atlantic, Benjamin Franklin spun out his own theory of the earth in a letter to the Abbé Soulavie in 1782, subsequently published in the *Transactions* of the American Philosophical Society. Assuming like everyone else that the earth had suffered some great revolution, Franklin tried to explain the catastrophe by assuming a change in the position of the earth's axis. Imagine, he suggested, that the earth contained a dense fluid consisting of highly condensed air. A sudden shift in the axis of the globe must then cause the internal fluid to burst the containing shell into a thousand fragments. A more gradual shift would cause the sea and land to change places in various parts of the world. "Such an operation as this possibly occasioned much of Europe, and among the rest, this mountain of Passy on which I live, and which is composed of limestone, rock, and sea shells, to be abandoned by the sea, and to change its ancient climate, which seems to have been a hot one."[1] The earth having become a permanent magnet, Franklin continued, further shifts in its axis were improbable, but its surface might still be disturbed by slight motions of the ponderous fluid within.

As the eighteenth century progressed, geological speculation began to be controlled to some extent by empirical investigation of the minerals, organic remains, and rock structures found in the crust of the earth. This was especially true in Germany, where interest in geology and mineralogy was closely linked to the development of mining, but it can also be seen in the "Analysis" that the Philadelphia mapmaker and geographer Lewis Evans appended to his *Map of the Middle British Colonies* (1755). In that essay Evans recognized the main physiographic and geologic regions of British North America, calling attention to what would now be called the Coastal Plain, the Fall Line, the Piedmont Plateau, the Blue Ridge, the Appalachians, the Allegheny Front, and the Allegheny Plateau. In his journal, parts of which were published by Thomas Pownall in 1776, Evans foreshadowed the principle of isostasy (equilibrium of the earth's crust) and recorded some rather bold ideas as to the significance of fossils. From observing a bank of shells undergoing fossilization near London, Md. — the upper layers mixed with loose sand, the next enclosed in a sandy clay, and the last embedded in a clay stratum hardened into "a loose kind of stone" — he concluded that the shell-bearing strata he had seen high in the mountains had been formed in a coastal region by a similar gradual process. But how had they subsequently become elevated high above sea level? Suppose, Evans suggested, that the Great Lakes were to be drained by some natural convulsion or by the gradual extension of the Niagara River gorge. The lake floor would rise as the weight of the water was removed, and the land on the other side of the earth would be correspondingly depressed. By the same reasoning a subsidence of the land in China, caused perhaps by a partial flood, would elevate parts of the American continent. "Some such changes may have come gradually, and advanced by such slow degrees, as that in a period of a few ages would not be perceptible; history, therefore, could take no notice of them."[2]

In the last quarter of the century still stronger links between theory and field work were forged. In 1775 Abraham G. Werner joined the faculty of the Bergakademie at Freiberg in Saxony and began the process of transforming the little mining academy into a world-famous center for the study of earth sciences. As a mineralogist, Werner invented a method of classifying minerals by their external characteristics and defended it against the rival claims of the practitioners of the new French chemistry, who insisted that chemical analysis was the proper basis of mineral classification, and against the founders of the science of crystallography, Romé de Lisle and René J. Haüy, who looked to the crystal structure of minerals to provide a sound classification. As a geologist, Werner proposed a classification of rocks based on a new science of earth history he called "geognosy." In Werner's view the rocks composing the earth's crust were mostly precipitates or deposits from a universal ocean that had once enveloped the earth, subsiding slowly and successively forming the primitive and floetz (sedimentary) rocks. To these were added in more recent times the volcanic and alluvial formations. To account for faults and fractures in the strata and for the fossil remains found in the floetz rocks, Werner was forced to postulate at least two general resurgences of the primitive ocean, although he was not able to assign any probable cause of these inundations. When it was discovered that some of the "primitive" rocks contained organic remains, Werner adopted a separate category called "transition" for them, supposing that they had been precipitated or deposited in the universal ocean during one or more stormy periods of earth history. Thus Werner became the leading exponent of the "Neptunist" theory that water had been the chief agent in forming the crust of the globe.[3] His ideas were carried to Scotland by John Murray and Robert Jameson and on to the United States by American students in Edinburgh like Benjamin Silliman and Samuel L. Mitchill and by Scottish emigrants like William Maclure.

Not all Scottish geologists were disciples of Werner, however. In 1788 James Hutton proposed a theory of the earth based on the idea that all geological phenomena have been produced by the uniform operation of processes similar to those now operating on the earth's surface. If this postulate were true—if the mountains and valleys had been formed by the slow erosion of running water and if the sediments deposited in the ocean by rivers and waves had been slowly transformed into rock strata by subterranean heat and pressure and gradually uplifted thousands of feet above sea level by the same forces, as Hutton asserted—it followed that the earth must be millions of years old and that both igneous and aqueous agencies were responsible for its present condition. These bold ideas, set forth in 1788 and expanded at book length in Hutton's *Theory of the Earth* (1795), were attacked vigorously in Richard Kirwan's *Geological Essays* (1795) and defended with equal vigor in John Playfair's *Illustrations of the Huttonian Theory of the Earth* (1802). When Silliman studied at Edinburgh in the winter of 1805–1806, he was confronted with the conflicting claims of the Huttonians and the Wernerians. Silliman agreed with the Huttonians that rocks like porphyry, trap, and granite could scarcely be of aqueous origin, but he could not stomach Hutton's general view of earth history with its vast time scheme and its indifference to the biblical history. "I held myself aloof from entire committal to either theory, or to any theory except one derived directly from the facts," he wrote to a correspondent.[4]

These ideas and controversies formed the backdrop for American contributions to mineralogy and geology in Thomas Jefferson's time. Jefferson himself, had little use for theories of the earth. Dissatisfied with all the various hypotheses that had been suggested to explain the presence of marine fossils in the Appalachian strata, he was content to leave the problem unresolved for the present.[5] Believers in the biblical revelation, however, could not bear to suspend judgment on a question involving the credibility of Scripture. To the geographer Jedidiah Morse the Deluge seemed a sufficient explanation of most of the geological phenomena on the earth's surface. In the second edition of his *American Geography* Morse scoffed at Evans's supposition that the present earth was made from the ruins of an earlier one. How much more likely, he declared, that the mountains were formed by the piling up of fragments of the earth's crust when the fountains of the deep were broken up in the days of the Flood. Morse conceded, however, that the Flood could not have produced the fossil-bearing strata in the mountains and elsewhere. All the evidence seemed to indicate that the creatures that inhabited the fossil shells had lived and died in the beds where their remains were preserved. "A very considerable period of time therefore must have elapsed between the chaotic state of the earth and the deluge, which agrees with the account of Moses, who makes it a little upwards of sixteen hundred years."[6]

Similar views were expressed in Samuel Miller's *Brief Retrospect of the Eighteenth Century.* In this work Miller summarized about twenty theories of the earth published between 1660 and 1800 and noted which could be reconciled with the biblical version of earth history and which could not. His own preference was for the system of the Irish chemist and geologist Richard Kirwan, who depicted the formation of the primitive, transition, and calcareous (floetz) formations in a manner reminiscent of Werner. Kirwan supposed "that the gradual retreat of the waters continued until a few centuries before the general deluge; that this event was occasioned by a miraculous effusion of water both from the clouds and from the great abyss, the latter originating in and proceeding from the great southern ocean below the equator, and thence rushing on to the northern hemisphere, spreading over the arctic region, and descending again southward; that during this elemental conflict the carbonic and bituminous matter ran into masses no longer suspensible in water, and formed *strata of coal;* and that other substances, by the combination or decomposition of their respective materials, formed various other kinds of mineral bodies, as *basaltic masses, chalcedonies, spars,* &c."[7]

Miller conceded that even writers like Buffon and Hutton who engaged in speculations irreconcilable with Scripture had made worthwhile contributions to the study of the earth's crust, but he gave chief credit for the growth of empirical knowledge to "enlightened mineralogists, practical miners, and patient chemical experimenters." In his section on mineralogy Miller gave a balanced account of the Wernerian, the chemical, and the crystallographic methods of studying and classifying minerals, praising Kirwan again for his contributions. The number of known mineral species, said Miller, had increased from the 500 described by Linnaeus to between 7,000 and 8,000. Huge mineral collections had been formed. Mineralogy, once considered "a low and trifling object of study," had begun to be viewed as "dignified in its nature, and most interesting in its relations."[8]

The mineral treasures of the United States, Miller observed, had been "but little explored." This was true, but his brief paragraph on American researches of this kind was totally inadequate compared to his much fuller treatment of European accomplishments. His usual patriotism failed him here, perhaps because it was easier to find out what was being done in Europe than in America. Although American scientists had not distinguished themselves as mineralogists or geologists by 1803, the groundwork had been laid for some useful contributions in both fields.

TEACHING AND RESEARCH TO 1800:
THE ROLE OF MEDICAL MEN

As in the case of chemistry, mineralogy made its way into the academic curriculum by way of the medical schools. Benjamin Waterhouse introduced the study of mineralogy at Harvard, aided by donations of mineral specimens by John C. Lettsom, James Bowdoin, and the government of France. The nature of Waterhouse's lectures on mineralogy and geology may be gathered from his "Heads of Lectures."

> XVII. MINERALOGY. The contents of the earth but little known: all below *three thousand* feet is dark conjecture. *Mountains* distinguished into *primaeval* and *alluvial.* The first are the "everlasting hills," which never contain metallic ores, nor petrefactions, nor any animal exuviae; of this kind are the *Alps* and *Pyrenees,* in Europe; . . . and the *Andes,* in America. These *preceded* the formation of vegetables and animals. The second are as evidently of *posterior* formation. They lie in *strata,* contain ores, petrefactions of vegetables, and vestiges of organic substances. These *Alluvial* mountains formed at, or since the deluge; the primaeval as old as the globe. *Kirwan* recommended.
>
> XVIII. THE MINERALOGICAL SCHOOLS of Sweden, Germany, and France . . . *Cronstedt* recommended.
>
> The LETTSOMIAN Cabinet of Minerals. Mineralogy of more importance, at present, to AMERICA, than Botany. We are dependent on foreign nations for riches that lie under our feet! . . . The extravagant price set on diamonds, and other glittering stones, ridiculous in the eyes of REPUBLICANS. . . .[9]

At the University of Pennsylvania, James Woodhouse introduced mineralogy into his chemical lectures. He organized the Chemical Society of Philadelphia and persuaded it to issue a circular urging Americans to send in mineral specimens for analysis free of charge. The specimens were to be analyzed by a committee including, among others, Woodhouse and Adam Seybert, whose three years of medical study and mineral collecting in Edinburgh, London, Paris, and Göttingen in 1793–1796 had made him one of Philadelphia's leading mineralogists. In New York City the example of the Chemical Society of Philadelphia was imitated by the American Mineralogical Society, organized by Samuel L. Mitchill and his colleagues on the staff of the *Medical Repository*. Like Waterhouse and Woodhouse, Mitchill included mineralogy in his college lectures.

The physician-naturalists of this early period produced two notable studies in geology, one by Mitchill and the other by a visiting German physician, Johann D. Schoepf, who spent eight months traveling from New York to Charleston and St. Augustine and westward as far as Pittsburgh in the summer, fall, and winter of 1783–1784. Schoepf subsequently published an interesting account of his travels and a remarkably perceptive geological description of the terrain he had traversed. Published in Erlangen, Germany, in 1787, his *Geology of Eastern North America* identified the fall line of the coastal rivers, recognized that the longitudinal valleys of the mountain regions had been cut into a vast plain sloping gently southeastward (now known as the Schooley peneplain), and outlined the main belts of bedrock from the coastal plain westward to the Allegheny plateau, including the crystalline rocks, the limestones, the mountain rocks composed of fragments of older rocks, and the sediments of the coastal plain. "Contemplated as an example of field work, Schoepf's accomplishment really shines," declared the late American geologist Edmund M. Spieker. Unfortunately, Schoepf's geological book, by contrast to his *Travels,* seems to have been little known in the United States. What might have been a foundation of American geology was destined instead to become a collector's item for rare book dealers.[10]

The same cannot be said of Mitchill's "Sketch of the Mineralogical and Geological History of the State of New York," which incorporated the results of the survey he undertook for the Society for the Promotion of Agriculture, Arts, and Manufactures. In this work Mitchill divided the terrain he had studied as follows:

I. Continent — subdivided into:
   1. The *granite country,* extending from the [Long Island] Sound to the termination of the Highlands, or first range of mountains.
   2. The *slate country,* beginning where the granite ends, and underlaying all other strata, further westward, and northward, than the researches of the Commissioner have extended.
   3. The *limestone country,* spreading in some places to considerable extent, and though scattered in various parts of the country,

in large bodies, yet always superficial, and bottomed upon shistic or granitical rock.

4. The *sand-stone country,* composing the Kaats-Kill, or Blue Mountains, and some smaller strata, but always resting upon some deep-laid and more ancient fossil bodies.

5. The *alluvial country,* consisting of horizontal layers of clay, loam, sand, turf, and generally speaking, all such matters as constitute intervale space between mountains and flat land, along creeks and rivers.

II. Islands: and these may be classed as they are,

1. *Primaeval,* as New York, Staten-Island, and the north side of Long-Island; or,

2. *Secondary,* like the beaches, hassocks, and marshes on the seacoast, and all that part of Long-Island lying south of the spine or ridge of hills which runs through it from east to west.[11]

Mitchill's performance in this pioneer survey was very creditable, proving useful to later writers such as the Count de Volney and William Maclure. That Mitchill was fully in touch with European publications on geology and by no means a dogmatic Wernerian is shown by his letter to Morse in April 1796 describing his discovery of basaltic rocks superimposed on sandstone strata on the west bank of the Hudson River between Bergen and Haverstraw. The significance of this conjunction had been made plain to him, he told Morse, by his having read Hutton's paper before the Royal Society of Edinburgh commenting on some similar appearances in Scotland "at the spot where the great Ridges of Granite break off and butt on the shistus near the Duke of Athol's at Blair." In both the Scottish and the New York locations, Mitchill wrote:

The Shistus is full of Cracks and fissures some strait, others crooked, passing lengthwise, crosswise. . . . And every one of these openings is *injected full of Quartz or Granitical Matter.* The inference from which is that the Shistic foundation was laid, its substance compacted to Rock, and this rent asunder by flaws & crevices at the Time when the Granite Rocks crystallized into their present forms, and at that Time became penetrated as thoroughly as we see them with their charge of a different and comparatively recent material. These facts clearly allow us to go back one revolution further in the Physical Chronology of the earth![12]

Apparently Mitchill had no fear that his findings would shock the Reverend Jedidiah Morse. He himself was evidently much impressed with this discovery, for he mentioned it again in a letter to Caspar Wistar, linking it once more with Hutton's earlier observations in Scotland.

GEOLOGY AND MINERALOGY IN PHILADELPHIA, 1800-1825

In Philadelphia the efforts of native American geologists and mineralogists were reinforced by the publications of foreign visitors, two of whom (William Maclure and Gerard Troost) became American citizens. The first such publication was Volney's *View of the Soil and Climate of the United States of America,* published in French in 1803 and translated for the American edition (1804) by the American novelist Charles Brockden Brown. Constantin-F. Chasseboeuf, Count de Volney was an experienced traveler with extensive scientific interests and contacts. His *Voyage to Egypt and Syria* (1787) had made his reputation and established a new standard of systematic observation and scientific accuracy in travel literature. His *The Ruins; or Meditations on the Revolutions of Empires* (1791) had won him literary fame and a place among the critics of Christianity and the champions of deism. Driven from France when the disciples of Robespierre overthrew the Girondins, Volney fled to the United States, where he was warmly received by the Jeffersonian Republicans. In the spring of 1796 he visited Jefferson at Monticello. He then pushed westward as far as Lexington, Ky., and Vincennes in the Indiana Territory, collecting geological specimens as he went; he returned to Philadelphia by way of Cincinnati, Detroit, Niagara Falls, Albany, and New York. Later he traveled extensively in New England and Virginia. Elected a member of the American Philosophical Society, he kept its president Thomas Jefferson informed of the progress of his travels and researches. Volney's stay in America was brought to an abrupt end by the rise of anti-French feeling among the Federalists and the passage of the Alien and Sedition Acts during John Adams's administration. Returning to France, Volney found a place of honor in the newly formed Institute of France and enlisted the aid of his scientific friends in preparing his book on the United States.[13]

Approximately one third of Volney's *View* was devoted to geography and geology, the remainder being concerned with climate, rainfall, winds, prevailing diseases, and aborigines. He seems not to have known of Schoepf's earlier book, but he drew freely on the works of Lewis Evans, Thomas Hutchins, Gilbert Imlay, Jedidiah Morse, Jeremy Belknap, Samuel Williams, and Samuel L. Mitchill, acknowledging his debt to some of these authors. He began by dividing the country physiographically into the Atlantic region, the mountainous district, and the western country. Proceeding to the geological structure of these regions, he followed Mitchill in distinguishing granitic, sandstone, calcareous, and alluvial formations and describing their relations to each other. By far his most original contribution was his description of the Mississippi valley area. It was, says the American geologist George W. White, "the first extensive, organized account of the physical features of the trans-Allegheny region."[14] In this section Volney gave a good description and analysis of the terraces of the Ohio and other river valleys and of the Niagara Falls and vicinity, illustrating his remarks with geological sections and presenting correct views as to the probable geological history of these features. The flat-lying beds of fossiliferous limestone in the Ohio and Mississippi valleys he recognized as evidence that the entire region had once been covered by water. Some of the fossils he collected there were examined by the great French naturalist Lamarck, who reported that "the districts of North America which furnished them have formerly been covered by the sea . . . for they all belong to the class of pelagia . . . , which . . . are found only at the greatest depths of the sea, and never near the shore."[15] Other organic remains, Volney speculated, had been deposited in great inland lakes, long since vanished.

To bring together as clearly and graphically as possible the information contained in his book, Volney included a map of eastern North America from the Mississippi River to the Atlantic Ocean and from the Gulf of St. Lawrence to Florida, indicating the location of various geological structures as well as the general geography of the region. He intended that the map should be colored to show the extent of each geological formation, but only a few copies of the French edition contained colored maps. Apparently Mitchill's copy was one of these, for he called particular attention to this innovation in his review of the book in the *Medical Repository*. "This is a beautiful and obvious manner of expressing the prevalence of any of the great strata of the earth, as overspreading or under-laying an extensive tract of country," wrote Mitchill. "It is a pity that map-makers have not more generally observed it." Modern historians of geology have agreed with Mitchill. "This extraordinary synthesis, the second geological map of the United States and one of the very earliest of published geological maps, surely deserves as much respect as [William] Maclure's attempt six years later," observed the Cornell geologist John W. Wells, adding that Volney's map first gave Europeans an idea of American geology.[16]

Volney's speculations concerning the causes of the phenomena he had observed were the least convincing and most controversial part of his book. Like Evans and Jefferson before him, Volney imagined that inland lakes or seas had broken through the Allegheny Mountains to form water gaps. For other purposes he resorted to earthquakes and volcanoes, citing William's study of New England earthquakes. Volney regarded these historically documented tremors as being produced by a line of subterranean fire "which pushes the earth upwards, in a line generally running from north-west to south-east, in the course of the Merrimack river, extending southward to the Potowmack, and north to the St. Laurence, particularly affecting the direction of Lake Ontario."[17] That lake, he conjectured, lay in the crater of an extinguished volcano. Apparently Volney's vulcanism in this connection did not preclude his accepting Werner's neptunism in other respects. After urging American scientific societies to devote themselves to systematic geological investigations, he predicted that these inquiries would tend to confirm the idea that the entire globe had once been completely covered with water and that North America had emerged from this universal ocean somewhat later than either South America or the Old World.

Volney's *View* was well received on both sides of the Atlantic. In Europe it was translated into German and Italian as well as English. The American translator, Charles Brockden Brown, considered it the best and most complete description of American physiography, geology, and climate yet published. Mitchill reviewed it at length in the *Medical Repository,* adding corrections and supplementary comments on the basis of his own researches. In Boston, where Volney's deism was suspect, the editors of the *Monthly Anthology and Boston Review* were less enthusiastic. How absurd, they declared, to explain the frequency of earthquakes in the United States "by placing a large volcano under Lake Ontario." President Jefferson praised Volney for his meteorological observations and conjectures but confessed that he had never taken much interest in geology "from a belief that the skin-deep scratches we can make or find on the surface of the earth, do not repay our time with as certain and useful deductions as our pursuits in some other branches." James Mease, a Philadelphia physician, thought well enough of Volney's book to capitalize on its popularity by issuing a similar work entitled *A Geological Account of the United States* (1807). He added little to what Volney and his predecessors had contributed, although he did include a catalogue of minerals. His chief objection to Volney's book was its suggestion that the American continent was less ancient than the European. Mease countered Volney's arguments on this score by attributing the formation of fossil-bearing strata to the agency of the Deluge.[18]

It was not Mease but an immigrant from Scotland, William Maclure, who made the next substantial contribution to the development of American geology. Born in Ayr in 1763, Maclure made a fortune as a partner in the London firm of Miller, Hart, and Company. He visited the United States several times in the course of business and in 1796 became an American citizen. Soon afterward he retired from business to devote himself to science and social reform. In 1803 Jefferson appointed him a member of a commission charged with settling the claims of American citizens against the French government. Meanwhile, Maclure had become deeply interested in geology. He traveled widely on the continent of Europe, from Scandinavia to the Mediterranean and as far east as Russia,

Two pioneers in the study and description of American geology: Constantin-François Chasseboeuf, Count de Volney (by Charles Willson Peale after a portrait by Gilbert Stuart) (left) and William Maclure (by Peale) (right). Courtesy of the Independence National Historical Park Collection (Volney) and the Frick Art Reference Library (Maclure)

studying rock structures and collecting specimens wherever he went. Returning to the United States in the summer of 1808, he commenced a geological tour through the eastern and northern states. "We understand," reported the editors of the *Medical Repository* in 1809, "that this enterprizing and able traveller has already examined, to a considerable extent, Maine, New-Hampshire, Vermont, Massachusetts, Rhode-Island, Connecticut, New-York, New-Jersey, Pennsylvania, Delaware, and Maryland. And for the purpose of employing himself more advantageously during the winter, he went to Georgia, where, with less inconvenience with frost and snow, he might pursue his scientific researches in the south. It is reported that this gentleman has already delineated on a map, the principal strata of rocks, mountains, and earths which he has observed during his journeys."[19]

Maclure is said to have crossed the Alleghenies at least fifty times in his travels, collecting specimens every half-mile and frightening the inhabitants of remote regions by his unfamiliar appearance and behavior. Upon his arrival in New Haven, Silliman called on him at his hotel room, which was littered with geological specimens that were being wrapped and packed by a servant. Outside the hotel, the horses that pulled Maclure's coach were lean and dull from transporting these treasures. As for Maclure himself, "his countenance had a ruddy glow and his manners were in a high degree winning and attractive. His language was pure and elevated; his mind being imbued with the love of science, he was successful in exciting similar aspirations in other and especially younger minds."[20]

In 1809 the results of Maclure's travels appeared in the *Transactions* of the American Philosophical Society under the title "Observations on the Geology of the United States, Explanatory of a Geological Map." In this paper Maclure outlined the geology of the country east of the Mississippi River, adopting the Wernerian classification of rocks as the most suitable.

> First, because it is the most perfect and extensive in its general outlines, and secondly, The nature and relative situation of the minerals in the United States, whilst they are certainly the most extensive of any field yet examined, may perhaps be found to be the most correct elucidation of the general exactitude of that theory, as respects the relative position of the different series of rocks.

Maclure warned his readers, however, that his use of the Wernerian terminology did not commit him to Werner's ideas about the origin of the various formations. So saying, Maclure proceeded to describe successively the location, composition, inclination, and mineral content of the primitive, transition, secondary, and alluvial rocks between the Mississippi River and the Atlantic Ocean. The boundaries of these regions, he declared, were delineated on the basis of "knowledge lately acquired by crossing the dividing line of the principal formations, in 15 or 20 different places, from the Hudson to the Flint river; as well as from the information of intelligent men, whose situation and experience, make the nature of the place near which they live familiar to them; nor has the information that could be acquired from specimens, when the locality was accurately marked, been neglected, nor the remarks of judicious travellers."[21] Many years later Charles Lyell was to declare, after his own extensive travels in the United States, that Maclure had traced the limits of these geological tracts "with no small accuracy."

To enable his readers to visualize the spatial relations between the various kinds of rocks he had described, Maclure included a map on which the formations were shown in various colors, as follows:

ORANGE . . . . . Primitive Rock [granite, gneiss, micaceous schist; other crystalline rocks]

PINK . . . . Transition Rocks [essentially steeply dipping stratified rocks, some slightly metamorphosed, as anthracite and some quartzite]

BLUE . . . . Floetz or Secondary Rocks [essentially horizontal
strata of limestone, sandstone, shale] and Old Red
Sandstone [with some greenstone trap gently dipping sandstone, now known to be Triassic]
GREEN . . . . Salt and Gypsum
YELLOW . . . . . Alluvial [unconsolidated or weakly consolidated
rock of Coastal Plain and lower Mississippi
valley] [22]

Modern geologists have differed considerably in their evaluations of
Maclure's memoir and map. Wells has criticized Maclure for fastening
the Wernerian nomenclature and classification on American geology and
suggested that Maclure drew heavily on Evans, Schoepf, Mitchill, and
Volney without acknowledgment ("his classification was hardly more than
Mitchill's put in Wernerian terminology"). Wells concedes that Maclure's
map was good — "the best of its time considering the large area covered" —
but suspects that Maclure prepared the map in Paris on the basis of information derived from Volney and subsequently perfected it during his
travels in the United States. [23] Edmund Spieker, translator of Schoepf's
Geology, considered Maclure's description of the geology of the eastern
United States less thorough and perceptive than Schoepf's but better organized and with the great advantage of a geological map. The
Wernerian terminology he forgave, especially since Maclure explicitly
disclaimed acceptance of Werner's theoretical conceptions. "There is
nothing inherently wrong with Werner's major division of the rocks that
make up the visible crust," declared Spieker. [24] Maclure certainly thought
so. The absence of volcanic rocks in the region east of the Mississippi River,
he declared, "is not the least of the many prominent features of distinction between the geology of this country and that of Europe, and may
perhaps be the reason why the Wernerian system, so nearly accords with
the general structure and stratification of this continent." In Maclure's view,
the geologist should try to "seize the great and prominent outlines of
nature" and "acquaint himself with her general laws." In the first of these
undertakings Maclure had succeeded brilliantly. To the second he had not
yet addressed himself.

The same volume of the *Transactions* that carried Maclure's "Observations" contained Benjamin H. Latrobe's "Memoir on the Sand-hills of
Cape Henry in Virginia," the Silliman-Kingsley account of the Weston
meteor, and Silvain Godon's "Observations to Serve for the Mineralogical
Map of the State of Maryland." Enthusiasm for mineralogy and geology
was reaching a high point in Philadelphia. One year earlier the *Philadelphia
Medical Museum* had published Adam Seybert's catalogue of forty
American minerals and Woodhouse's analyses of titanium and zinc ores
and his memoir on the Weston meteor. In 1808 Godon, a Parisian

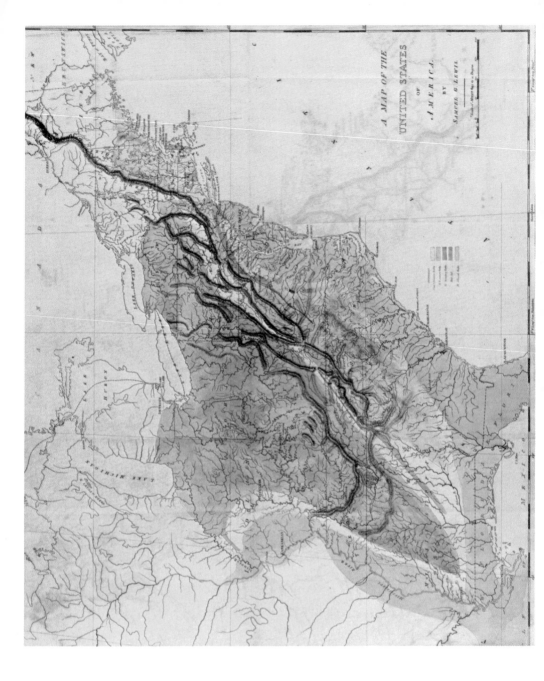

William Maclure's geological map, designed to illustrate his "Observations on the Geology of the United States . . ." (1809). The colors show the extent of the different geological structures. Courtesy of Special Collections, University of Connecticut Library

mineralogist who had emigrated to the United States and settled in Philadelphia after spending a year lecturing and geologizing in the vicinity of Boston, put the American Philosophical Society's cabinet of minerals in scientific order, announced a course of public lectures on mineralogy and geology, and began the manufacture of yellow pigments from a combination of lead with chromic acid. Then came a series of setbacks for the earth sciences. Woodhouse died suddenly of a stroke in 1809, Seybert deserted science for politics and business, and Godon was thrown in prison for debt. The new incumbent of the chemical chair at the University of Pennsylvania, John R. Coxe, had none of Woodhouse's enthusiasm for mineralogy, and the American Philosophical Society, which had managed to publish six volumes of its *Transactions* since 1771, allowed nine years to elapse before the next volume appeared.

Fortunately for the progress of the earth sciences, the slack created by these reverses was taken up by the founding and rapid growth of the Academy of Natural Sciences of Philadelphia. Under the successive presidencies of Gerard Troost (1812–1817) and William Maclure (1817–1840) the academy became an important center for mineralogical and geological studies and for natural history generally. Troost, who had been trained at the universities of Amsterdam and Leyden and at Paris under the Abbé Haüy, introduced the members of the academy to Haüy's methods and principles in the crystallographic analysis of minerals and contributed many articles to the academy's *Journal,* which soon became a leading outlet for publications on American mineralogy and geology. Troost wrote on zircon, phosphate of lime, pyroxene, yenite, chrysoberyl, andalusite, and other minerals. Isaac Lea described the minerals of Philadelphia and vicinity; Lardner Vanuxem and William Keating, those of the mineral-rich area around Franklin, N.J.; Thomas Nuttall, those of the rocks at Hoboken, N.J. Nuttall also wrote on the geological structure of the Mississippi valley, as did Edwin James. The collections of the academy grew rapidly, beginning with the acquisition of Adam Seybert's collection in 1812. Several of the younger members, including Lea, Henry Seybert, Vanuxem, and Keating, emulated the example of Adam Seybert and Troost by going to Paris to pursue their studies.[25]

Meanwhile, the president of the Academy of Natural Sciences, William Maclure, continued to supply the institution liberally with books and specimens while pursuing his researches and publications. In 1815–1816 he studied the geology of several of the West Indian islands in company with Charles A. Lesueur. In 1816 and 1817 he made his first excursion into the Mississippi valley, accompanied by Lesueur, and in the latter year published a revised version of his "Observations" and the accompanying geological map, adding an introductory chapter on methods of geological investigation and two concluding chapters on the relations between geology and agriculture. In 1818 this expanded version of the "Observations" was republished in the *Transactions* of the Philosophical Society, and Maclure's "Essay on the Formation of Rocks" appeared in the academy's *Journal.*

In these publications Maclure indicated his general views about geology and its applications more fully than ever before. As usual he insisted strenuously on the limits of human knowledge, the dangers of theoretical speculation, and the necessity of extensive and thorough research, but he was now more willing than he had been in 1809 to discuss modes of rock formation and the ways in which geological processes had altered the surface of the earth. The divergent views of the neptunians and the vulcanists, he was convinced, were dictated largely by the character of the rocks they observed in the countries where they lived and worked. In southern Europe, where there were active volcanoes, geologists argued for an igneous origin of many rocks. In northern Europe, where no volcanoes were found, neptunism reigned supreme. To settle the issue between these parties, said Maclure, it would be necessary to begin with rocks whose mode of formation was open to observation, as in the case of volcanic products and sediments deposited by rivers, and then to conjecture by rational analogy the probable mode of formation of other rocks. Adopting this procedure, Maclure divided rocks into those indubitably formed in water by deposition or chemical precipitation; those probably so formed; those indubitably formed by igneous agency; those (like basalt) probably so formed; and those whose mode of origin was doubtful, bearing traces of both aqueous and igneous agency.[26]

In the fourth category Maclure placed rocks such as pitchstone, greenstone, pearlstone, porphyry, and clinkstone, which Werner and his followers had classified as aqueous in origin. Maclure had encountered rocks of this kind in his travels in the Crimea, Bohemia, the Alps, Saxony, Thuringia, Hesse Cassel, central and southern France, and Spain, usually in connection with basalt formations, but he had found no trace of them in the United States. Since volcanic products and basalt were also absent in America, the inference seemed plain that rocks of the fourth category were of igneous origin. In this as in many other cases the range of Maclure's observations was truly impressive. Probably no other geologist of his time had traveled so widely and collected so assiduously throughout Europe and America.[27]

Comparing the mountain ranges of eastern North America with those of Europe, Maclure found the latter much more difficult to interpret geologically. He attributed the difficulty to "the accidents and changes that have been effected in the stratifications of the different classes of rocks on the European continent, since their original formation; by the effects of water, during the immensity of time, partially washing away the superincumbent stata . . . and leaving the more hard and durable parts of the same stratification in their original positions; or by the long and continual action of rivers wearing deep beds, and exposing to view the subordinate strata, giving the whole the present appearance of a confused and interrupted stratification, though it might have been uniform and regular in its original state."[28] Despite the greater disorder in the European mountains, Maclure was convinced that the "natural structure and arrangement of the original stratification" was the same in Europe and America.

Maclure's habit of distinguishing between the "natural structure" of the "original stratification" and the "accidental changes" that had distorted it conflicted sharply with his earlier criticism of Werner for using terms like "primitive" and "secondary," implying succession in time. Likewise, his conjectures in his "Essay on the Formation of Rocks" clashed with his declared intention not to speculate on the mode of origin of the various rocks, the agencies that had subsequently modified them, or the time required for these modifications. By 1818 Maclure had begun to form opinions on these subjects, though he was tentative in expressing them. Observing in his "Essay" that granite seemed to form the basement rock in the crust of the earth, he nevertheless went on to point out that geologists had no way of knowing what lay beneath the granite.

> Is it the nucleus of the earth, from which, and on which all changes and formations emanate and rest, as an eternal foundation? or is it only a link of those changes, the circle and recurrence of whose action is lost in the immensity of time? We know nothing; we may form theories and systems without end, and perhaps one system is as good as another; but still we must recur to the humiliating truth, we know nothing.[29]

In this passage Maclure seems to vacillate between the Wernerian idea of an original structure of the globe modified by subsequent accidental changes and the Huttonian conception of uniform geological processes operating endlessly throughout vast eons of time. There is a strong Wernerian note in Maclure's reiterated distinction between the original order of the mountain ranges of Europe and North America and the "accidental and subsequent changes" that altered that arrangement. But Maclure had no sympathy for the catastrophism of many Wernerians. "In all speculations on the origin, or agents that have produced the changes on this globe," he declared in 1818, " we ought to keep within the boundaries of the probable effects resulting from the regular operations of the great laws of nature which our experience and observation has brought within the sphere of our knowledge. When we overlap those limits, and suppose a total change in nature's laws, we embark on the sea of uncertainty, where one conjecture is perhaps as probable as another; for none of them have any support, or derive any authority from the practical facts wherewith our experience has brought us acquainted."[30]

When he did permit himself to speculate on the geological origins of the North American continent, Maclure found himself unable to imagine a time when "that primitive chain of mountains called the Alleghany" did not exist, but he thought it likely that there had been a time when that range was not overlaid by transition, secondary, or alluvial deposits and the chain ran unbroken the length of the continent, impounding, with the aid of the Rocky Mountains on the west, a great inland lake whose waters eventually spilled over the mountain range at various places, wearing the channels now occupied by the coastal rivers.[31] Apparently Maclure had advanced but little on the conjectures of Evans, Jefferson, Mitchill, and Volney in attempting to envisage the formation of the American continent.

Basically, however, Maclure was a utilitarian and an empiricist, impatient of abstract inquiries and carefully elaborated systems, whether Wernerian or Huttonian. He believed that Werner's classification and nomenclature were the soundest, most useful, and most widely known, but he did not hesitate to depart from them when observation seemed to require it. Some of the Wernerian names he never used, he told Silliman, because he had never encountered the rocks they were supposed to denote.

> Transition trap is a rock that I have not met with, and may perhaps be a part of the floetz trap that happened to cover the transition, without any immediate connexion, but like a current of lava, overlying all the classes of rocks it meets with. This misapplication of names naturally arises from the system of neptunian origin, on which the nomenclature of Werner is founded.

The most useful classification, Maclure added, would be "to retain Werner's five classes as being well defined, that is, as well as the graduated variety of nature will permit, . . . and to make some subdivisions in each class, without deranging the system already best known, or the ideas of those who follow it." The surest method of advancing geology as a science, Maclure concluded, was to collect facts without committing oneself to any system. "The thirty or forty cases of geological specimens, which I thought necessary to support my sketch of our geology, (now lodged in the collection of the academy of Natural Sciences in Philadelphia,) will, I presume, remain unchanged by the mutation of names introduced into theories."[32]

True to his utilitarian ideals, Maclure devoted two chapters of the second printing of his "Observations" to the subject of the decomposition of rocks in relation to the quality of the soil in different parts of the United States. He illustrated his remarks with five geological sections of the region between the Appalachians and the Atlantic Ocean at five different latitudes. Like his fellow deists Franklin and Jefferson, he was sustained by a cheerful faith that the processes of nature had been arranged in such a way as to promote human happiness. The earth, he declared, was "every day moulding down into a form more capable of producing and increasing vegetable matter, the food of animals, and consequently progressing towards a state of amelioration and accumulation of those materials, of which the moderate and rational enjoyment constitutes [a] great part of our comfort and happiness. . . . On the surface of such an extensive and perpetual progression, let us hope that mankind will not, nay cannot, remain stationary."[33]

Maclure was to continue his geological researches and efforts at social reform until his death in 1840, growing gradually more radical in his political and scientific opinions as time went on. In geology his influence was conservative, however, partly because of his adoption of the Wernerian terminology but also because, like Werner, he identified and traced rock strata by lithological characters rather than by their fossil content. The latter, he declared in 1818, had "not yet been examined with that accuracy of discrimination necessary to form just conclusions."

Maclure's best days as a geologist were over, but he had done much for American geology and would continue to support it throughout his life. He had traced in the field the rock structures of the United States, compared them to those in Europe, and delineated them in one of the earliest geological maps. He had given freely of his time, money, books, and specimens to American scientific societies, especially the Academy of Natural Sciences of Philadelphia, and had subsidized and inspired many young scientists. Not without reason has he been called "the father of American geology."

FURTHER DEVELOPMENTS IN NEW YORK CITY

In New York City the second decade of the nineteenth century brought rapid progress in the earth sciences. In 1810 Archibald Bruce founded *The American Mineralogical Journal,* the first specialized scientific periodical in the United States apart from medical journals. Bruce was well equipped for his task as editor. After graduating from Columbia he had spent three years in Europe studying medicine and cultivating the acquaintance of leading mineralogists such as the Count de Bournon, the Abbé Haüy, and Heinrich Struve. Returning to the United States in 1803 with a substantial collection of minerals, he continued his mineralogical researches. He established a laboratory for chemical analysis of minerals, gave lectures on mineralogy at the College of Physicians and Surgeons, and maintained an extensive correspondence with mineralogists and geologists in both the United States and Europe.

No sooner was Bruce's *Journal* begun than it became a rallying point for the devotees of the earth sciences throughout the United States. In New York City it drew support from Samuel L. Mitchill, David Hosack (Bruce's former teacher), Samuel Akerly (Mitchill's son-in-law), the chemist John Griscom, and many others. Benjamin Silliman was a regular correspondent and a frequent visitor to New York. So were George Gibbs of Newport, R.I., and Robert Gilmor, Jr., of Baltimore, both of whom had acquired handsome mineral cabinets during their European travels and had formed many of the same acquaintances with European mineralogists that Bruce had. Farther afield, Bruce maintained contact with Zacchaeus Collins, Adam Seybert, Charles Wister, and Solomon Conrad in Philadelphia, Parker Cleaveland at Bowdoin College, and correspondents in Georgia, Kentucky, and Ohio.

Filling the pages of the *Journal* with American contributions was no easy task, but Bruce managed to bring out four issues in four years until failing health and professional difficulties forced him to discontinue the project. Altogether, twenty-one Americans contributed forty-two articles to the journal. Mitchill and Silliman each contributed five papers, Gibbs, William Meade, and Bruce four each, and Akerly three. Most of the articles were descriptive accounts, either of minerals, mineral waters, the geological features of particular localities, or metallurgical operations. Several were reports of chemical analyses of minerals. Of these the most outstanding were Bruce's descriptions and analyses of two new mineral species, "native magnesia [magnesium hydroxide, subsequently named *brucite* in his honor]" and "red oxide of zinc [zincite]." "So thoroughly was the work done by Bruce," wrote the Yale mineralogist George Brush in 1882, "that these species remain today essentially as he described them, and his papers may well be studied by mineralogists now as models of accuracy and clearness of statement."[34]

Bruce's *Journal* was well received in Europe, especially in Paris, where Haüy called it to the attention of the readers of the *Journal des Mines* and published translations of some of the leading articles. Above all, Bruce's *Journal* was a source of great pride to the American scientific community. Its demise after 1814 caused great concern in their ranks and led eventually to the founding of the *American Journal of Science and Arts* as its successor and, in a sense, its continuation.

Meanwhile, interest in the earth sciences continued to gather momentum in New York and its vicinity. In 1814 Gibbs established a magnificent country estate on Long Island and entered actively into the scientific life of the city. With Mitchill, Bruce, Hosack, Griscom, and others he promoted mineralogy and geology in the New-York Historical Society, the Society for the Promotion of Useful Arts, and the newly formed Literary and Philosophical Society of New York. Mitchill continued to foster the earth sciences in his lectures and in the *Medical Repository,* and in 1817 he took the lead in founding the Lyceum of Natural History of New York, which eventually acquired Mitchill's own collections and those of the New York Historical Society. In 1818 he published an American edition of William Phillips's *Elementary Introduction to the Knowledge of Mineralogy,* with extensive notes on American mineralogy, and in the same year brought out an American version of Robert Jameson's translation of Georges Cuvier's *Essay on the Theory of the Earth,* appending his own "Observations on the Geology of North America." The members of the Lyceum of Natural History were adding rapidly to the Lyceum's cabinet of natural history and publishing memoirs in Silliman's journal, Mitchill's *Medical Repository,* Biglow's *American Monthly Magazine and Critical Review,* and *The New York Medical and Physical Journal.* "Our young men have formed a Lyceum of Natural History," Mitchill wrote to Reuben Haines of Philadelphia, "and hold weekly meetings. They are influenced by a zeal unknown among us until now."[35]

TEACHING AND RESEARCH IN NEW ENGLAND
AND THE ALBANY REGION

In New England, Parker Cleaveland and Benjamin Silliman, both teachers at the undergraduate level, led the way in developing the earth sciences.

Parker Cleaveland is remembered chiefly for his *Elementary Treatise on Mineralogy and Geology,* first published in 1816. Educated at Harvard, Cleaveland became professor of mathematics and natural philosophy at Bowdoin College in 1805. His interest in minerals was aroused when workmen excavating near the Androscoggin River brought him pieces of quartz and iron pyrites they had mistaken for diamonds and gold. Cleaveland was unable to identify the substances, but his curiosity was excited. Soon he was teaching mineralogy to his students and working hard on a textbook. By his own exertions and with the help of Benjamin Vaughan of Hallowell, Maine (son of the Samuel Vaughan who had played a prominent role in Philadelphia science in the 1780s and brother of John Vaughan, treasurer of the American Philosophical Society), Cleaveland built up the mineralogical collection and library at Bowdoin and began to solicit descriptions and specimens of minerals from all parts of the United States. Maclure, Gibbs, Silliman, Bruce, Mitchill, Gilmor, and correspondents from the remoter regions of the country sent him their mineral

View of Bowdoin College (by John G. Brown) as it appeared when Parker Cleaveland went there to teach in 1805. Courtesy of Bowdoin College Museum of Art, Brunswick, Maine

localities. Silliman performed chemical analyses for Cleaveland and advised him on the structure of his textbook, promising to use it in his courses at Yale. Cleaveland's heavy teaching load and the invasion of the coast of Maine by British troops in 1814 delayed the progress of the work; but finally, in December 1816 the book appeared.[36]

The main body of the *Treatise* was mineralogical, although Cleaveland did reproduce Maclure's geological map with the addition of some information about the transition rocks of southern and western Maine. In the "Introduction," the Wernerian, chemical, and crystallographic methods of studying and classifying minerals were described and evaluated. Rocks were distinguished from minerals as "minerals which fall under the cognizance of geology." The classification of minerals, presented in a tabular view, was a compromise between the systems of Alexandre Brongniart and Haüy. The nomenclature was eclectic, stressing the use of well-known names. Of special interest to both American and European readers was Cleaveland's list of American mineral localities, compiled from the descriptions and specimens sent to him by his correspondents.

The *Treatise* sold well and was favorably reviewed, especially in Europe. Robert Jameson at Edinburgh and William T. Brande at Cambridge praised it and made use of it in their lectures. Haüy described it to his American visitors as "the best elementary work on the science extant. . . ." "It was highly approved by all who saw it in France, England, Scotland, & Ireland," Robert Gilmor, Jr., wrote to Cleaveland, "and was constantly out of my possession. I lent it to Hauy, Brongniart, Brochant, Bournon, & Lucas, the last of whom made copious extracts from it, probably with a view to improve the next edition of his work on mineralogy."[37]

On the strength of this book Cleaveland was elected to the American Philosophical Society and the Geological Society of London and offered a position at Harvard. After long consideration he declined the Harvard offer, preferring to stay at Bowdoin and work on a new edition of his *Treatise*. Descriptions of mineral localities poured in from all parts of the United States in response to his appeals. From Cincinnati, Daniel Drake solicited specimens for the newly formed Western Museum Society, adding that the society had adopted Cleaveland's classificatory system and would be glad to send specimens for his collections. Benjamin Vaughan continued to assist Cleaveland in obtaining the latest foreign books, and Silliman again provided chemical analyses and general advice. In 1822 the second edition of the *Treatise* appeared, expanded to two volumes to include the new mineral localities Cleaveland's correspondents had supplied. Like Bruce's *American Mineralogical Journal,* Cleaveland's *Treatise* did much to unify, inspire, and invigorate the cultivators of the earth sciences in the new republic.

Yale rather than Bowdoin, however, was to exert the greatest influence on the development of mineralogy and geology in the early nineteenth century. Benjamin Silliman, strongly supported by President Timothy Dwight and his successor Jeremiah Day, established Yale's preeminence in these sciences. On returning from his studies at the University of Edinburgh in 1806, Silliman began introducing lectures on geology and mineralogy in his chemical course and took advantage of every opportunity to expand the college's mineral collections. Within a year he had acquired an excellent mineral cabinet of 2,000 specimens belonging to a former Yale student, Benjamin D. Perkins, and had begun a private course of lectures devoted entirely to mineralogy. Soon afterward he made the acquaintance of George Gibbs, who had just returned from his travels in Europe, bringing with him several magnificent collections of minerals purchased at no small price. At Lausanne, Gibbs had studied under the well-known

Four important contributors to American geology and mineralogy: George Gibbs (after Gilbert Stuart, artist unknown) (upper left), Archibald Bruce (upper right) and Parker Cleaveland (lower left) (after Thomas Badger by Joseph Rodefer de Camp), and Amos Eaton (by A. H. Ritchie) (lower right). Courtesy of the Yale University Art Gallery (Gibbs), Bowdoin College Museum of Art, Brunswick, Maine (Cleaveland), Rensselaer Polytechnic Institute Archives (Eaton), American Philosophical Society (Bruce)

mineralogist and chemist Heinrich Struve and had bought the mineral cabinet of Count Gregorii Razumovsky. In Paris he became a friend of the Count de Bournon, François Gillet-Laumont, J. F. d'Aubuisson, and other French mineralogists and acquired the cabinet of Gigot d'Orcy when the collections of that victim of the guillotine were placed at auction. Through Bournon he made further acquisitions from Saxony, Dauphiny, Vesuvius, Padua, Verona, and the extinct volcanoes of the Rhine.[38]

Soon after Gibbs's return to Newport in 1807, he and Silliman became fast friends. In 1810 Gibbs agreed to place some of his collections on loan at Yale if Silliman could arrange adequately for their display. With Dwight's approval, the second floor of South Middle Hall was converted into a mineral gallery, and Gibbs and his servant came over from Newport to assist in arranging the specimens according to Haüy's system of classification. In June 1812 the Razumovsky specimens were added to the collection, and the gallery was opened to the public despite the fear of invasion occasioned by the outbreak of the War of 1812—"Mr. Madison's War," as it was known by staunch Federalists like Silliman and Gibbs. "The fame of the cabinet was now blazoned through the land, and attracted increasing numbers of visitors," Silliman later recalled. "I had become a zealous student of mineralogy and geology, and now felt that the time had come to present them with more strength and fulness than in former years."[39]

Silliman now began to offer separate courses in mineralogy and geology instead of interspersing these subjects in his chemical lectures. The Perkins cabinet was joined to the Gibbs collection, and the mineral gallery was fitted as a lecture hall. Gibbs supplied Silliman with the latest mineralogical and geological books from Europe and established prizes for students in the junior and senior classes. Meanwhile, Silliman was preparing to give instruction in the chemical analysis of minerals. With the erection of a new building in 1820, the second floor of which was reserved for the chemical laboratory and the mineral collections, chemistry and the earth sciences were firmly and handsomely established in the undergraduate curriculum at Yale. Silliman was in the process of training a generation of chemists, geologists, and mineralogists. His student assistants, among whom were James D. Dana, Chester Dewey, Charles U. Shepard, Amos Eaton, George T. Bowen, Edward Hitchcock, and Benjamin Silliman, Jr., would soon carry Silliman's methods and message to colleges throughout the nation. Many of them would surpass their teacher in substantive contributions to chemistry, geology, and mineralogy.[40]

Silliman had launched two other projects of importance for the earth sciences. The first of these, the American Geological Society, was no great success. Parker Cleaveland had urged the formation of a national geological society for some years, "to stimulate us to unite and concentrate our efforts." Gibbs took up the idea, and in 1819 he and Silliman persuaded the Connecticut legislature to pass an act constituting Gibbs, Silliman, Cleaveland, Robert Hare, Robert Gilmor, Jr., John W. Webster of Harvard, and their associates as the American Geological Society. The list of officers elected at the society's first meeting, held in the "philosophical room" of Yale on September 7, 1819, read like an honor roll of American geologists and mineralogists except for the unaccountable omission of Mitchill's name. William Maclure was elected president. Gibbs, Silliman, Cleaveland, Stephen Elliott of South Carolina, Robert Gilmor, Jr., Samuel Brown of Kentucky, and Robert Hare were made vice-presidents, and F. C. Schaeffer, John W. Webster, and Edward Hitchcock corresponding secretaries. Invitations to membership were sent to mineralogists and geologists in every part of the country, but the society never became an effective national organization. The time was not yet ripe for scientific meetings on a national scale. The society became an appendage of Silliman's program at Yale, receiving donations of books and specimens from various members and holding occasional meetings that were thinly attended.

The society's last meeting appears to have taken place in November 1828. William Maclure was there, Silliman reported. "The brilliant man whom I first saw twenty years before, had now hoary locks; he stooped as he walked, and an ulcer on his leg made him lame." Maclure's friend Thomas Cooper, now president of South Carolina College, was also there—"patriarchal and venerable, mild and conciliating." These two veterans of American science attended one of Silliman's lectures and called on him at his house, where the conversation turned to "the moral relations of science and the expositions it gives of the mind and thoughts of the Creator." Maclure and Cooper then left New Haven, never to return.[41]

Silliman's other project, the *American Journal of Science* was much more successful. Ever since Bruce's *American Mineralogical Journal* had begun to falter, Silliman had been urged by Gibbs, Cleaveland, Griscom, and others to take over the journal from Bruce or start a new one in its place. In October 1817, a few months before Bruce's untimely death, Silliman drafted a proposal for a journal to be called *The American Journal of Science, More Especially of Mineralogy, Geology, and the Other Branches of Natural History; Including Also Agriculture as Well as Useful Arts* and secured Bruce's approval of the project. "I saw Dr. Bruce in November," Silliman wrote to Cleaveland in March 1818; "he was obviously a broken man but he approved of my plan & authorized me to use his name; but he has since paid the debt of nature as you have doubtless seen in the papers."[42]

The first number of the *American Journal of Science* appeared in July 1818, with a preface announcing that the journal would feature contributions by American scientists in natural history (including mineralogy, geology, botany, and zoology), chemistry, natural philosophy, and mathematics. The *Journal* also invited articles on agriculture, manufactures, 'and domestic economy; music, sculpture, engraving, and painting; and comparative anatomy, physiology, and other branches of medicine. Mineralogical and geological articles constituted a third to one half of each volume in the early years of the journal. Silliman himself contributed many of these. Others came from his former students Edward Hitchcock, Amos Eaton, Denison Olmsted, George Bowen, and Chester Dewey, still others from Gibbs, William Maclure, James Pierce, and Jeremiah Van Rensselaer. From the western country came John H. Kain's description of the mineralogy and geology of northwestern Virginia (now West Virginia) and eastern Tennessee, W. B. Stilson's sketch of the geology and mineralogy of part of Indiana, Caleb Atwater's essay "On the Prairies and Barrens of the West," and Henry R. Schoolcraft's account of the geology of the lower Mississippi valley. There were also numerous descriptions of mineral localities. The *Journal* gave every sign of accomplishing its purpose of stimulating and providing an outlet for American scientific endeavor, if only it could pay its way.

The financial issue hung in the balance for at least a decade. The original publishers soon withdrew their support; other arrangements were made, but by 1826 Silliman found himself forced to assume complete responsibility for the *Journal*. It now became in fact what it had long been in popular parlance — "Silliman's Journal." From that time on the enterprise gradually became self-supporting, although the number of subscribers never reached 1,000 and rarely exceeded three quarters of that figure. In Europe it was the best known and most highly respected American scientific journal for several decades.[43]

Silliman's contributions to the progress of the earth sciences in the United States were many and varied. His original researches were of a very modest order (see Chap. 7), but he probably did as much to advance the earth sciences through his teaching, textbooks, training of graduate assistants, work in building up the mineralogical collections at Yale, public lectures, founding of the American Geological Society, and statesmanlike editorship of the *American Journal of Science* as any man of his generation. So testified Charles Lyell in a letter to Silliman written in April 1842.

> Now that I have travelled from Niagara to Georgia, and have met a great number of your countrymen on the Continent of Europe and heard the manner in which they ascribe the taste they have for science to your tuition, I may congratulate you, for I never heard as many of the rising generation in England refer as often to any one individual teacher as having given a direction to their taste.[44]

If Benjamin Silliman had a rival in his own generation, it was Amos Eaton, one of his own students. Educated at Williams College, Eaton studied law and set up practice in Catskill, N.Y. In his spare time he began the study of natural history and laid plans for opening a botanical school. In August 1811, however, he was convicted of forgery and sentenced to life imprisonment. Pardoned by Governor De Witt Clinton in 1815, Eaton decided to devote his life to science. He was thirty-nine years old, three years older than Silliman, when he went to Yale in 1816 to study chemistry, mineralogy, geology, and botany. Something of the contrast between his personality and Silliman's comes out in Eaton's evaluation of his teacher in a letter to John Torrey written in 1819.

> I can tell you the good and the bad of him in a few words. He is an excellent practical chemist and a good *cabinet* mineralogist. With very little knowledge of geology, he affects much. He is impatient when his opinions are questioned, and has very lofty conceptions attached to the stupendous title of Professor of Yale College; and expects us to be ever mindful of his honorable marriage into the family of Governor Trumbull. Still he is a pretty good fellow; though quite too formal for a man of science.[45]

After learning all he could from Silliman and Yale's botany professor Eli Ives and from careful study of the Gibbs collection of minerals ("You will never be perfect in mineralogy until you are familiar with Gibbs' cabinet," Eaton wrote Torrey), Eaton spent a year teaching at Williams College, followed by seven years of itinerant lecturing and field work in western Massachusetts, Vermont, and the vicinity of Albany and Troy. In 1819, through his connections with De Witt Clinton, Stephen Van Rensselaer, and Theodoric R. Beck, principal of the Albany Academy, he was invited to lecture before the New York state legislature on the applications of chemistry and geology to agriculture. With Van Rensselaer's financial support, Eaton then undertook geological surveys of Albany and Rensselaer counties and the entire route of the proposed Erie Canal. In 1824, the year in which the first part of the canal route survey was published, Stephen Van Rensselaer founded the Rensselaer School (now the Rensselaer Polytechnic Institute) and chose Eaton to plan and direct its program in pure and applied science. Eaton was now established in a career that would rival Silliman's in its influence on American science.

Like Silliman, Eaton was a transitional figure, linking the science of the Jeffersonian era with that of the Jacksonian. He was likewise a gifted teacher whose pupils (James Hall, Ebenezer Emmons, George H. Cook, Douglas Houghton, and others) played important roles in the development of American geology. Unlike Silliman, however, Eaton was an indefatigable field geologist and a prolific writer. Beginning with his *Index to the Geology of the Northern States,* first published in 1818, he set out to describe the stratigraphy of the northern states and to establish a nomenclature for American geology. Maclure had already "struck out the grand outlines of North American geographical geology," Eaton acknowledged. Gibbs had placed American mineralogists in his debt with his mineral collections and Silliman had used these collections skillfully in teaching mineralogy and geology. Cleaveland had contributed his textbook, Mitchill a mass of information. "But the drudgery of climbing cliffs and descending into fissures and caverns, and of traversing, in all directions, our most rugged mountainous districts, to ascertain the distinctive characters, number and order of our strata, has devolved upon me," Eaton proclaimed. "I make no pretensions to any peculiar qualifications, other than that bodily health and constitutional fitness for labor and fatigue, which such an employment requires."[46]

Like Werner and Maclure, Eaton recognized primitive, transition, secondary, and alluvial rocks, but he followed the English geologist Robert Bakewell in introducing another class, "superincumbent rocks," between the secondary and the alluvial. Of the new class he said, "These rocks overlay others in a non-conformable position. In our district they generally rest upon breccia or red sandstone; and are always above primitive hornblende rocks. . . . Bakewell and many others consider them of volcanic origin."[47] Eaton then proceeded to divide each class of rocks into its constituent strata, as follows:

Secondary Class	Rock salt	
	Gypsum	Gypsum (sometimes with rock salt)
	Compact limestone	Compact limestone (often contains organic relics)
	Breccia	Breccia
	Red sandstone	Red sandstone
Transition Class	Rubblestone	
	Graywacke slate	Graywacke slate
	Argillaceous and siliceous slate	Argillaceous and siliceous slate
	Metalliferous and limestone	Metalliferous and limestone
	Sienite	Granular limestone
	Calcareous and granular quartz	
Primary Class	Soapstone rocks	Talcose rocks
	Mica slate	
	Gneiss	Sienite
	Granular limestone and quartz	Granite

[While Eaton's 1818 classification (left column) was being printed, he revised it (right column) on a separate sheet that was distributed with some copies of the *Index.*]

Geological section showing the main types of rocks from the Susquehanna River (upper left) to Boston, Mass. (lower right), from Amos Eaton's Index to the Geology of the Northern States, 2nd ed. The numbers on the plate refer to the following types of rocks: (1) granite, (2) gneiss, (3) hornblende rock, (4) mica-slate, (5) talcose rock, (6) granular limestone, (7) argillite, (8) metalliferous limestone, (9) graywacke, (10) red sandstone, (11) breccia, (12) compact limestone, (13) gypsum, (14) secondary sandstone, (15) basalt, (16) greenstone trap

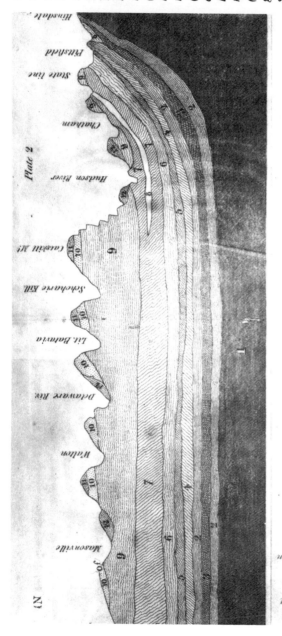

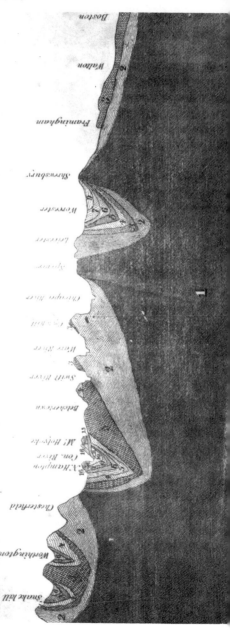

These formations were displayed graphically in a section extending from Boston to the eastern part of the Catskill Mountains. For the first time the transition and secondary rocks were shown in their relations to each other, for the most part correctly. Both types were characterized mineralogically and lithologically without reference to their fossil contents, but Eaton did list and name six kinds of organic remains found in the region he had traversed, calling them *mammodolite, ornitholite, amphibolite,* and the like according to their resemblance to living animals or plants.

When he came to explain how the various kinds of rocks had been formed, Eaton proved to be a good Wernerian. He pictured the primeval earth as a globe enveloped in a fluid mass in which the solid materials that were to form the various strata of rocks were mixed together, gradually separating from each other as they settled in concentric circles around the nucleus of the earth. But these concentric sheaths were eventually disrupted by the force of subterranean steam generated in the bowels of the earth, "whether . . . from the admixture and combustion of substances then abounding beneath the granite [or] . . . by the concentric layers of metallic plates serving as a vast galvanic battery."

> The projecting edges of granite, together with the uplifted strata of transition and secondary rocks, formed the first islands and continents of dry land. Alluvial deposits had already commenced under water, and therefore parts of the raised islands and continents were prepared for the reception and support of plants and animals of the more perfectly organized structure.[48]

These continents, however, were sunk beneath the oceans and the ocean beds were elevated in their place "when the wickedness of man drew down the vengeance of the Almighty" and the entire human population except those immured in Noah's ark was destroyed.

Like Silliman, Hitchcock, and many other American geologists of the period, Eaton was a devout Christian, firmly convinced that the findings of geological science would eventually vindicate the doctrine of the plenary inscription of the Bible. In the "Reflections on the History and Structure of the Earth," which concluded the second edition of the *Index,* Eaton traced a general concordance between the sequence of events in earth history recorded in Genesis and the sequence discovered by geologists. Conceding that six days was a rather short period of time for the deposition and alteration of the earth's strata, he noted that certain learned divines had suggested that each day might have been a thousand years, since a thousand years were but as a day in the sight of God. "We know," said Eaton, "that all operations were hastened in effecting so much in so short a time; and we are only to believe, what is certainly most rational, that though these operations were hastened, the Almighty established laws for

Amos Eaton's conception of the original arrangement of the strata composing the earth's crust (upper left) before "the grand explosion which first upturned and disfigured the rocky crust of the globe" and produced the mountain ranges shown at the upper right. The lower figure represents a section of "secondary" country, "where both transition and secondary rocks appear." The numbers refer to the same types of rocks listed below Eaton's section of the country between the Susquehanna and Boston (see previous figure). The numbers of the concentric strata surrounding the primitive earth correspond to the same sequence up through (10) red sandstone, after which is (11) "muddy waters surrounding the earth." From Eaton's Index, 2nd ed., Plate I governing his work at the very first, and endowed the first created atom with its present properties."[49]

With Van Rensselaer as patron of his geological researches, Eaton was able to push his stratigraphical inquiries still further. When Van Rensselaer agreed to finance a survey of the proposed Erie Canal route, Eaton wrote to Silliman to enlist his aid in identifying doubtful specimens and unraveling the stratigraphy of western New York.

> I have hitherto found much difficulty with all the secondary rocks of the west. . . . I shall be very attentive to that subject, collect large specimens in various localities and study their relative positions with minute attention. I shall enter upon that subject without any prepossessions, and cautiously reserve all opinions until I bring you a load of specimens. . . . It wants all your weight of character to bring geologists together in relation to our western rocks. I will be very particular about the proportion of territory which each rock covers, and to collect all other facts in relation to them.[50]

Silliman agreed to collaborate, suggesting that duplicate specimens be sent to Cuvier and Brongniart in Paris, and on November 11, 1822, Eaton and his assistant Matthew H. Webster set off in a one-horse wagon on the first of several surveys of the canal route. "I challenge all correct geologists upon the accuracy of the survey," Eaton wrote to Torrey in February 1824. "I have become acquainted with almost every individual fragment of rock, creek, and hillock on the canal line. I travelled with my one horse waggon and boy more than three thousand miles last summer."[51]

In 1824 the results of Eaton's labors were published under the title *A Geological and Agricultural Survey of the District Adjoining the Erie Canal.* They were summed up in a geological section of the region between Boston and Lake Erie. In the dedication to his patron Eaton announced:

> This section is about five hundred and fifty miles in extent, stretching across nine degrees of longitude, and crossing both the great primitive ranges of New-England (the White Mountain and Green Mountain ranges,) also the broad transition and secondary ranges of the eastern and western parts of the state of New-York. Therefore if your directions have been faithfully executed, you will have presented to the science more *elementary facts in a connected view,* than any other individual.[52]

Eaton's representation of the stratigraphy of this region was an important contribution. Maclure had presented geological profiles of the eastern United States at various latitudes in the second edition of his "Observations" (1817–1818), but these had indicated only whether the successive rock structures were primitive, transition, secondary, or alluvial. In Eaton's sections the rock strata within these general categories were shown.

In working out the stratigraphy of the northern states, Eaton was forced to make important decisions concerning geological nomenclature and classification. After studying the relevant European and American works, he decided to adopt the European names of the strata wherever they were applicable and to invent his own descriptive names, based on some lithological characteristic of the rocks in question, where no European name was available. The European names he took primarily from Werner, but he also drew on English writers like Robert Bakewell, William Conybeare, and William Phillips, preferring Bakewell's "Superincumbent Class" to the "Tertiary" of Conybeare and Phillips. He also followed Bakewell in transferring the Old Red Sandstone from the secondary, where Werner had placed it, to the transition formations. "The few names which I have added," wrote Eaton, "were admitted with extreme reluctance; and, as it were, forced upon me for want of established names which appeared to be appropriate." As to European rocks that were absent in the United States and American rocks not found in Europe, Eaton concluded:

> If transition trap, primitive silicious slate, primitive gypsum, and the chalk basins of Europe, are not to be found in America, shall we make ourselves ridiculous in our own estimation, by insisting on their presence, and by giving them their allotted places in our systems? Or shall we exclude from our list of rocks our vast iron formation, because similar rocks have never been found in the eastern continent, excepting a few limited patches or beds? . . . Would not a student in geology be as much embarrassed at such an arrangement, as the student in geography at seeing the continent of America set down as a petty island, included in one of the sea-board counties of England?[53]

After listing the various types of rocks found in the primitive, transition, secondary, and superincumbent classes and indicating their mineral and fossil content and localities, Eaton described these rocks as they were encountered successively along the Erie Canal route. The *Survey* concluded with Hitchcock's commentary on his own profile of the strata between Boston and Plainfield, Mass. This profile, constructed for Eaton's use at the request of Van Rensselaer, differed from Eaton's profile of the same region in some respects.

As time went on, Eaton continued to revise his nomenclature and picture of the stratigraphy of the northern states and its relation to European stratigraphy. Like Maclure before him, he was hampered by his insistence on trying to fit American rocks into the Wernerian categories and his reliance on lithological characters instead of fossil contents in correlating American and European strata. In dealing with the rocks of western New York, moreover, he was entering unknown territory as far as geological science was concerned. Since the rocks abounded in fossils and lay in regular strata with only a slight dip, he assumed they must belong to Werner's secondary class, the presumption being that transition rocks were invariably disturbed and metamorphosed and relatively deficient in fossils. Actually, however, the New York strata were much older than the secondary formations of Europe, being correlated with the Silurian, Ordovician, and Devonian strata that had not yet been brought within the domain of science by the European geologists. Eaton noted the vast extent of the undisturbed strata in New York, observing that "we have at least five distinct and continuous strata, neither of which can with any propriety take any name hitherto given and defined in any European treatise which has reached this country." Nevertheless he did his best to correlate these strata with the European secondary. The true correlation would not be discovered for another two decades and would require the combined efforts of the New York geologists and their European colleagues. But the *sequence* of the rocks of western New York had been ascertained with relative accuracy by Eaton in his *Survey,* as James Hall, geologist and paleontologist for the western district of the New York survey and a former pupil of Eaton, testified in 1843:

> Nearly all the rocks of western New-York are enumerated in the order of succession; and, with some exceptions and omissions, the order is correct, and the subdivisions will always hold good in the science. It is a remarkable fact, that at this early period, Mr. Eaton should have recognized the sandstone of the Catskill mountains as the Old Red [Sandstone] of Europe; which, now that we have identified its characteristic fossils, is proved to be true. Had he seized this grand idea, and confined himself to the elucidating of the strata below the Catskills, he would have brought to light the most interesting series of rocks yet known in any part of the globe. . . . The great source of error throughout seems to have been the prevailing desire to identify, within the limits of New-York, all the rocks and systems published in Europe from the Tertiary downwards.[54]

The twentieth-century New York geologist John W. Wells is equally generous in evaluating Eaton's work. Conceding that Eaton was led into error in attempting to correlate the New York rocks with the succession of English strata delineated in the *Outlines of the Geology of England and Wales* by Conybeare and Phillips (1822), Wells concludes nevertheless that Eaton's picture of the rock succession in central and western New York was "essentially as it is understood today and has formed the ultimate basis for all subsequent work." Wells adds that, although Eaton relied on lithological rather than fossil evidence in correlating strata, he did eventually include figures and descriptions of organized remains in his *Geological Text-book* (1832). "Eaton's descriptions are very brief and in many cases, even with the poor figures, the species are impossible to recognize," says Wells. "Nevertheless, it was an earnest first attempt to analyze the vast array of fossils in the New York Devonian."[55]

CONCLUSION

The progress of the earth sciences in the United States in the Jeffersonian era was the result of the combined efforts of all the main types of investigators involved in American science in this period. The efforts of medical men like Waterhouse, Woodhouse, Mitchill, and Bruce were reinforced by those of visiting or immigrant geologists and mineralogists like Schoepf, Volney, Maclure, Godon, and Troost, by the contributions of gentleman-scientists like Gibbs, Gilmor, and Vaughan, and by the work of outstanding teachers like Cleaveland, Silliman, and Eaton. In mineralogy, all three modes of mineral analysis — petrological, chemical, and crystallographic — were introduced and cultivated successfully. In geology, William Maclure was the dominant figure until the 1820s, by which time Amos Eaton had begun to make his bid for fame. With characteristic immodesty Eaton characterized his *Index to the Geology of the Northern States* as having inauguarated the second, or Eatonian, epoch in American geology, distinguished from the first, or Maclurean, by its attention to stratigraphy. There was some truth in Eaton's generalization: he was indeed the first to explore the succession of American rock strata in detail. But 1816–1820 marked a transition to a new period in other ways as well. With the publication of Cleaveland's *Elementary Treatise*, Eaton's *Index*, and the expanded version of Maclure's "Observations" and with the founding of the *American Journal of Science, Journal of the Academy of Natural Sciences of Philadelphia*, Lyceum of Natural History of New York (soon to have its own journal), and American Geological Society, the earth sciences had acquired a solid foundation. The Jeffersonian era with its strong Wernerian orientation was drawing to a close. The Jacksonian era of state surveys, stratigraphy based on paleontology, resurgent geological uniformitarianism, and countless reconciliations of Genesis and geology lay just ahead. The "theory of the earth" was being transmuted into earth science.

10

Natural History in a New World: Botany

LIKE Adam in the garden of Eden, the naturalists of the infant American republic faced the exhilarating task of naming, classifying, and describing the plants and animals of a new world. The American continent was theirs, and theirs the duty and high privilege of giving a scientific account of its productions. "What a field have we at our doors to signalize ourselves in!" Jefferson wrote to Joseph Willard, president of Harvard College.

> The botany of America is far from being exhausted: it's Mineralogy is untouched, and it's Natural history or Zoology totally mistaken and misrepresented. As far as I have seen there is not one single species of terrestrial birds common to Europe and America, and I question if there be a single species of quadrupeds. . . . It is the work to which the young men, whom you are forming, should lay their hands. We have spent the prime of our lives in procuring them the precious blessing of liberty. Let them spend theirs in shewing that it is the great parent of science and virtue.[1]

Enthusiasm ran high, but the ranks of American naturalists were thin and their resources meager. A few physician-naturalists burdened with the practice and teaching of medicine; a handful of clergymen with scientific interests; an occasional merchant, planter, or nurseryman; an assortment of wandering naturalists from Europe; a few fledgling botanic gardens and natural history museums; a small number of libraries in private hands or associated with colleges or learned societies; several medical journals; and a number of relatively isolated scientific societies — these were the personnel and resources available for the vast work of exploring the natural history of a continent.

The work had been going on since the first colonists arrived in America. From the seventeenth century on, men like John Banister, John

253

Clayton, and John Mitchell in Virginia; John Bartram and his son William in Pennsylvania; John Brickell in North Carolina; Cadwallader Colden (aided by his daughter Jane) in New York; and Cotton Mather and Manasseh Cutler in Massachusetts sent specimens and descriptions of plants, animals, and fossils to Europe. These helped to make possible such basic works in natural history as John Ray's *History of Plants* and Linnaeus's *Systema Naturae*. Other American contributions appeared in the journals of European learned societies. A few were published separately in London or Leyden. Botany was the most popular subject of investigation, but the animals of the New World attracted some attention as well.[2]

After the revolutionary war, botany continued to be the favorite study of most American naturalists. From his parish in Lancaster, Pa., Henry Muhlenberg, youngest of three distinguished sons of the patriarch of the Lutheran Church in America, corresponded tirelessly in an effort to organize a brotherhood of American botanists dedicated to producing an American flora. In November 1792 he wrote to Manasseh Cutler at Ipswich, Mass.

> Let each one of our American Botanists do something, and soon the riches of America will be known. Let Mechaux [André Michaux] describe South Carolina and Georgia; [Samuel] Kramsch, North Carolina; [James] Greenway, parts of Pennsylvania; Bartram, Marshall, Muhlenberg, their Neighborhood; Mitchell [Samuel L. Mitchill], New York; and You, with the northern Botanists, your states — how much could be done! If then, one of our younger Companions (I mention Dr. Barton, in particular, whose business it is) would collect the different Floras in one, how pleasing to the botanical world. . . . I have written to pretty near all the mentioned Gentlemen on this Head, and hope to receive their concurrence. Pray let me have your opinion on this subject.[3]

Cutler was Muhlenberg's most likely prospect for a botanical correspondent in the New England states, since Benjamin Waterhouse was not a field naturalist. Cutler had taken a first step toward a flora of New England with his "Account of Some of the Vegetable Productions, Naturally Growing in This Part of America, Botanically Arranged," published in the *Memoirs* of the American Academy of Arts and Sciences in 1785. A few years later he had supplied botanical and zoological information for Jeremy Belknap's *History of New Hampshire* and (only a few months before Muhlenberg wrote to enlist his aid) had sent Jedidiah Morse several catalogues of New England trees and plants for inclusion in his *American Geography,* stating:

> I have made considerable progress, & altho I have had very little leisure for this favourite pursuit, yet I believe I have examined nearly all ye

vegetables of N. England. But I have not yet found time for arranging & digesting my minutes. . . . I have it in contemplation to publish a botanical work of considerable magnitude, though it will be principally confined to ye productions of N. England.[4]

In the ensuing years this "gentleman so mild in his manner and so ardent in his researches," as Samuel L. Mitchill described Cutler, continued to gather materials for a flora of New England, although hampered considerably by the lack of suitable reference works. He corresponded widely in both Europe and America. One of his British correspondents, Jonathan Stokes, sent him "a very fine botanical microscope (an instrument I very much wanted)." Andrea Murray, professor of botany and materia medica at the University of Göttingen, sent him a variety of publications and arranged a correspondence for him with some of the Parisian botanists. In Sweden he corresponded with Olaf Swartz and Gustav Paykull. Meanwhile, manuscript volumes of plant descriptions accumulated on the shelves of Cutler's study while he served in Congress, helped organize the Ohio Company and its settlement at Marietta, Ohio, maintained a school for boys, and ministered to the needs of his congregation. On a cold snowy day in January 1812, however, fire broke out in his study while he was at dinner. Many of the botanical manuscripts were destroyed; Cutler, now seventy years old, sorrowfully abandoned his long-cherished dream of publishing a botany of New England.[5]

Like Cutler and Muhlenberg, most of the other members of Muhlenberg's projected botanical network were botanists by avocation only. André Michaux was the closest to being a professional botanist. Samuel Kramsch was a Moravian missionary who botanized in the vicinity of Salem, N.C., in his leisure hours. James Greenway, a landed gentleman in Dinwiddie County, Va., hoped eventually to publish a flora of Virginia, but he did not live to complete it. Mitchill was interested in botany (he was interested in everything!) but found relatively little time for it in his busy round of activities. Only in Philadelphia and its vicinity was there a sufficient concentration of botanical enthusiasts and botanical libraries and gardens to warrant entertaining the idea of producing an American flora.

John and William Bartram had paved the way for such an enterprise by their travels and collections in the middle states and in the Carolinas, Georgia, and Florida. Unfortunately their specimens, descriptions, and drawings remained unpublished in the libraries of their English patrons while the species they had discovered were made known to the scientific world in works such as Humphry Marshall's *Arbustum Americanum* (1785), Thomas Walter's *Flora Caroliniana* (1788), William Aiton's *Hortus Kewensis* (1789), and by Continental botanists like Lamarck, L'Héritier, and Gmelin. But the knowledge acquired by the Bartrams in

*Title page of the
famous* Travels *of
William Bartram and
his portrait (by Charles
Willson Peale). Portrait
by courtesy of the
Independence National
Historical Park
Collection*

their travels lingered in William's mind after his father's death in 1777, and many of the trees and plants they had collected grew in the Bartram garden on the banks of the Schuylkill. A broadside *Catalogue* listing the plants in the garden was published in 1783, and in 1791 many of the plants listed therein were described in William Bartram's *Travels through North & South Carolina, Georgia, East & West Florida,* although not in the systematic manner of a botanical treatise.[6] Bartram, his garden, and his artistic talents as an illustrator were to remain of central importance in the development of American natural history for another three decades, but largely as resources to be drawn on and sometimes pilfered by other naturalists, foreign and domestic.

Of those who drew on Bartram's knowledge and talents none was more assiduous than Benjamin S. Barton, a young Philadelphian to whom Muhlenberg had tentatively assigned the task of combining the researches of American devotees of botany into an American flora. With the gardens of William Bartram, William Hamilton, and Humphry Marshall and the libraries of Philadelphia as well as his own at his disposal and with the willing assistance of such nearby enthusiasts as Bartram, Muhlenberg, Marshall, and Zacchaeus Collins and the support of the American Philosophical Society, it seemed quite possible that Barton would accomplish the task that Muhlenberg had envisioned for him and that he himself had placed on his scientific agenda.

Unfortunately Barton, like Mitchill, had so many items awaiting his attention that he found it difficult to concentrate on any one of them. Full of energy and ambition, he projected a series of comprehensive works dealing with American botany, zoology, materia medica, and Indian linguistics. But his teaching duties, combined with medical practice and service on the staff of the Pennsylvania Hospital, made field work difficult, and his voracious appetite for knowledge of all kinds far outran his capacity to assimilate and publish in coherent form the information he accumulated. He did manage to make a trip up the Hudson River and across from Albany to Niagara Falls in 1797. A few years later he collected plants in Virginia, where his brother lived, and around Passaic Falls and Perth Amboy, N.J.; but most of the specimens in his herbarium were collected by others, including former students, foreign travelers with whom Barton became acquainted, correspondents like Muhlenberg, and Philadelphia friends like Collins and Bartram. He acquired specimens and descriptions wherever he could, but he lacked Muhlenberg's openhandedness in sharing his acquisitions with others. "With Dr. Barton I correspond but seldom except when he puts some queries to me," Muhlenberg wrote to William D. Peck in 1812. "I could never persuade him to let me see his Herbarium although he has seen mine twice. His Principle seems to be 'it is more blessed to receive than give' however he is indefatigable in Collecting and has done much for Botany."[7]

Barton's only substantial botanical publication was a textbook entitled *Elements of Botany: Or Outlines of the Natural History of Vegetables.* It was published in 1803 and dedicated "To the Students of Medicine in the University of Pennsylvania, and to the Lovers and Cultivators of Natural History, in Every Part of the United States." The book was handsomely illustrated with thirty plates, some of them made from drawings by William Bartram, to whom Barton made warm acknowledgment. The work was divided into three parts, the first devoted to a description of the parts of plants, the second to vegetable anatomy and physiology, and the third to systematic botany. But Barton had so much to say about the presumed functions of the various plant organs in the first and third parts that Part 2 was reduced to a mere thirty-two pages. In all three parts the text was a running commentary on the ideas of Linnaeus, with frequent criticisms of the great Swede's opinions and copious quotations from other

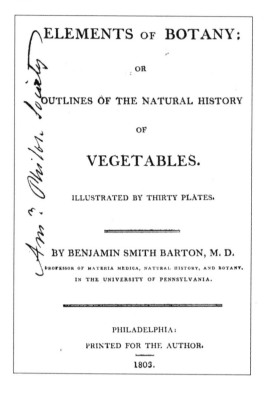

ELEMENTS OF BOTANY;

OR

OUTLINES OF THE NATURAL HISTORY

OF

VEGETABLES.

ILLUSTRATED BY THIRTY PLATES.

BY BENJAMIN SMITH BARTON, M. D.

PROFESSOR OF MATERIA MEDICA, NATURAL HISTORY, AND BOTANY,
IN THE UNIVERSITY OF PENNSYLVANIA.

PHILADELPHIA:

PRINTED FOR THE AUTHOR.

1803.

Title page of the first American botanical textbook and portrait of the author, Benjamin Smith Barton (engraving). Portrait by courtesy of the American Philosophical Society

authorities, ancient and modern, and the Roman poets as well. There were also frequent digressions on the medical uses of different kinds of plants, a subject Barton had developed more fully in an earlier textbook, *Collections for an Essay toward a Materia Medica of the United States* (1788).[8]

The result was a lengthy and rather diffuse discourse more notable for general erudition, independence of thought, and wide-ranging curiosity about plants than for clarity and rigor. Like Barton's teaching, his textbook was admirably adapted to acquaint the student with a wide variety of perspectives, issues, and opinions without indoctrination in any. Barton was skeptical of Linnaeus's idea that the country of origin of plants could be told from their appearance. He pointed out that many species were common to Siberia, Kamchatka, northern Japan, and North America. He was equally skeptical of Christian Sprengel's suggestion, later to be used by Darwin in developing his theory of natural selection, that the bright petals of plants served to attract insects and hence to promote pollination. Barton also rejected the opinion of Darwin's grandfather, Erasmus Darwin, that the corolla was part of the plant's pulmonary system. "Many experiments

*William Bartram's
drawing of* Xanthorhiza
simplicissima *(yellow
root) as published in
Benjamin Smith
Barton's* Elements of
Botany. *Courtesy of the
American Philosophical
Society*

remain to be made, before the uses of the corolla can be completely demonstrated, to the satisfaction of naturalists and philosophers," Barton concluded. "I am disposed, in the meanwhile, to believe, that both this part and the calyx are essentially concerned in the office of vegetable respiration."[9]

In regard to plant classification Barton expounded the "sexual system" of Linnaeus (so-called because it was based on the reproductive organs of the plant) but with many reservations and criticisms. A truly natural system of classification, Barton declared, would be "an arrangement that should bring together, under particular classes or orders, vegetables agreeing not only in their structure and appearance, but also in their properties." Whether such a system could ever be discovered he rather doubted, but he thought it important to search for one. To this end he outlined in an appendix seventeen different systems of classification proposed by botanists from the time of Andrea Cesalpino to his own. Oddly enough he omitted the system (that of Antoine L. de Jussieu) that was to dominate European botany in the next several decades, though it was not accepted in England,

Germany, and the United States until the 1830s. Of one thing only was Barton sure: "THE SEXUAL SYSTEM OF LINNAEUS CANNOT BE IMMORTAL."[10]

Barton's ambitious plans for a systematic work on American botany never bore fruit. He announced a *Prodromus of a Flora of the States of New-York, New-Jersey, Pennsylvania, Delaware, Maryland, and Virginia* in 1805 and published a preliminary sample in his *Philadelphia Medical and Physical Journal* in March of the following year. Apparently 500 copies of the work were printed, but Barton never allowed them to come before the public, perhaps because he had begun to suspect that the book would do him no great credit. He continued to collect materials for an American flora, however. In 1806–1807 he subsidized the visiting German botanist Frederick Pursh in two collecting trips, the first into Maryland, Virginia, and northern North Carolina, the second into northern Pennsylvania, central New York, and as far north as Rutland, Vt. Barton was not always prompt with his payments, partly because he distrusted Pursh's bibulous habits. Drinking, Barton warned his brother in Virginia, was Pursh's greatest failing "(and God knows it is a big one), but the poor fellow, who has been well educated, has merit, and I verily believe would not steal anything in the world, rum, gin, whisky, or the like excepted."[11]

When Pursh left Philadelphia for New York, taking with him descriptions and scissored pieces of specimens of the Lewis and Clark plants Jefferson had entrusted to Barton's care (see Chap. 8), Barton found a new assistant in the young English printer Thomas Nuttall, who had arrived in Philadelphia in 1808. At Barton's expense Nuttall traveled first into Delaware and then northward to Niagara Falls and the western end of Lake Ontario. Then in April 1810 Barton sent him on a trip into the Great Lakes region with the idea that he would continue on to Lake Winnipeg in Canada, making observations on Indians, plants, animals, and minerals and returning by way of the Missouri River. Nuttall was to receive eight dollars a month and a share of his collections, but Barton was to have exclusive rights to publication of the material. After making his way alone to Michilimackinac on Lake Huron, Nuttall joined the Astoria fur-trading expedition and accompanied it to St. Louis and thence up the Missouri River to the Dakota country in the spring of 1811. He returned by way of New Orleans, where the outbreak of the War of 1812 prompted him to take ship for England after sending specimens, seeds, and a manuscript book to Barton.[12]

Barton's botanical collections were growing by leaps and bounds, but he never found time to produce the North American flora he had long dreamed of writing. When Benjamin Rush died in 1813, Barton applied for and received the chair in the theory and practice of medicine his old enemy had left vacant. But the strain of preparing lectures for a field in

which he was poorly qualified proved too much for Barton's physique, long
weakened by intermittent attacks of gout. Abandoning his North American
flora, Barton attempted instead a new edition of the *Flora Virginica*
published by Gronovius in 1762 on the basis of specimens sent to him by
John Clayton of Virginia. Seventy-four pages of the new edition (a mere
beginning) were published in Philadelphia in 1812, but Barton carried all
the copies off to Europe when he went there in 1815 in an unsuccessful attempt to save his failing health. Returning to America in November, he
made last-minute arrangements for the disposition of his collections and
manuscripts and the reprinting of some of his papers. Three days before
his death he wrote a paper for the American Philosophical Society entitled "Miscellaneous Facts Concerning the First Discovery of the Genus of
Bartonia, Accompanied with a Drawing of One of the Species." The genus
had been named and handsomely illustrated in the *Botanical Magazine* in
London by Pursh; but Nuttall, who was then in England, had written to
Barton pointing out that in the manuscript book he had sent from New-
Orleans he had proposed the name *Bartonia superba* for the species Pursh
now proposed to name *Bartonia ornata*. Barton was anxious that Nuttall's
name be adopted. "I might," Barton added, "say something more . . . concerning my claims to the collections of these two travellers; but this is rather
a matter which respects the individuals concerned than the Public."
Ironically and with a kind of poetic justice, it turned out that Muhlenberg,
whose generous assistance Barton had never reciprocated, had named a
small-flowered genus of the gentian family in honor of Barton many years
earlier. It is by this genus, rather than the handsomer one proposed by
Pursh and Nuttall, that Barton is commemorated in modern botanical
nomenclature.[13]

A more striking contrast than that between Barton and Muhlenberg,
both of whom aspired to bring about the production of an American flora,
can scarcely be imagined. Muhlenberg was a modest, generous, deliberate
man, a true lover of botany and natural history but devoid of ambition
to cut a great figure in the world of science. Educated at Halle in Prussia,
he was ordained a Lutheran minister shortly after his return to Philadelphia
in 1770. He commenced his botanical researches when the British occupation of Philadelphia drove him to New Hanover, Pa., and deprived him
temporarily of clerical duties. In 1780 he accepted an appointment to the
Lutheran church in Lancaster, where he spent the remainder of his life busi-
ly engaged as pastor, college president, and part-time botanist. He began
his herbarium by collecting the plants in his own vicinity, with special attention to grasses and sedges, and rapidly expanded it by exchange of
specimens with correspondents on both sides of the Atlantic. Meanwhile,
he developed his own botanical garden, in which he grew plants from the

seeds sent to him. As early as the spring of 1791 Muhlenberg wrote to Cutler that he had collected more than 1,100 different plants in the Lancaster area.

The task of identifying the plants that came into his hands was not an easy one, although he did his best to obtain the latest botanical works from Europe. Thus in September 1792 he invited William Bartram to visit him, promising to show him copies of such valuable and inaccessible works as Joseph Gaertner's *De Fructibus* and the latest editions of the Linnaean *Amoenitates Academicae* and the *Systema Naturae.* In the same year he notified Moses Marshall of the arrival of a new edition of Linnaeus's *Genera Plantarum,* in which the editor, Johann von Schreber of Erlangen, had named a plant in Marshall's honor at Muhlenberg's request.[14] Linnaeus had named, classified, and described about 2,000 American plants in his works, and the posthumous editions by German editors continued to add to the list, but Muhlenberg repeatedly encountered difficulties in identifying the plants he collected. In December 1792 he appealed for help to James E. Smith, owner of Linnaeus's herbarium and library and president of the Linnean Society of London. Muhlenberg had made considerable progress in describing the plants in the vicinity of Lancaster, he told Smith.

> But I met so many *adversaria, nova genera et novas species* [adversaries, new genera and new species], which I am not able to class according to the present editions of Linnaeus, that I long ago and earnestly looked for a friend, who would kindly assist me to find out which plants are already described by Linnaeus, and which are nondescripts. Some of my doubts have been cleared up by my worthy friend, Dr. Schreber, . . . but very many remain.[15]

Soon Smith was added to the long list of Muhlenberg's European correspondents, which included Jonathan Stokes, Kurt Sprengel, Johann D. Schoepf, Christian Persoon, Carl Willdenow, Johann von Schreber, Olaf Swartz, and many others. To these correspondents, especially Willdenow and Schreber, Muhlenberg sent specimens and descriptions that were published to the botanical world. In 1801 Willdenow published notes on twenty species of plants growing near Lancaster, and in 1803 notes on ten new species of willows; these were subsequently republished in London in an English translation.

Muhlenberg sent hundreds of specimens to his European correspondents for identification, accompanying them with his own descriptions when he considered them to be new species. He prepared a considerable number of botanical manuscripts, but only a fraction of these were published. Before 1800 he had nearly completed a descriptive flora of the plants growing in the vicinity of Lancaster, arranged according to the Linnaean system and including both flowering plants and cryptogams; yet he published only an "Index" to this flora, a mere list of names without descriptions or synonyms.[16] Presumably he wished to complete his study of the cellular cryptogams before publishing a descriptive work. Thus was lost to science a work that might have been an important contribution to the botany of the eastern United States and a convincing demonstration of Muhlenberg's high competence as a systematic botanist.

Muhlenberg had not given up the more ambitious project of a North American flora. His zeal to see this accomplished by an American was raised to a new pitch by the realization that some foreign botanist might carry off these laurels if the Americans did not work together to prevent it. In 1801 André Michaux, having returned to France from his travels in America (see Chap. 5), published his *History of the Oaks of America* in Paris before meeting an untimely death on an expedition to Australia. In 1803 his son, François Michaux, saw through the press his father's *North American Flora,* the first general account of North American botany. This work, based entirely on André Michaux's observations and collections over the vast territory extending from northern Florida to Hudson's Bay and westward to the Mississippi River, owed many of its handsome illustrations to the master hand of Pierre-Joseph Redouté, illustrator of the *Jardin de la Malmaison* for Napoleon's wife, Josephine de Beauharnais. Tradition has it that the characters of the species and genera were the work of the leading French botanist of the time, Louis C. Richard. Michaux's son was to continue his father's botanical travels in the United States and publish a handsome three-volume *History of the Forest Trees of North America.*[17]

Undaunted by the inroads of foreign botanists on the botanical treasures of the New World, Muhlenberg continued to urge his American correspondents to work together to produce an American flora. His New England correspondents were slow to respond, but he was more successful in the South. Writing to John Brickell of Savannah in 1803, Muhlenberg recommended to him the wandering plant collectors John Lyon and Matthias Kinn and plied him for information that would supplement Thomas Walter's *Flora Caroliniana* (1788). The sight of some plants not listed by Walter would be a feast to his eyes, he wrote Brickell. "Michaux has figured a few Oaks from the southern States, which I have not seen. . . . Could you get me a specimen of each. . . . Have you any new Mosses or Lichens. Walter has but little. . . . Do You, dear Sir, give them a touch and bring them to light."[18]

In 1811 Muhlenberg found a valuable correspondent in William Baldwin, a Pennsylvanian who had begun his botanizing at Humphry Marshall's garden and continued it under Benjamin S. Barton's instruction during his medical studies at the University of Pennsylvania. Baldwin was settled in Wilmington, Del., when Muhlenberg first wrote him to open a botanical correspondence. Of Baldwin's wife, Muhlenberg wrote: "Dr. Perlee informs me she is a great Botanist and Entomologist." In September 1811, when Baldwin took a southern trip in the hope of improving his consumptive condition, Muhlenberg wished him Godspeed and gave him the names of some of his correspondents in the South. He must be sure to visit "that excellent Botanist and Entomologist" Stephen Elliott at Beaufort, S.C., and Oemler and Abbot in Savannah. At Natchez, Miss., if he traveled that far, he would find Henry Moore, "a native of Lancaster, who . . . is in trade, — but likewise attends to Botany, and every branch of natural history. . . . Without a doubt you will find many other valuable gentlemen, who are unknown to me," Muhlenberg added. "Remember me to all such; and try to open communion and correspondence with such. By joining hands, we may do something clever for the Science. Mine, indeed, begin to get old and stiff, — but the heart still beats high, and wishes that others may do what was left undone. Away with all jealousy."[19]

Baldwin kept in touch with Muhlenberg as he went. He noted that the Michaux garden near Charleston was in ruins and the Botanic Garden in the city in little better condition. John Shecut, he reported, had been forced to leave his *Flora Carolinaeensis* unfinished for lack of subscribers. Baldwin's own botanizing had been hampered by the lack of any systematic work on botany, and he had lost his copy of Bartram's *Travels*. Somewhat later, however, Willdenow's fourth edition of Linnaeus's *Species Plantarum* and André Michaux's *History of the Oaks of America* reached him by land. Meanwhile, in June 1813 Baldwin's specimens for Muhlenberg's herbarium finally reached Lancaster by way of New York through the kind offices of Samuel L. Mitchill. Altogether, Baldwin contributed more than 1,550 specimens to Muhlenberg's herbarium between 1811 and 1815.

In October 1813 the chief product of Muhlenberg's long and arduous labors on American botany issued from the press at Lancaster, bearing in Latin and English the title: *A Catalogue of the Hitherto Known Native and Naturalized Plants of North America.* This work had been substantially completed in 1809, but Muhlenberg, wishing to have it published locally so that he could correct the proof sheets, had trouble in finding a printer in Lancaster willing and able to undertake the task. He eventually found one, a Mr. Hamilton, but the printing went very slowly. "My Catalogue goes very slow," Muhlenberg reported to Elliott on January 14, 1813; "the five first Classes to the Umbellatae are printed from Nov. 1811, until now, and all my begging and driving is of no use." The rough paper and crude typography of the book bespoke the difficulties that had attended the publication of this simple list of American plants. The brief preface acknowledged by name the assistance Muhlenberg had received from his many correspondents and closed with a cheerful salute: "HEALTH AND PROSPERITY TO THEM AND ALL LOVERS OF BOTANY."[20] Publication had come none

Three pioneers of North American botany: Thomas Nuttall (engraving) (upper left), François-André Michaux (by Rembrandt Peale) (above), and Gotthilf Henry Ernst Muhlenberg (by Charles Willson Peale) (left). Courtesy of the American Philosophical Society

too soon. Two months later Muhlenberg suffered a severe stroke. He lived another fifteen months, until May 23, 1815, his ardor for botany unquenched to the end.

Muhlenberg's *Catalogue* listed 853 genera and about 3,670 species. For each of the flowering plants it gave a Latin name; an English name; the common name when known; and descriptive information concerning the calyx, corolla, growth habit, habitat, longevity, time of flowering, and sometimes also a locality such as New York or Pennsylvania. Much of the information was in abbreviated form to save space. In the case of the cryptogams (including the ferns, mosses, lichens, and fungi) Muhlenberg gave only the Latin names. Many of the species listed were new, but because Muhlenberg did not give detailed descriptions and listed very few synonyms, his names were rarely adopted by later botanists. Some present-day botanists consider Muhlenberg's brief descriptions as providing sufficient information to render some of his Latin names for new species valid according to the rules laid down in the *International Code of Botanical Nomenclature;* other botanists disagree and hence do not accept these names.[21]

Muhlenberg seems to have cared little about priority. He failed to indicate which of the names he had originated himself or even which of the species were new, except for a few lichens marked "N.S." In a second, posthumous edition of the *Catalogue,* published in 1818 from Muhlenberg's corrected manuscripts, the number of genera was increased to 877 and species to about 3,810, again without synonyms or detailed descriptions.[22] Descriptions were provided, however, in another posthumous work of 1817 devoted to the grasses and sedges. Here again Muhlenberg showed little concern to gain credit for describing new species, much to the amazement of Constantine Rafinesque, that passionate seeker after botanical glory, who reviewed the work for the *American Monthly Magazine and Critical Review.* Modern botanists agree in wishing that Muhlenberg had indicated clearly the names he originated and that he had cited standard abbreviations of the names assigned by botanists whose specific names he accepted when he redescribed their species. By doing so, he would have simplified the task of establishing a correct botanical nomenclature according to modern rules.

As to the general caliber of Muhlenberg's work in systematic botany, modern authorities agree in giving him high marks. A. S. Hitchcock notes that Muhlenberg's interpretations of the species of grasses described by his predecessors were usually correct and that in about sixty cases Muhlenberg originated the specific names he used, although he claimed credit for first naming them in only five of these cases.[23] The main regret of modern commentators is that Muhlenberg published so few of the manuscripts he prepared. "Seldom in the history of systematic botany has any author written so much of which actually so little was published," wrote E. D. Merrill and Shiu-Ying Hu, adding that two of the three new generic names Muhlenberg proposed (*Bartonia* and *Elliottia*) are still accepted by botanists and more than 100 of his specific names are still used in the sense that he proposed them.[24] Joseph Ewan agrees as to the quality of Muhlenberg's work but notes that his influence was relatively small "owing to the small editions of his works, the German contacts that soaked up his time, and [the fact] that he did not write in English for the American audience. . . . Foreigners were . . . more important [in American botany] than Americans."[25]

Fortunately for botanical science, Muhlenberg's herbarium of more than 5,000 specimens, the largest in the United States at that time, was acquired by the American Philosophical Society in two installments shortly after Muhlenberg's death. Since the specimens were unmounted and not in the best order, the herbarium was of little immediate use to American botanists. Eventually, however, it was placed on loan at the Academy of Natural Sciences of Philadelphia and put into serviceable condition. In this way it has become an important resource for researchers in botanical nomenclature in our own time.[26]

If Muhlenberg and Barton, both of whom died in 1815, had lived a few years longer, they would have been astonished at the sudden increase in the quantity of literature on North American botany as the second decade of the nineteenth century drew to a close. The coming of this new and more affluent era was heralded by the publication of Pursh's two-volume *Flora Americae Septentrionalis: Or, A Systematic Arrangement and Description of the Plants of North America*. Pursh had acquired a first-hand acquaintance with American botany through his experience with the Philadelphia gardens, travels at Barton's expense, employment on the plants collected by Lewis and Clark, tour of duty at the Elgin Botanic Garden in New York, and subsequent brief travels in the West Indies and New England. He then had the good sense to go to England, where he would have access to the Linnaean herbarium and English plant collections that had been receiving American plants since the seventeenth century. With the cooperation of Aylmer B. Lambert, vice-president of the Linnean Society of London, who supplied him with liberal quantities of beer as well as information, advice, and injunctions to keep his nose to the grindstone, Pursh made rapid progress in his work. It was published in January 1814.

Pursh was fairly generous in acknowledging the assistance he had received from American botanists and gardeners and from wandering foreign botanists like himself. In particular, he paid his respects to Muhlenberg, "a gentleman whose industry and zeal for the science can only be surpassed by the accuracy and acuteness of his observations"; Humphry Marshall;

William Bartram, "a very intelligent, agreeable, and communicative gentleman"; William Hamilton and his gardener John Lyon; his former patron Benjamin S. Barton; Meriwether Lewis and Bernard M'Mahon; Aloysius Enslen, plant collector for Prince Liechtenstein of Austria; David Hosack; John Le Conte of Georgia; and William D. Peck of Harvard. He also listed the sources of information concerning American plants that had been opened to him in England through the courtesy of Joseph Banks, Aylmer B. Lambert, William Roscoe, and others. These sources included the herbaria of earlier collectors such as John Bartram, John Clayton of Virginia, Mark Catesby, Thomas Walter of South Carolina, the British botanical explorer Archibald Menzies, William Bradbury (Nuttall's companion on the fur-trading expedition up the Missouri), and others. Together with his own collections, these American and European sources gave Pursh a knowledge of twice as many North American plants as had been described in André Michaux's *North American Flora.* With this information and the assistance of England's leading botanists, Pursh was able to publish a work of great importance for American botany.

Adopting the Linnaean system of classification with some modifications, Pursh modeled his plant names on those found in Willdenow's edition of Linnaeus's *Species Plantarum* and in André Michaux's *Flora,* avoiding the penchant for proposing new names that had characterized the younger Michaux's *History of the Forest Trees of North America* and carefully indicating any change made in the description of the various species and genera. He also indicated the localities where the plants were found, their range, situation, soil, times of flowering, color and size of flowers, uses, and whether he had seen them in the living state or only as dried specimens and, if the latter, in whose herbarium. Lengthy descriptions were given only in the case of interesting new species such as those discovered by Lewis and Clark. Many of these were shown in the plates. Among the cryptogams only the ferns were included, the rest being reserved for a later work.

In all there were 3,076 species in 740 genera, embracing the plants of four regions: the Columbia River valley and Pacific Northwest, the Missouri River country, the Southeast as far as New Orleans, and the best known middle Atlantic states. Pursh's "Index of Authors" listed 190 botanical writers on whom he had drawn. It constituted, says Joseph Ewan, editor of the modern reprint of Pursh's *Flora,* "the first extensive list of references relating to the North American flora."[27] By his exertions Pursh had brought the combined resources of the botanical world to a focus on the flora of North America. Rafinesque might complain that Pursh had omitted many plants, changed some names unnecessarily, named some genera badly, neglected to mention many relevant publications (including those of Rafinesque himself), and taken credit that probably should have been assigned to his patron and adviser Lambert, but he had to concede that Pursh's *Flora* was the best manual of North American botany that had yet been produced. Later botanists have concurred in this judgment. "Pursh's *Flora*," says Ewan, "complements the later manuals of Nuttall, Elliott, Torrey and Gray, indeed it is fundamental for hundreds of taxonomic decisions for North American plants." The Canadian botanist David P. Penhallow is of the same opinion. Pursh, he declared, "not only laid the foundations for the development of systematic botany on this continent, but placed all future botanists under obligations which they have been proud to acknowledge."[28]

Among those who were not happy with Pursh's performance was the other beneficiary of Barton's patronage, Thomas Nuttall. To Nuttall's dismay, Pursh had published some of the species from the Missouri country that Nuttall had shown him or that Pursh had seen preserved in the herbarium Nuttall presented to Lambert or growing from seeds Nuttall had brought back and planted in John Fraser's nursery. On his return to Philadelphia in 1815, Nuttall began to formulate plants for a botanical work of his own. But first he wanted to round out his knowledge of American plants by exploring the southern and southwestern regions of the country. He went first to Savannah, where he learned much concerning the flora of Florida from William Baldwin, whom he described as "better acquainted with the plants of America than any other person I have yet met with." Through Baldwin he made contact with several other botanists: Stephen Elliott, Lewis and John E. Le Conte, Thomas Wray of Augusta, Ga., and others. Making his way back to Philadelphia by way of Wilmington, N.C., Nuttall soon embarked on a much longer journey that took him through Pittsburgh to Cincinnati, Lexington, and Tennessee and then eastward through the Cumberland Gap into the Carolinas and down the Catawba valley to Camden and on to Charleston, where Elliott received him cordially and showed him the first section of his *Sketch of the Botany of South Carolina and Georgia.*

Barton had died before Nuttall set out on this tour, hence he was free to publish whatever he chose. Philadelphia, to which he now returned, provided the best resources for undertaking a major botanical publication of any city in the United States. At the Academy of Natural Sciences, Nuttall found a congenial place to work, surrounded and encouraged by eager naturalists, geologists, and mineralogists like Thomas Say, Joseph Correa da Serra, Zacchaeus Collins, George Ord, William Maclure, and Gerard Troost. Besides his own collections and those of the academy, Nuttall had access to Collins's and Muhlenberg's herbaria. For nearly two years he worked at his plants, "often remaining up all night, and when tired lying down under the bones of the great Mastodon for repose."[29] He had no publisher. His book was "printed for the author," and tradition records that Nuttall, a typesetter by trade, set much of the type himself. He had at first thought of producing a revised edition of Pursh's *Flora,* but the final product was an original work in two volumes, *Genera of North American Plants.* It was dedicated, not to Barton, but to Correa da Serra, who may possibly have contributed something toward its publication. Pursh had also passed over Barton in framing a dedication for his book but had at least acknowledged his indebtedness to his former patron.

Like Pursh's work, Nuttall's *Genera* followed the Linnaean system of classification, but only because he believed his American readers would be confused if he attempted to introduce them to the new "natural" system of Jussieu. Each genus of plants was described in English rather than in the traditional Latin, and its component species were listed briefly; only new species or species of special interest were described. The main purpose, said Nuttall, was "a Synopsis of the Genera," with a full description of each, including the mode of vegetation, the geographic distribution, and remarks intended to indicate their natural relationships. The list of species was designed to supplement those described by Pursh by providing additional information based on Nuttall's own observations and by adding several hundred new species discovered and described since Pursh's work appeared. Many of these were Nuttall's own discoveries, and he took pleasure in paying tribute to his botanical friends by naming genera or species in their honor with brief dedicatory footnotes. Among the Philadelphians so honored were Barton, Collins, Wistar, and M'Mahon. Genus No. 588 was named *Balduina* "as a just tribute of respect for the talents and industry of William Baldwyn, late of Savannah in Georgia; a gentleman whose botanical zeal and knowledge has rarely been excelled in America." Unfortunately, Baldwin thought that Nuttall had plagiarized some of his species, and the two men were still estranged when Baldwin, serving as botanist on the Long expedition to the Rocky Mountains, succumbed to tuberculosis at the age of forty-one.[30]

Nuttall's *Genera* was well received on both sides of the Atlantic. Rafinesque had none of the charges of plagiarism he had leveled against Pursh's *Flora.* Nuttall's work, he declared, was "superior in many respects to any other yet published on either side of the Atlantic." The Moravian mycologist Lewis D. von Schweinitz, whose important publications on fungi had just begun to appear, thought Nuttall's observations "uncommonly excellent," and the great British botanist William J. Hooker praised the *Genera* as forming "an era in the history of American botany." Looking back on Nuttall's work twenty-four years after its publication, Asa Gray declared, "No botanist has visited so large a portion of the United States, or made such an amount of observations in field and forest. Probably few naturalists have ever excelled him in aptitude for such observations, in quickness of eye, tact, in discrimination and tenacity of memory."[31]

THE

GENERA

OF

NORTH AMERICAN PLANTS,

AND

A CATALOGUE

OF THE

SPECIES,

TO THE YEAR 1817.

———◆———

BY THOMAS NUTTALL, F. L. S.

FELLOW OF THE AMERICAN PHILOSOPHICAL SOCIETY,
AND OF THE ACADEMY OF NATURAL SCIENCES
OF PHILADELPHIA, &c.

———◆———

VOLUME I.

———

PHILADELPHIA:

PRINTED FOR THE AUTHOR BY D. HEARTT.
1818.

Title page of Thomas Nuttall's important contribution to North American botany. Courtesy of the American Philosophical Society

Nuttall's *Genera* was the most distinguished of a series of botanical publications that issued from American presses toward the end of the Jeffersonian era. In 1821 Nuttall's *Journal of Travels into the Arkansas Territory* appeared, reporting his botanical and other discoveries on a new exploration subsidized by Correa da Serra, Collins, Maclure, and John Vaughan, and Nuttall became the chief botanical contributor to the newly established *Journal* of the Academy of Natural Sciences of Philadelphia. William P. C. Barton, drawing on the combined resources of his own herbarium and that of his uncle Benjamin S. Barton, produced a flora of the Philadelphia area in 1818 and a three-volume North American flora in 1820–1824.[32] Meanwhile, Stephen Elliott was bringing out the successive numbers of his *Sketch of the Botany of South Carolina and Georgia* (1816–1824). "No one in Europe can . . . appreciate correctly the difficulty of the task in which I have engaged," Elliott wrote to William J. Hooker. "The want of books, the want of opportunities for examining living collections or good herbaria, the want of coadjutors, have all served to render my task arduous, and to multiply its imperfections."[33] Yet Correa da Serra considered Elliott the ablest botanist in the United States, and Hooker had nothing but praise for his *Sketch*.

> There are many new species, described with care and fidelity, and the grasses, which are accompanied with some neat plates, have particularly attracted the author's attention. . . . [We only regret that] the work cannot . . . take in the Cryptogamia; . . . we consider Mr. Elliott's talent for minute description admirably calculated for such plants as that class embraces.[34]

Elliott's treatise was dedicated to Henry Muhlenberg. With what delight would the reverend botanist have read the dedication and the statement in the preface that "in a science like botany, depending upon fact and observation, the progress must necessarily be slow and gradual. It is only by the cooperation and contributions of many individuals that it can ultimately attain any degree of perfection." For his own part, Elliott acknowledged his debt not only to Muhlenberg and his informants in South Carolina and Georgia but also to Jacob Bigelow and Francis Boott in New England, Torrey and Rafinesque in New York, Collins and Nuttall in Philadelphia, and Schweinitz in Salem, N.C. With the help of these correspondents Elliott had been able to add 180 genera and 1,000 species to those listed in Thomas Walter's pioneering *Flora Caroliniana* (1788).[35]

In Boston, Bigelow found public interest in botany so much increased since the days of Cutler's catalogue of New England plants that in 1814 he published a *Florula Bostoniensis* and an American edition of James E. Smith's *Introduction to Physiological and Systematical Botany.* Three years later he began publication of his *American Medical Botany,* a substantial contribution to a field of inquiry pioneered by his teacher Benjamin S. Barton and William P. C. Barton.[36] Meanwhile, the formation of the Linnaean Society of New England had provided an institutional setting for the pursuit of botany and other branches of natural history in eastern Massachusetts.

In New York the enthusiasm of Mitchill, Hosack, Torrey, Eaton, Rafinesque, and the other members of the newly formed Lyceum of Natural History produced a spate of publications in natural history that rivaled the output of the Philadelphia naturalists. Torrey, Eaton, and Rafinesque were the most prolific writers on botany, each with his own special scientific personality. Eaton's main contribution to American botany was his *Manual of Botany,* the first edition of which appeared in 1817 while he was still teaching at Williams College. As he explained in a letter to Torrey, he made no pretense of being a great botanist. "You are made for the highest walks of science — nice accurate investigation — new discoveries and improvements — to correct the blunders of others and to keep the ship of science in trim. . . . I am made for noise and bustle. My forte is among the rabble. And if I have any merit, it consists in the art of simplifying."[37]

As a botanical textbook, Eaton's *Manual* was a great success. It went through eight editions between 1817 and 1840 and became the standard field reference book for students in American academies and colleges. Written in English rather than Latin, it first described the characters of the genera systematically arranged and then described the species at length under their appropriate generic headings. The reviewer in the *North American Review* observed:

> The work . . . is wholly popular in its arrangements, design, and nomenclature, embracing not only indigenous vegetables, but even all those exotics which are cultivated in the interior parts of the country. Taking [Christian] Persoon as the basis of his descriptions of phaenogamous plants, . . . the author has corrected and enlarged them, either from personal examination, or by reference to Pursh, — adding also twenty or thirty new species chiefly communicated by Dr. Bigelow and Dr. Torrey. . . . As a convenient and accurate guide in botanical inquiries . . . compressing a great variety of matter into a narrow compass, the Manual is deserving of great praise. And . . . whilst Pursh and Nuttall have omitted all cryptogamous plants excepting the ferns, Eaton . . . has given quite a full account of the cryptogamous species, . . . amounting to four hundred and twenty more than Michaux describes in the Flora Boreali-Americana.[38]

Unfortunately Eaton remained committed to the Linnaean system of classification at a time when English and American botanists were gradually shifting over to the natural system of Jussieu.

As for young John Torrey, he lost little time in achieving the prowess and distinction as a botanist that Eaton had prophesied for him. Soon after his graduation from the College of Physicians and Surgeons in 1818, he began building up his herbarium through extensive correspondence in Europe and America. Beginning with a catalogue of plants growing in the vicinity of New York, published in 1819, he went on to describe the plants collected by Edwin James on the Long expedition (an expedition Torrey himself had been invited to join) and by David B. Douglas on the Cass expedition to the Great Lakes and upper Mississippi valley.[39] In 1824 Torrey's first major work appeared, *A Flora of the Northern and Middle Sections of the United States: Or A Systematic Arrangement and Description of All the Plants Hitherto Discovered in the United States North of Virginia.* With this publication Torrey was well on his way to becoming one of the two leading American botanists of the mid-nineteenth century, the other being his early friend and later collaborator Asa Gray.

Unlike Torrey, Rafinesque was primarily a field naturalist, "the best naturalist I am acquainted with, but . . . too fond of novelty," Torrey wrote to Eaton. In 1815 Rafinesque settled for a time in New York City, where he helped to launch the Lyceum of Natural History and published voluminously in Biglow's *American Monthly Magazine and Critical Review.* Then in 1819 after an exploratory trip down the Ohio River, he accepted an appointment as professor of botany, natural history, and modern languages at Transylvania University (see Chap. 5) and threw himself passionately into the study of the natural history and antiquities of the Ohio country. A flood of manuscripts describing new species and genera of plants and animals poured from his pen. When Benjamin Silliman, editor of the *American Journal of Science,* and other eastern editors balked at publishing such a deluge of communications, Rafinesque sent them to Paris, Brussels, or London or published them in short-lived periodicals of his own devising or in the *Western Review and Miscellaneous Magazine,* or the *Kentucky Gazette.* Modern botanists are still trying to clear up the confusion caused by Rafinesque's shotgun mode of publication. In 1943 the Harvard botanist E. D. Merrill estimated that Rafinesque originated about 3,000 new generic and subgeneric plant names. Of the generic names, said Merrill, only about five percent are acceptable today, and Rafinesque's names for species have fared but little better.

> While Rafinesque described a great many new genera and new species *de novo* on the basis of actual specimens, he based an extraordinarily large number of his "new" entities on the published work of other authors. . . .
> He proposed his own laws of nomenclature, and many of the changes in both generic and specific names were made because of his confidence in his own rules — rules that other botanists never accepted.[40]

Merrill conceded, however, that Rafinesque's penchant for subdividing large, complex genera into smaller ones anticipated the practice of many modern botanists, and he urged botanists to adopt Rafinesque's names for these smaller genera when possible. Francis W. Pennell, curator of the botanical collections at the Academy of Natural Sciences of Philadelphia, was more favorably disposed toward Rafinesque, calling him "the best field-botanist of his time." Pennell agreed that Rafinesque's species and genera were loosely and scantily described, often on the basis of inadequate evidence, but noted that Rafinesque's ideas about plant classification, including his belief in the mutability of species, anticipated those of later botanists.[41] "Brilliant, but erratic and eccentric" seems to be the botanical consensus about Rafinesque.

In 1826, having failed to persuade Thomas Jefferson to arrange an appointment for him at the projected University of **Virginia**, Rafinesque resigned his position at Transylvania and returned to **Philadelphia**, there to spend the remainder of his life eking out a living by selling a cure for tuberculosis, adding continually to his herbarium, and publishing as always on an amazing variety of subjects, mostly at his own expense. He died in 1840 from cancer of the liver and stomach. His large herbarium, containing forty or fifty thousand specimens from all parts of the world, was acquired by Elias Durand, then curator of plants at the Academy of Natural Sciences. Unfortunately Durand destroyed most of the specimens as "worthless" before returning to his native country (France) in 1868.[42] Rafinesque was dead, but his multifarious, fugitive publications lived on, destined to excite as much controversy and comment in the twentieth century as they had in his own day. His remains now repose in a place of honor at Transylvania University. His name is commemorated in a sign over the entrance to the student refreshment lounge, the "Rafskeller."

CONCLUSION

In botany as in geology the progress of scientific research depended upon the combined efforts of native Americans and immigrant or visiting scientists. Attracted to America by the lure of an unexplored continent, European botanists like Pursh, Nuttall, the Michaux, and Rafinesque moved from place to place as fancy or scientific opportunity dictated, ranging the country from Canada to Florida and westward to the Mississippi River and even beyond in some cases. Wherever they went, they added to their botanical knowledge not only by their own researches but also by their contacts with American botanists. In Philadelphia they were assured of a cordial welcome at Bartram's and Hamilton's gardens or (after 1812) at the Academy of Natural Sciences; in Lancaster, Pa., at the home of Henry Muhlenberg; in New York, at the Elgin Botanic Garden or the Lyceum of Natural History; in Boston, at Harvard's botanical garden; in South Carolina and Georgia at the plantations of Elliott and other planter-naturalists. Small wonder, then, that these botanical wanderers from Europe produced the first extensive North American floras. Meanwhile, the American botanists, tied down to professional callings, did what field research they could in their localities and built up their knowledge by correspondence abroad and throughout the United States. One of them, Henry Muhlenberg, managed to produce a useful catalogue of North American plants. Others like Benjamin S. Barton and Amos Eaton advanced the cause of American botany by their textbooks and teaching, and Barton made an important additional contribution by subsidizing the researches of Pursh and Nuttall. Still others, like Bartram, Hamilton,

M'Mahon, and Hosack, maintained botanical gardens. In one of these, M'Mahon's garden, plants from the Lewis and Clark expedition were grown and preserved and made available to botanical science through Frederick Pursh. By the 1820s American botanists like Eaton, Torrey, and Gray were ready to take over from their European brethren the task of naming, classifying, and describing the flora of the United States.

11

Natural History in a New World: Zoology and Paleontology

![decorative oak leaf illustration] IN zoology and paleontology, as in botany, Philadelphia naturalists led the way in exploring, collecting, and describing the riches of the North American continent. Some of these had been named, illustrated, and described in eighteenth-century works like Mark Catesby's *Natural History of Carolina, Florida and the Bahama Islands,* Buffon's *Natural History, General and Particular,* Linnaeus's *Systema Naturae,* Johann R. Forster's *Catalogue of the Animals of North America,* George Edwards's *Natural History of Birds* and *Gleanings of Natural History,* and Thomas Pennant's *History of Quadrupeds* and *Arctic Zoology;* and Jefferson had included a list of American animals in his *Notes on the State of Virginia.* But these were only beginnings. The main task lay ahead, and Philadelphia afforded the best resources for a concerted effort at producing a book on the fauna of North America. William Bartram had a better field knowledge of American animals than any of his compatriots, Charles W. Peale had begun building a first-class museum of natural history, the American Philosophical Society had been revitalized under the leadership of Benjamin Franklin and David Rittenhouse, and Benjamin S. Barton had returned from his studies in Europe burning with ambition to become the Thomas Pennant of America.

ZOOLOGICAL PIONEERS IN PHILADELPHIA

William Bartram and his father had supplied information about American plants and animals to English devotees of natural history for many years, and some of their contributions had been published. The London merchant-naturalist Peter Collinson published two of William Bartram's drawings of the "Horne-tailed Turtle" in the *Gentleman's Magazine,* Pennant quoted both Bartrams in his *Arctic Zoology* (1784–1787), and

277

Falco. Eagle and Hawk.

¶ Falco regalis, the great grey eagle.
¶ F. leucocephalus, the bald eagle.
* F. pifcatorius, the fifhing eagle.
¶ F. Aquilinus, cauda ferrug. great eagle hawk.
¶ F. gallinarius, the hen hawk.
¶ F. pullarius, the chicken hawk.
* F. columbarius, the pidgeon hawk.
¶ F. niger, the black hawk.
* F. ranivorus, the marfh hawk.
* F. fparverius, the leaft hawk or fparrow hawk.

a Milvus. Kite Hawk.

‖ Falco furcatus, the forked tail hawk, or kite.
‖ F. glaucus, the fharp winged hawk, of a pale fky-blue colour, the tip of the wings black.
‖ F. fubcerulius, the fharp winged hawk, of a dark or dufky blue colour.
‖ Pfitticus Carolinienfis, the parrot of Carolina, or parrakeet.

Corvus. The Crow kind.

* Corvus carnivorus, the raven.
‖ C. maritimus, the great fea-fi le crow, or rook.
¶ C. frugivorus, the common crow.
¶ C. criftatus, f. pica glandaria, the blue joy.
¶ C. Floridanus, pica glandaria minor, the little joy of Florida.
¶ Gracula quifcula, the purple jackdaw of the fea coaft.

‡ Kite hawks   Thefe are characterifed by having long fharp pointed wings, being of fwift flight, failing without flapping their wings, let light bodies, and feeding out of their claws on the wing, as they gently fail round and round.

Pica glandaria cerulea non criftata, the little jay of Eaft Florida.

* Gracula purpurea, the leffer purple Jackdaw, or crow blackbird.
• Cuculus Carolinienfis, the cuckoo of Carolina.

Picus. Woodpeckers.

‖ Picus principalis, the greateft crefted woodpecker, having a white back.
* P. pileatus, the great red crefted black woodpecker.
* P. erythrocephalus, red headed woodpecker.
⊦ P. auratus, the gold winged woodpecker.
¶ P. Carolinus, the red bellied woodpecker.
¶ P. pubefcens, the leaft fpotted woodpecker.
¶ P. villofus, the hairy, fpeckled and crefted woodpecker.
¶ P. varius, yellow bellied woodpecker.
¶ Sitta Europea, grey black capped nuthatch.
† S. varia, ventre rubro, the black capped, red bellied nuthatch.
† Certhia rufa, little brown variegated creeper.
* C. pinus, the pine creeper.
* C. picta, blue and white ftriped or pied creeper.
* Alcedo alcyon, the great crefted king-fifher.
* Trochilus colubris, the humming bird.
* Lanius grifcus, the little grey butcher-bird of Pennfylvania.
* L. garrulus, the little black capped butcher-bird of Florida.
⟨ L. tyrannus, the king bird.
* Mufcitapa nunciola, the pewit, or black cap flycatcher.
⟨ M. criftata, the great crefted yellow bellied flycatcher.
* M. rapax, the leffer pewit, or brown and greenifh flycatcher.
⟨ M. fubviridis, the little olive cold. flycatcher.

P p

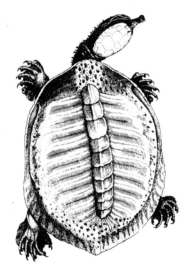

Pages from William Bartram's list of 215 American birds in his Travels, *"the starting-point of a distinctively American school of ornithology" (Elliott Coues). The "great soft-shelled tortoise" depicted by Bartram (left) is similar in most respects to the Southern soft-shelled turtle* Amyda ferox. *Courtesy of the American Philosophical Society*

George Edwards, in the sixth volume of his *Gleanings* (1758–1764), described the spotted tringa and the pine creeper from Bartram specimens. "These birds, with many others," said Edwards, "were shot near Philadelphia, in Pennsylvania, by my friend, Mr. William Bartram, who sent them to London, for me to publish the figures and natural history of them." Unfortunately the many drawings and descriptions Bartram sent to John Fothergill during his travels in the Carolinas, Georgia, and Florida in the 1770s lay unpublished until 1968, when the modern botanist and historian of science Joseph Ewan brought them out in a handsome folio volume.[1] Had these been published in Bartram's time, says Ewan, many errors and crudities in subsequent publications on North American fauna might have been avoided.

Fortunately Bartram did name and describe many animals in his *Travels* (1791), although some of his names have been disallowed by modern zoologists because he sometimes used polynomials instead of binomials in identifying them. Especially important was his list of 215 birds, described by the late nineteenth-century ornithologist Elliott Coues as "the starting-point of a distinctively American school of ornithology" and by another modern writer, Witmer Stone, as "a landmark in the progress of American ornithology . . . and the first ornithological contribution, worthy of the name, written by a native American." Of Bartram's 215 birds of the eastern United States, Coues identified all but 52. He considered that most of the bird names in Bartram's list were valid binomials.[2] It should be added that Bartram did not confine himself to listing birds. Some were described in considerable detail, with information as to their migrations and habits.

Bartram also described a good many mammals, reptiles, and amphibia, many of them for the first time, but he did not bother to give them scientific names. According to Francis Harper, Bartram was the most notable of the eighteenth-century commentators on North American amphibians and reptiles: "He was apparently the original discoverer and describer of the Southern Bull-frog . . . , the Georgia Tree-Frog . . . , the Southern Cricket Frog . . . , the Little Tree-Frog . . . , and the Southern Leopard Frog. . . . He neglected a golden opportunity in not providing them with technical names."[3] Bartram's description of the alligator became as famous in literary as in scientific circles. After describing its egg-laying habits and general appearance, Bartram went on to evoke the scenes he had witnessed:

> An old champion . . . darts forth . . . on the surface of the waters, in a right line; at first seemingly as rapid as lightning, but gradually more slowly until he arrives at the centre of the lake. . . . He now swells himself by drawing in wind and water through his mouth, which causes a loud sonorous rattling in the throat for near a minute, but it is immediately forced out again through his mouth and nostrils, with

a loud noise, brandishing his tail in the air, and the vapor ascending from his nostrils like smoke. At other times, when swollen to an extent ready to burst, his head and tail lifted up, he spins or twirls round on the surface of the water. He acts his part like an Indian chief when rehearsing his feats of war; and then retiring, the exhibition is continued by others who dare to step forth, and strive to excel each other, to gain the attention of the favourite female.[4]

Bartram's *Travels* attracted far more attention in Europe than in America. It was translated into French and German, and several of the animals described were named, classified, and commented on by European naturalists in these editions. In 1794 Frederick A. Meyer, who had provided the zoological commentary for the German edition of the *Travels* (1793), published an "Alphabetical Sketch of all Living Zoologists" in the *Zoologisches Annalen*. William Bartram was the only American listed, and he was described as "English"!

If William Bartram was the most knowledgeable zoologist in Philadelphia, Benjamin S. Barton was the most ambitious. During his studies in Britain he made the acquaintance of several leading figures in the world of science and medicine, including John Walker of the University of Edinburgh, the London physicians John C. Lettsom and John Hunter, and the zoologist and literary figure Thomas Pennant. Pennant in particular welcomed Barton's offer of a scientific correspondence. In October 1790 he wrote to Barton inviting him to send drawings and descriptions for inclusion in his *Arctic Zoology* and offering to delay publication until Barton's contributions had been received. "I wish to usher a rising genius of the new world to the literati of the old. . . . Write soon & write much," Pennant urged.[5]

Despite Pennant's pleas for haste, nearly three years elapsed before Barton's manuscript "Notes on the Animals of North America" reached London, by which time the new edition of *Arctic Zoology* had already been published. Nevertheless, Pennant was grateful for the information contained in Barton's notes on thirty-five mammals, three kinds of birds, and four fish, which was based partly on Barton's own observations but even more on those of his informants, including "my ingenious friend William Bartram." The notes indicated not only the appearance and habits of the various animals but sometimes also their Indian names and Indian lore about them. The Delaware Indians called the beaver *amochk*, Barton reported: "The badge, or *Armorial Bearing* of the Lenopi [Delaware Indians] . . . is the Beaver. In assuming this badge, it is not improbable that these Indians associated the idea of their own importance . . . with the sagacity of the animal."[6]

Apparently Barton sent other information to Pennant as well, for in

July 1793 Pennant wrote to thank him for valuable topographical and geographical data and for his drawings and descriptions of various animals, including the caribou, a new species of finch related to the red-headed linnets, and a black-cheeked nuthatch. Among the drawings were some by William Bartram. Pennant promised to return these and continued to importune Barton for information, promising to sponsor a memoir on the elk, which Barton proposed to submit to the Royal Society of London. Pennant was now seventy-one years old and still vigorous despite a severe injury to his knee in April 1795. "I am now nearly as active as ever," he wrote Barton. "Can walk almost to any distance & ride as much. I cannot indeed vault on my horse as usual: that is a trifle."[7]

The indomitable Pennant died in 1798, having commissioned Barton to carry on the scientific study of North American animals. Barton certainly aspired to do so, but it was no easy task given his professional duties and his tendency to scatter his energies in many fields of research. He collected information assiduously, but he never succeeded in completing a work of substantial importance. His *Fragments of the Natural History of Pennsylvania* (1799), "a rude and imperfect sketch of the natural History-Picture in the neighborhood of Philadelphia" (as Barton described it), contained much useful information about bird migrations in relation to the progress of vegetation, much of it derived from Bartram, and introduced terms such as "resident species," "occasional visitants," and the like that are still in use. Besides this work, Barton published occasional articles and pamphlets on a variety of animals, including the rattlesnake, a new species of salamander, the jerboa, the mole, the alligator, the opossum, the Indian dog, and the wild sheep brought back by Lewis and Clark.

In these researches, as well as in his teaching, Barton was greatly aided by the collections in Peale's Museum, which by 1800 included (according to Claypoole's *American Daily Advertiser*) over 100 quadrupeds, about 700 birds, 150 amphibious animals, many thousands of insects, a number of fishes, nearly 1,000 minerals and fossils, and 11 wax figures of human racial types. Peale was anxious that his museum should be a place of education as well as enjoyment. With the aid of a visiting French naturalist, Ambroise P. de Beauvois, he prepared a *Scientific and Descriptive Catalogue* of the exhibits and undertook to give public lectures. In two series of lectures in 1799–1800 and 1800–1801 Peale commented on the museum specimens as each was exhibited, beginning with a live orangutan he had recently acquired. He continued with the other quadrupeds, of which nearly 200 were described and 122 illustrated, after which Peale presented 765 specimens from his splendid collection of birds. In his commentary on the specimens Peale exhibited a considerable familiarity with the literature of natural history and a shrewd independence of mind based on his own observations of these animals, not only those he had kept in a living state but also those he had dissected. "We should not adopt too hastily the opinions of any great Man," he warned his auditors, "but . . . judge for ourselves whether they deserve credit at all times, and always . . . distrust that author who shall presume to treat lightly of the works of Creation."[8]

Like Buffon, Peale believed that the description of an animal should indicate not only its appearance but also "its manners, disposition and general character." Thus, "the Cat, . . . although an animal of prey, is a useful domestic. It is neither wanting in sagacity nor sentiment; but its attachments are stronger to *places* than to *persons*. The Cat is handsome, light, adroit, cleanly and voluptuous." Doubtless Peale's disquisitions fell short of the high standards set by the public lectures at the Museum of Natural History in Paris, but they were, as Keir Sterling has said, "the most comprehensive effort by an American up to that time to summarize all that was known on mammals and birds."[9]

THE BEGINNINGS OF PALEONTOLOGY IN PHILADELPHIA

In 1797 paleontology made its way into the American Philosophical Society when Thomas Jefferson presented his memoir on the huge clawed animal whose bones and teeth had been found in western Virginia and Peale was asked to "cause those bones to be put in the best order, for the Society's use" (see Chap. 2). The discovery and description of these remains was important for paleontology generally and for the stimulus it gave to paleontological researches at the American Philosophical Society.[10] The remains were found in July 1796 by workers digging for saltpeter in "one of those caves beyond the blue ridge," Jefferson reported to David Rittenhouse. They seemed to belong to "an animal of the family of the lion, tyger, panther &c. but as preeminent over the lion in size as the mammoth is over the elephant. . . . I have now in my possession the principal bones of a leg, the claws, and other phalanges, and hope soon to receive some others. . . . The whole of them shall be deposited with the [American Philosophical] society."[11]

In March 1797 Jefferson arrived in Philadelphia to assume his new duties as vice president of the United States and as president of the American Philosophical Society, succeeding the lately deceased Rittenhouse in the latter office. He brought with him his memoir on the megalonyx, or "great claw," still convinced that it was a member of the cat family. On visiting a Philadelphia bookstore, however, he came across the September 1796 issue of the *Monthly Magazine,* a British journal, in which was an engraving of the skeleton of a great clawed animal discovered in Paraguay in 1788 and subsequently mounted in the Royal Cabinet of

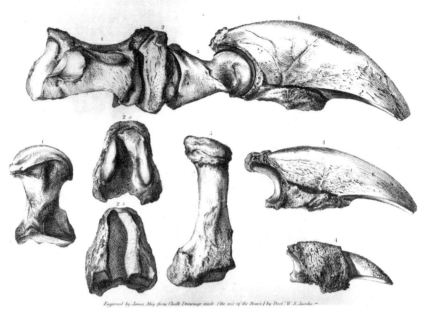

Engraved by James Akin from Chalk Drawings made (the size of the Bones) by Doct.^r W. S. Jacobs.—

Bones and claws of the Megalonyx jeffersoni, *a fossil edentate described by Thomas Jefferson and Caspar Wistar in 1799. Courtesy of the American Philosophical Society*

Natural History in Madrid. Copies of a Spanish engraving of the skeleton had found their way to the rising French comparative anatomist Georges Cuvier in Paris, who had promptly published a pamphlet in which he named the animal the megatherium and classified it among the edentates (toothless animals) as a relative of the sloth. This pamphlet formed the basis of the notice in the *Monthly Magazine.*[12] Sensing that his original guess as to the nature of the Virginia *incognitum* might be wrong, Jefferson hastily revised his memoir for the *Transactions* of the American Philosophical Society before presenting it to the society at their meeting on March 10. The megalonyx was now described as an animal "of the clawed kind" rather than as belonging to "the family of the lion, tyger, panther &c.," and an addendum was added in which Jefferson alluded to the notice in the *Monthly Magazine.* He noted, however, that the skeleton of the Paraguayan animal was "not so done as to be relied on, and the account . . . only an abstract" and left the question of the identity or nonidentity of the megalonyx and the megatherium for future determination when further facts should be available.

Strange to say, Jefferson had in his own library a drawing of the Paraguayan skeleton with exact measurements of some of the bones. The drawing and description had been sent to him in Paris in the spring of 1789 by William Carmichael, American chargé d'affaires at Madrid, shortly before Jefferson's departure for America. Jefferson wrote to thank Carmichael, stowed the papers in his luggage, and promptly forgot about them. Had he remembered them when the megalonyx bones came to his attention, he might have stolen the march on Georges Cuvier and the Spanish naturalists. As it was, he suspended judgment concerning the relations between the Paraguay and the Virginia *incognita,* still retaining some vestige of his original conception of the nature of the latter. Even in its revised form, Jefferson's memoir on the megalonyx, published in the *Transactions* in 1799 with an accompanying memoir on the same subject by the Philadelphia anatomist Caspar Wistar, held out the possibility that the megalonyx was a member of the cat family, three times the size of any lion described by Buffon and doubtless still extant in some remote corner of the world. ("If one link in nature's chain might be lost, another and another might be lost, till this whole system of things should vanish piecemeal.")[13]

Caspar Wistar had a different idea. He concluded that the megalonyx was similar in some respects to the bradypus or sloth illustrated by Daubenton in Buffon's *Natural History* and in other respects to the Paraguayan creature depicted in the *Monthly Magazine,* but not of the same species with either. His guess was verified in 1804 when Cuvier published memoirs on the megalonyx and the megatherium in the *Annales* of the Museum of Natural History, paying tribute to Jefferson as the discoverer and first describer of the megalonyx. In writing these memoirs Cuvier had at his disposal the casts of the megalonyx bones sent to him by Jefferson and a copy of José Garriga's handsome *Descriptión del esqueletto de un quadrupedo muy corpulento y raro que se conserva en el Real Gabinete de Historia Natural de Madrid,* published at Madrid in 1796 shortly after Cuvier's first memoir on the megatherium appeared. In 1822 the French naturalist Anselme Desmarest gave the Virginia *incognitum* the scientific name *Megalonyx jeffersoni,* by which it is known today.[14]

Stimulated by the discovery and description of the megalonyx, the American Philosophical Society established a committee to promote research on American antiquities and natural history and in its 1799 *Transactions* published the committee's circular urging the importance of obtaining "one or more entire skeletons of the Mammoth, so called, and of such other unknown animals as either have been, or hereafter may be discovered in America. . . . The committee suggest, to gentlemen who may be in the way of inquiries of that kind, that the Great Bone Lick on the Ohio, and other places where there may be mineral salt, are the most eligible spots for the purpose, because animals are known to resort to such places."[15]

One of the committee's members, Charles W. Peale, was especially eager to obtain a complete skeleton of the mammoth, for it would be sure to be of popular as well as scientific interest when exhibited in his museum. His opportunity came, however, not at Big Bone Lick, but in Orange and Ulster counties in New York. Giant bones and teeth had been turning up in this area for some years. In 1793 Robert Annan had described in the *Memoirs* of the American Academy of Arts and Sciences four huge teeth dug up on his farm in Orange County during swamp-draining operations in 1780. Further discoveries were made soon after and reported in Mitchill's *Medical Repository* in 1801. According to one of Mitchill's informants, James G. Graham of Shawangunk, N.Y., huge bones in various states of preservation had been unearthed by farmers digging for marl in several places located within a ten-mile radius of Ward's Bridge in the town of Montgomery, the latest discovery being only three miles from Graham's house.[16]

The news of these multiple discoveries in an easily accessible area raised to a fever pitch the hopes of Jefferson, Wistar, Peale, and others that a complete skeleton of the mammoth might be obtained. At Wistar's request, Jefferson wrote to Robert Livingston in New York asking his assistance in obtaining some of the bones found at Shawangunk. Livingston replied that he had already tried and failed, "the whole town having joined in digging for them till they were stopped by the autumnal rains."

To Peale the prospect of obtaining a complete skeleton of the American *incognitum* was as alluring financially as it was scientifically. Such an exhibit would make his museum famous throughout the world. In June 1801 Peale took the stagecoach to New York, procured letters of introduction to Graham, and sailed up the Hudson River to West Point. At Newburgh, Graham welcomed him cordially and drove him to the farm of John Masten. There on the floor of Masten's granary lay an assortment of huge bones — ribs, vertebrae, pelvis, shoulder blades, a femur, a five-foot section of a tusk, and many others — which the inhabitants of the region had dug out of Masten's marl pit with reckless enthusiasm until the flow of water had put an end to the search. Exhilarated by the sight of these scientific treasures, Peale obtained permission to sketch the bones and ended by paying $300 for the entire collection and the right to return to dig for the missing parts of the skeleton. A double-barrelled gun for Masten's son and some New York gowns for his daughters clinched the bargain.

Returning to Philadelphia with his booty, Peale lost no time in mounting an expedition to begin the search. Having obtained an interest-free loan of $500 from the American Philosophical Society at a special meeting on

July 24, 1801, he set off for the Masten farm in company with his son Rembrandt, his museum assistant Jotham Fenton, and James Woodhouse of the University of Pennsylvania. The problem of draining the marl pit challenged Peale's practical ingenuity to the limit, but he finally succeeded in constructing a huge bailing machine operated by three or four persons walking side by side in a wheel twenty feet in diameter. Onlookers and passersby, of which there were a great number, were only too glad to lend their services at the wheel. Cheered on by the plaudits of the crowd and the animating influence of a plentiful supply of rum, the workers toiled mightily in the muck, turning up many of the missing parts but without discovering the top of the head, some of the tail and toe bones, or the very essential lower jaw. Sounding operations using pointed iron rods were then tried at other locations until finally, in a marshy woodland on the farm of one Peter Millspaw, new discoveries were made. There at last, when all hope of success seemed to have evaporated, Rembrandt Peale's sounding iron struck a succession of solid objects underneath the surface of the marsh. The shovels bit the muck again, and the weary workers soon turned up a thigh bone, two foreleg bones, a scapula, several toe bones, and then — hallelujah! — a complete lower jaw.

> The unconscious woods echoed with repeated huzzas, which could not have been more animated if every tree had participated in the joy. 'Gracious God, what a jaw! how many animals have been crushed between it!' was the exclamation of all: a fresh supply of grog went round, and the hearty fellows, covered with mud, continued the search with encreasing vigor.[17]

Peale had now acquired parts sufficient to mount two nearly complete skeletons of the mammoth, but the task of assembling them in correct order was not easy. Wistar's knowledge of anatomy proved very helpful, but trial and error had to be resorted to in piecing together many of the broken bones.

> The tusk was . . . broke into three pieces, but the fractures being put together gave the true curve and twist of this enormous tusk, 11 feet in length. The fractures also of the several pieces of the head, fitting nicely together, produced the true form, but they could not be found out but [by] numberless tryals of puting first one piece, then another, together.[18]

The work of reconstruction went on for three months in the Peale family parlor in Philosophical Hall. As finally mounted, the skeleton stood 11 feet high at the shoulder; 9 feet at the hips; 5 feet, 8 inches wide at the chest; and 17 feet, 6 inches from tusk to tail. The lower jaw was of wood, copied from the jaw found at the Millspaw farm. Since no original existed for the top of the head, it was modeled in papier-mâché from an elephant's skull, with a horizontal red line to indicate its conjectural nature. Soon after the reconstruction was completed, the southeast chamber of Philosophical Hall was converted into a "Mammoth Room." The members of the American Philosophical Society were invited to a special viewing on December 24, 1801, after which the general public were admitted to behold the new scientific sensation at fifty cents per head.

The mammoth was a brilliant financial success, but the scientific task of naming, classifying, and describing the animal remained to be accomplished. One wonders why Caspar Wistar, professor of anatomy at the University of Pennsylvania and author of an earlier memoir on the megalonyx, published nothing on the huge skeleton he had helped to mount in Philosophical Hall. Perhaps he felt that Peale or one of his sons should have the honor of describing it. Rembrandt and Rubens Peale were dispatched to London with the second skeleton with instructions to proceed to the Continent after exhibiting it in England and to dispose of it to the highest bidder before returning home. Unfortunately, the resumption of hostilities between England and France prevented a Continental tour and distracted the attention of the British public from scientific curiosities. The exhibit in Pall Mall drew disappointingly small crowds and the skeleton proved unsalable. Nevertheless, the tour was not without important scientific results. Rembrandt Peale published two accounts of the skeleton, the first a brief brochure that was reprinted in the *Philosophical Magazine,* the second a lengthy *Historical Disquisition on the Mammoth,* containing a drawing of the skeleton. In this work he made detailed comparisons between the anatomy of the elephant and that of the mammoth and concluded, as Jefferson had in his *Notes,* that the American *incognitum* was an animal *sui generis* adapted to cold climates. It was, said Peale, a carnivorous beast different from both the elephant and the woolly mammoth of Siberia. Peale was mistaken in his belief that the teeth of the mammoth were those of a carnivore, but he was on safer ground in predicting that similar teeth and accompanying bones would eventually be found in Siberia, since remains like those of the woolly mammoth had been found in North America.[19]

Peale's *Disquisition,* although a creditable performance for a twenty-four-year-old artist, left much to be desired as a treatise on comparative anatomy and vertebrate paleontology; but it proved highly useful to Georges Cuvier, especially when taken in conjunction with the drawing of the mammoth skeleton made at Pall Mall by the British anatomist Everard Home and with the casts of the mammoth bones sent to Cuvier by Peale's father. On the basis of these materials and others that had been accumulating at the Jardin des Plantes for nearly a century Cuvier was able to publish a series of memoirs in 1806 in which he named, classified, and described the various elephantlike animals whose bones and teeth had been found in various parts of the earth. By careful anatomical comparisons he was able to show that the American mammoth, though herbivorous, belonged to a different genus from that which included the Asian and African elephants and the woolly mammoth. This new genus, which he proposed to call the *mastodonte,* would embrace four known species, two from the Old World and two from the New, the largest of them being *le grand mastodonte* unearthed and described by the Peales. Thanks to the Peales, wrote Cuvier, the complete osteology of this largest mastodon was now known except the top part of the cranium.[20]

While the Peales were making their dramatic contribution to the new science of vertebrate paleontology, Jefferson, Wistar, Barton, and others continued their search for new evidence of *animalia incognita* on the North American continent. As usual, Barton was active in gathering information and relaying it to his correspondents in Europe, including Cuvier, Eberhard von Zimmerman, J. H. Reimarus, W. T. von Tilenau, the Count de Lacépède, and others. Meanwhile, he kept the readers of his *Philadelphia Medical and Physical Journal* informed of "miscellaneous Facts and Observations" that came to his attention, such as Cuvier's discovery of an extinct species of opossum near Paris. It was now clear, he told the members of the Linnaean Society of Philadelphia in 1807, that "the continent of North-America was formerly inhabited by several species of animals, which are now entirely unknown to us, except by their bones, and which, there is reason to believe, now no longer exist." The old idea of the great chain of nature, of the absolutely necessary dependence of one species upon another, was exploded. A new science was arising, which Barton proposed to name Elminthology, that took as its province the domain between geology on the one hand and zoology and botany on the other. Only by close attention to the fossils found in the earth's crust, said Barton, would it be possible to form a correct theory of the earth.

> I possess specimens of slate found in Pennsylvania, upon which are distinctly impressed several of Filices, or Ferns, of South America and Jamaica. . . . In the neighbourhood of Chambersburg and Winchester, the limestone contains Cornua-ammonis, of different sizes. And the impressions of Sertulariae, Madrepores, Celepores, Tubipores, Echini, and many other marine animals, are abundantly distributed in the calcareous pavement of the valley, all the way from the neighbourhood of Easton, in Pennsylvania, to the first streams of the Roanoke in Virginia.[21]

Barton entertained hopes of publishing a work on the extinct flora and fauna of North America, but this project like many others he dreamed of had to be abandoned when he succeeded to Benjamin Rush's chair in the theory and practice of medicine in 1813. The most he could do was to publish in 1814, one year before his death, a brief collection of his letters to Cuvier, Jefferson, J. H. Reimarus, and others on the subject of the remains of the mammoth and other extinct animals under the grandiloquent Latin-English title, *Collections, with Specimens, for a Series of Memoirs on Certain Extinct Animals and Vegetables of North-America. Together with Facts and Conjectures Relative to the Ancient Condition of the Lands and Waters of the Continent.* The publication added nothing to Barton's reputation, being much heavier on conjectures than facts.

Meanwhile, Jefferson and Wistar were making further contributions to vertebrate paleontology. In 1805, in response to an appeal from Jefferson and Peale, Samuel Brown of Lexington, Ky., had searched for a specimen of the missing frontal bone of the mammoth at Big Bone Lick and had sent as a result a skull fragment that, when examined by Wistar for the American Philosophical Society, turned out to resemble the skull of the South American peccary described by Daubenton in Buffon's *Natural History.* "This first identification of a fossil peccary was correct," wrote the twentieth-century paleontologist George G. Simpson, "and it long antedated the generally accepted time of this discovery."[22]

In December 1806 Wistar wrote to William Goforth of Cincinnati on behalf of the American Philosophical Society, requesting his aid in procuring a frontal bone of the mammoth and accounts and specimens of "all other bones" as well. When these negotiations failed to produce any tangible result, Jefferson asked William Clark to stop at Big Bone Lick on his trip to Kentucky and to collect at Jefferson's expense the kinds of bones particularly wanted by the American Philosophical Society. Clark seems to have executed his mission with dispatch, for on December 19, 1807, President Jefferson wrote to thank him for the bones already received and to ask that he also send certain other specimens he had given to his brother so that Jefferson could send a collection of duplicates to the National Institute of France, "for whom I am bound to do something."[23]

The bones sent by Clark were spread out in a room of the White House, and Wistar was invited by Jefferson to come and select specimens for the collections of the American Philosophical Society.

> It is a precious collection, consisting of upwards of three hundred bones, few of them of the large kinds which are already possessed. There are four pieces of the head, one very clear, and distinctly presenting the whole face of the animal. . . . There are four jaw-bones tolerably entire, with several teeth in them, and some fragments; three tusks like elephants. . . ; an abundance of teeth studded and also of those of the striated or ribbed kind; a foreleg complete; and then about two hundred small bones, chiefly of the foot. . . . There is one horn of a colossal animal.[24]

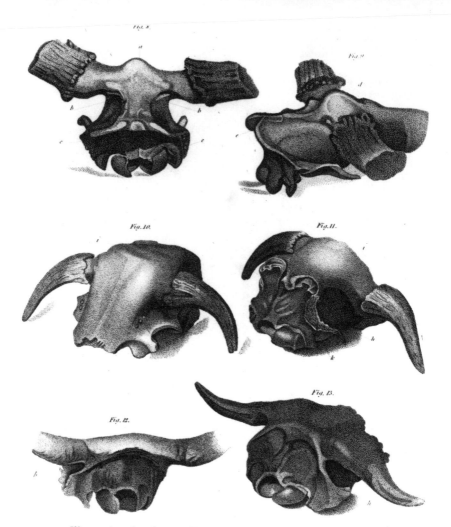

Illustration for Caspar Wistar's description of the remains of a fossil ox found at Big Bone Lick by Capt. William Clark. Courtesy of the American Philosophical Society

Of Wistar's visit in response to Jefferson's invitation we have no detailed account, but it must have been a great pleasure for both men to examine the bones, speculate on their significance, and select those most needed by the society they both served. When Wistar had made his choices and Jefferson had reserved a few specimens for his little museum at Monticello, the rest were shipped to the National Institute of France, where in due course they were acknowledged in the *procès verbaux* of the class of mathematical and physical sciences. The official report, signed by Lacépède and Cuvier, stated that most of the bones and teeth were those of the mastodon and served to confirm the conclusions Cuvier had drawn in his memoir of 1806 on the basis of much less evidence. "The gift of M. Jefferson," said the report, "not only enables us to verify by immediate observation that which we knew only on the testimony of others but adds new and precious details to those we already possessed."[25]

Meanwhile, Wistar was making new discoveries from his examination of the smaller bones in the Clark collection. Unfortunately his memoir on these bones was lost by the committee appointed to review it in 1809. As a result, his "Account of Two Heads Found in the Morass, Called Big Bone Lick" did not appear in the society's *Transactions* until 1818 and then only in truncated form. The first of the two incomplete skulls was shown to be that of an extinct deer different in important respects from the wapiti and the moose, its nearest living relatives. The second skull, now known to be similar to that of a musk ox, was placed by Wistar in the genus *Bos,* which then included all the bovines, and was shown to be distinct from, yet allied to, that of the bison. "Although the identification by Wistar was less brilliant than his other work," writes Simpson, "the error, if it can be called such, was one of degree and not of kind."[26]

Unfortunately for the society and for the progress of American paleontology, Wistar, who had succeeded Jefferson as president of the organization in 1815, died of a severe attack of typhus shortly before this last memoir was published. But the paleontological tradition he had established at the medical school of the University of Pennsylvania and the American Philosophical Society was to find a brilliant continuation in the work of Richard Harlan, William Horner, Samuel G. Morton, and Joseph Leidy. Philadelphia remained the chief center of American vertebrate paleontology for the rest of the century.

ALEXANDER WILSON AND THE
DEVELOPMENT OF AMERICAN ORNITHOLOGY

At the same time that paleontology was securing a foothold in Philadelphia the older natural history tradition of the Bartrams was gaining new adherents. About 1803 William Bartram, then sixty-two years old, became acquainted with a middle-aged schoolmaster of Scottish origins named Alexander Wilson. The son of a poor Scottish distiller, Wilson had arrived in Philadelphia in 1794 at the age of twenty-eight with nothing but his wits, his health, a slight reputation as a humorous poet, and a knowledge of the trade of weaving. He soon took to school teaching to make a living and in 1804, having been invited to teach at Union School in Kingsessing Township on the outskirts of Philadelphia, settled at Gray's Ferry just down the river from Bartram's garden. By this time Wilson had fallen in love with the American landscape and its flora and fauna, especially its birds. After a walking trip to Niagara Falls with his friend William Duncan in the autumn of 1804, Wilson made up his mind that he would write a history of American birds illustrated by his own hand. He was poorly equipped for this huge undertaking. He had no money, no training in drawing or engraving, no command of science or foreign languages, no books, nothing but enthusiasm and dogged determination. But he persevered, making the most of the opportunities the Philadelphia area provided.

At Peale's Museum there was a splendid collection of American birds open to Wilson's inspection. Bartram introduced him to the works of Catesby and Edwards, taught him something of botanical and zoological nomenclature and classification, and examined his descriptions and drawings with a critical eye. Bartram even sent some of Wilson's drawings to President Jefferson, who responded with cordial interest. From a Philadelphia engraver, Alexander Lawson, Wilson borrowed engraving tools and taught himself the art of engraving and coloring the plates. Then in April 1806 Samuel Bradford, a leading Philadelphia publisher, offered him a job as assistant editor of a new American edition of *Rees's Cyclopaedia,* and Wilson soon persuaded his employer to back his projected multivolume work on American ornithology. Wilson was to furnish at his own expense all the drawings and descriptions and to supervise the whole work, meanwhile performing his duties as assistant editor of the *Cyclopaedia.* The publisher would underwrite the cost of the first volume, an edition of 200 copies. If Wilson could secure that many subscribers, the work would continue; otherwise it would be dropped. An illustrated prospectus promising ten volumes with ten color plates each at a total price of $120 was then mailed to 2,500 potential subscribers, among them President Jefferson. Jefferson's subscription arrived in due course, with the message: "He salutes Mr. Wilson with great respect."

Two years elapsed before the first volume of the *American Ornithology* issued from the press in September 1808. During all that time Wilson worked feverishly procuring new birds, preparing drawings, writing descriptions, and superintending the difficult work of coloring the plates. One of the colorists, a gifted seventeen-year-old boy named Charles R. Leslie who later became court painter to Queen Victoria, has left a word portrait of Wilson, his first mentor:

> He looked like a bird. His eyes were piercing, dark and luminous, and his nose shaped like a beak. He was of spare, bony frame, very erect in his carriage, inclining to be tall; and with a light, elastic step, he seemed perfectly qualified by nature for his extraordinary pedestrian achievements. . . . I assisted him to color some of his first plates. We worked from birds which he had shot and stuffed, and I remember

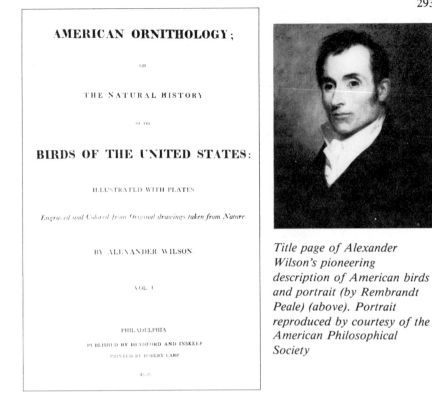

AMERICAN ORNITHOLOGY;

OR

THE NATURAL HISTORY

OF THE

BIRDS OF THE UNITED STATES:

ILLUSTRATED WITH PLATES

Engraved and Colored from Original drawings taken from Nature.

BY ALEXANDER WILSON.

VOL. I

PHILADELPHIA

PUBLISHED BY BRADFORD AND INSKEEP
PRINTED BY ROBERT CARR

1808

Title page of Alexander Wilson's pioneering description of American birds and portrait (by Rembrandt Peale) (above). Portrait reproduced by courtesy of the American Philosophical Society

the extreme accuracy of his drawings, and how carefully he had counted the number of scales on the tiny legs and feet of his subject.[27]

Wilson was doubly fortunate in receiving from Meriwether Lewis the birds collected on the Lewis and Clark expedition. Lewis also gave him considerable information about birds seen on the expedition and put him in touch with Sergeant Ordway, from whom Wilson secured further information.

Shortly after the first volume appeared, Wilson set out on a trip northward in search of subscribers, carrying his book and prospectus under his arm. In Princeton, Newark, New York, New Haven, Springfield, Worcester, Boston, Portland, Hanover, Albany, and many smaller towns along the way he searched out the wealthy and learned or the merely curious and showed them his book. One hundred and twenty dollars for a single set of volumes was more than most Americans could manage or even imagine, but Wilson eventually procured forty-one subscriptions, including one from that aged champion of the rights of man Thomas Paine, then living in Greenwich, Conn. Wilson wrote to Alexander Lawson,

> I found this extraordinary man, sitting wrapt in a night gown, the table before him covered with newspapers, with pen and ink beside them. Paine's face would have excellently suited the character of Bardolph; but the penetration and intelligence of his eye bespeak the man of genius. . . . He complained to me of his inability to walk . . . — examined the book, leaf by leaf, with great attention — desired me to put his name as subscriber.[28]

Besides subscribers, Wilson found a considerable number of potential correspondents who promised to keep him informed concerning the birds in their vicinities.

Soon after his return to Philadelphia, Wilson set off for the South, traveling by stagecoach. He obtained sixteen subscriptions in Baltimore but none in Annapolis. In Washington he found seventeen new subscribers and visited President Jefferson, who gave him a letter of introduction that served him well in Virginia. Continuing on his way, Wilson passed through Norfolk, Va.; New Bern and Wilmington, N.C.; Charleston, S.C.; and on to Beaufort, S.C., where Stephen Elliott welcomed him cordially and invited him to explore the woods for birds both at Beaufort and at his plantation on the Ogeechee River. In March 1809 Wilson arrived in Savannah, Ga., exhausted but comforted in the knowledge that he had obtained a total of 250 subscriptions, enough to ensure the continuation of the work in its original style and to warrant striking off 300 more copies of the first volume. Besides this, he had discovered several new birds and established a chain of correspondents all along the eastern seaboard, so that "scarcely a wren or tit" should escape his notice.

Of these correspondents the most valuable was John Abbot of Savannah, whose plantation on the Ogeechee River was not far from Elliott's. Born in London, Abbot had achieved a considerable reputation as a painter of landscapes, butterflies, and rare insects before leaving England in 1773 to try his hand at painting American birds and insects, as Catesby had done before him. He migrated from Virginia to Georgia as the revolutionary crisis developed, eventually sided with the American patriots, and received a substantial land grant for his services in the revolutionary war. He then married, settled down as a planter, and devoted his leisure hours to painting insects and birds. He made several hundred bird paintings for English collectors, sent them innumerable stuffed birds and mounted insects, and made thousands of drawings and paintings of American insects. In 1797 James E. Smith, president of the Linnean Society of London, published a two-volume collection of Abbot's butterflies under the title *The Rarer Lepidopterous Insects of Georgia,* taking the main credit to himself, however. Abbot seems to have been discouraged from publishing anything further, but he continued to paint birds and insects for his own pleasure and for such small financial rewards as were sent to him from time to time by the English scientists with whom he corresponded. Delighted to find a fellow enthusiast in Wilson, Abbot promised to send him southern birds for the *American Ornithology*. He was as good as his word, and Wilson paid him due credit as well as a substantial amount of cash for his invaluable contributions to the success of the work.[29]

Returning to Philadelphia by ship, Wilson threw himself into the work of preparing the second volume of the *Ornithology*. When it finally appeared in January 1810, he set off on a western trip dedicated to finding new subscribers and birds. Passing through Lancaster, Harrisburg, and Carlisle, he arrived in Pittsburgh, where he garnered nineteen subscriptions in four days. Then in the last weeks of winter he embarked on the Ohio River in a skiff, surrounded by swirling ice cakes. Three weeks later he landed at Bear Grass Creek, Ky., after a voyage of 720 miles. He visited Cincinnati and Louisville, where he met John J. Audubon, and pushed on overland to Lexington and Nashville. Ignoring all warnings against traveling alone, he set out for Natchez and New Orleans by the hazardous Natchez Trace, where he nearly perished of dysentery. At Natchez he found a note from William Dunbar, scientist friend of Jefferson and Andrew Ellicott, inviting him to spend some time at Dunbar's plantation nine miles to the south.

> It is unfortunate that I should be so much indisposed as to be confined to my bedroom; nevertheless, I cannot give up the idea of having the pleasure of seeing you. . . . The perusal of your first volume of *Ornithology,* lent me by General Wilkinson, has produced in me a very great desire of making your acquaintance. . . . My house stands literally in the forest, and your beautiful orioles, with other elegant birds, are our courtyard companions.[30]

The invitation was accepted, and it is hard to say which of the two Scotsmen enjoyed the visit more. Wilson wounded and captured a new bird, the Mississippi kite, in his rambles about the plantation, though not until the kite had scored Wilson's hand deeply with its claw. Dunbar and his wife and seven children found Wilson delightful company. Dunbar's days were numbered, but his letters of introduction procured a warm reception for Wilson as he made his way slowly to New Orleans. There he was lionized and rewarded with sixty new subscriptions, almost as many as the Philadelphians ever gave him. On June 24, 1810, Wilson took ship for Philadelphia, taking with him his new subscriptions, specimens of a dozen new birds, two tame parrots, six painted buntings in a cage, and a wealth of firsthand information about the birds of the southwestern United States.

The task that confronted Wilson when he returned to Philadelphia late in the summer of 1810 was a staggering one. Two volumes of the *Ornithology* had been published and had been widely praised, but little financial reward could accrue to either Wilson or his publisher until the remaining volumes were produced. Meanwhile, the packet of bird drawings and data Wilson had mailed to Philadelphia from Nashville had been lost in the mail, and the third and fourth volumes could not be published until Wilson reproduced the lost material. Working feverishly throughout the year of 1811 with constant encouragement from the Bartrams at whose garden he stayed during the summer months, Wilson finished the work on the third, fourth, and fifth volumes. By June 1812 he could report to François Michaux that he had completed the sixth volume and was hard at work on the seventh. Since each volume contained ten colored plates, with from two to four birds on each, it is apparent that Wilson was turning out descriptions and illustrations of more than 100 birds per year — and all this with, as yet, little financial recompense. Having given up his assistant editorship with Bradford to devote more time to the *Ornithology,* he was now dependent on what little income he derived from helping to color the plates for that work. "I have sacrificed everything to publish my Ornithology," he wrote to Michaux, "yet I have never yet received a single *cent* of its proceeds."[31] Nor would he ever do so. At the publisher's insistence, he made a trip to New England in the autumn of 1812 to collect from delinquent subscribers for volumes already delivered. He returned to Philadelphia exhausted, only to find that the colorists had stopped work on the plates for the sixth volume and that he must color the plates himself if the project was to continue.

Wilson was pushing himself to the limits of his endurance, working late into the night coloring plates and devoting his daylight hours to gathering and preparing materials for the volumes on water birds. In the latter enterprise he had considerable assistance from a newly found friend, George Ord, a well-to-do Philadelphia businessman-sportsman-scholar-scientist who had become fascinated with Wilson's passion for ornithology and had begun to share it himself. Ord accompanied Wilson on some of his ornithological expeditions to the New Jersey coast and brought Wilson the fruits of his forays. Late in April 1813 the seventh volume of the *Ornithology* was published. Elated by this and the news of his election to the American Philosophical Society, Wilson set off with Ord for Cape May to procure materials for the last two volumes, the total number having now been reduced from ten to nine. It was to be Wilson's last trip. On his return to Philadelphia he drove himself relentlessly as if aware that he was in a race with death. "Intense application to study has hurt me much," he wrote to a friend in Scotland. "My 8th volume is now in the press and will be published in November. One volume more will complete the whole which I hope to finish in April next."

It was not to be. On August 23, 1813, at the age of forty-seven, Wilson died of acute dysentery, the illness that had nearly ended his life on the Natchez Trace several years earlier. But the *American Ornithology* did not remain incomplete, thanks to the friendship of Ord, who saw Wilson's eighth volume through the press within the next five months and himself gathered the materials for and wrote the ninth volume. It contained descriptions of sixteen species, thirteen of them illustrated on four plates, and a sketch of Wilson's life. Within a year of Wilson's death his *Ornithology* was finished.

Judged by any appropriate standard, the *American Ornithology* was a heroic achievement. It was not, like Cuvier's *Animal Kingdom,* a masterpiece of systematic natural history devoted to naming, classifying, and describing animals in terse scientific prose primarily on the basis of museum specimens and the published works of other naturalists. Wilson had indeed studied the specimens in Peale's Museum, many of which he had himself contributed; and he had examined "with an eager and inquisitive eye" all the works of European naturalists he could lay hands on, only to turn away from them in disappointment. They contained vague and formal particulars about the size and specific marks of the birds they described and contained illustrations that were often caricatures of the birds they represented. From these "barren and musty records" Wilson turned with relief to "the magnificent repository of the woods and fields, the *Grand Aviary of Nature,*" there to learn from the "GREAT MASTER OF CREATION lessons of his wisdom, his goodness and his love, in the conformation, the habitudes, melody and migrations of this beautiful portion of the work of his hands." To communicate these lessons to the public was, Wilson declared, his heart's wish.[32]

One should not conclude from this rapturous declaration of purpose that Wilson paid no attention to systematic natural history. He was familiar with the main systems of bird classification. Some writers, he noted, "increased the number of orders to an unnecessary extent, multiplied the genera, and, out of mere varieties, produced what they supposed to be entire new species," while others, seeking to simplify the science, lumped together birds whose habits, food, and other distinguishing characteristics were widely different, thereby confounding "that beautiful gradation of affinity and resemblance, which Nature herself seems to have been studious of preserving throughout the whole." These errors in classification arose, said Wilson, primarily from lack of firsthand knowledge of the birds concerned. Only by repeated excursions into the woods and fields could the naturalist learn the particular haunts, the modes of nest construction, manner of flight, seasons of migration, favorite food, and other characteristic features of the various kinds of birds.[33]

As a case in point Wilson adduced the orchard oriole, called the Bastard Baltimore by Catesby and Buffon and the *Oriolus spurious* by Gmelin. Buffon had confused this bird with the Baltimore oriole, supposing it to be "a variety of a more generous race, degenerated by the influence of climate, or some other accidental cause." John Latham had also been led astray, suggesting that the color changes in the Baltimore oriole had led naturalists to imagine two species where in fact there was only one. Wilson, for his part, rejected the very notion of a "bastard" species and showed that the orchard oriole differed from the Baltimore in size, colors, form of bill and tail, song, mode of nest building, and color and shape of eggs. "If all these circumstances . . . be not sufficient to designate this as a distinct species," he wrote, "by what criterion . . . are we to discriminate between a *variety* and an *original* species, or to assure ourselves, that the Great horned Owl is not in fact a *bastard* Goose, or the carrion Crow a mere *variety* of the Humming-bird?"[34] The principal cause of the confusion, he went on to explain, was the multiplicity of changes of color to which both the male and the female orchard oriole were subject. "I may add," said Wilson after describing these changes of plumage, "that Mr. Charles W. Peale . . . has expressed to me his perfect conviction of the changes which these birds pass thro; having himself examined them both in spring and towards the latter part of summer, and having at the present time in his possession thirty or forty individuals of this species, in almost every gradation of change."[35]

Since some system of classification was necessary, Wilson chose the one set forth in Latham's *General Synopsis of Birds* (1781–1785) and *Index Ornithologicus* (1790) as being the least objectionable. This system was outlined in the "Introduction" to Wilson's first volume. In a later volume the birds described in the *American Ornithology* were listed under their appropriate orders and genera with page references to the volumes in which they were described and illustrated. The presentation of the birds in the successive volumes followed no special order, except for a division between land and water birds. Each description was preceded with a Linnaean binomial followed by synonyms taken from the works of other naturalists such as Brisson, Klein, Buffon, Catesby, Edwards, Pennant, and Bartram and by the specimen number in Peale's Museum. Then came the description in Wilson's spirited and often eloquent prose. That of the blue jay begins: "This elegant bird, which, as far as I can learn is peculiar to North America, is distinguished as a kind of beau among the feathered tenants of our woods, by the brilliancy of his dress; and, like most other coxcombs,

Icterus spurius, Orchard Oriole. 1. *Female*. 2 and 3. *Males of the second and third years.*
4. *Male in complete plumage.* a. *Egg of the Orchard Oriole.* b. *Egg of the Baltimore Oriole.*

The orchard oriole in various plumages as depicted in Alexan-
der Wilson's American Ornithology. Courtesy of the
American Philosophical Society

*The goldfinch, bluejay, and Baltimore oriole as depicted in
Alexander Wilson's* American Ornithology. *Courtesy of the
American Philosophical Society*

makes himself still more conspicuous by his loquacity, and the oddness of his tones and gestures."

Wilson's performance was inevitably deficient in some respects. As a delineator of birds he fell short of his rival and successor Audubon, the perfection of whose artistry Wilson had glimpsed when he visited him in Louisville. Nor was he well trained in systematic natural history and the intricacies of nomenclature. As a field naturalist, however, he was a worthy successor to his friend William Bartram and equal if not superior to Audubon. "No other work in ornithology is equally free of error," wrote Elliott Coues, dean of late nineteenth-century American ornithologists. "He has treated of American birds better than those of Europe have yet been treated," Cuvier declared. Charles L. Bonaparte, whose own *American Ornithology* was a continuation of Wilson's, concurred.

> Placed where he could derive little or no aid from scientific books or men, Wilson's ardent and perspicacious mind triumphed over circumstances, and enabled him to exhibit the truths he discovered in that warm, lucid, and captivating language, which never fails to reach the heart of his reader, because it flowed direct from his own; whilst his clearness of arrangement, accuracy of description, and faithfulness of delineation, show most advantageously the soundness of his judgment and the excellence of his observation. We may add, without hesitation, that such a work as he has published in a new country, is still a desideratum in any part of Europe.[36]

By the reckoning of Emerson Stringham, in his *Alexander Wilson,* Wilson described 264 of the 343 species of birds found in the United States in his time, added 48 new species to those previously known, prepared good life histories of 94 species, and gave 40 species the vernacular names by which they are presently known. "He missed only 79 species in the immense wilderness," adds his biographer Robert Cantwell, "and scarcely more than a score of errors, most of them minor, have been found in the *Ornithology* in a century and a half."[37]

Wilson would have asked for no higher praise. Writing in the preface to his fifth volume, he had asked only that it be remembered of him "that in the period in which three-fourths of our feathered tribes were altogether unknown even to the proprietors of the woods which they frequented — that without patron, fortune or recompence, he brought the greater part of these from the obscurity of ages, gave to each 'a local habitation and a name' — collected from personal observation whatever of their characters and manners seemed deserving of attention; and delineated their forms and features, in their native colors, as faithfully as he could, as records, at least, of their existence."

RESEARCHES IN NEW ENGLAND, NEW YORK,
AND KENTUCKY

In zoology as in botany the decade following the War of 1812 was a period of rapid expansion. In the Boston area the formation of the Linnaean Society of New England and the quiet but effective teaching of William D. Peck at Harvard laid the groundwork for the flowering of natural history after 1830. Peck, who had first acquired a reputation as a naturalist with his "Natural History of the Slug Worm" in the 1790s, subsequently published a series of memoirs on insects injurious to agriculture in the *Massachusetts Agricultural Repository and Journal.* In 1816 his observations on the life history and control of the cankerworm appeared, in 1817 his memoir on insects that attack the pear tree and the Weymouth pine, in 1818 his account of the enemies of the locust tree, and in 1819 a memoir on those of the oak and cherry. These researches on economic entomology were to be continued brilliantly by his pupil Thaddeus W. Harris after Peck's death in 1822. A modest man, Peck sent many insect specimens and descriptions to the English entomologist William Kirby, who published some of them in the *Transactions* of the Linnean Society of London. Kirby named one of these, a curious insect that inhabits the joints in the abdomen of the wasp, *Xenos peckii* in honor of his American correspondent.[38]

Among the active zoologists at the Lyceum of Natural History of New York were its president Samuel L. Mitchill, his brother-in-law Samuel Akerly, Constantine Rafinesque, James De Kay, and John L. Le Conte. The latter two made their main contributions after 1825. Mitchill published memoirs on insects, polyps, parasites, reptiles, rodents, cetaceans, and fossil vertebrates, but he is best remembered for his work on fishes. From the day in 1789 when he sent Jedidiah Morse a list of the fishes of New York for inclusion in his *American Geography* Mitchill had continued to collect information on this subject. In 1814 he published a "Report . . . on the Fishes of New-York," in which he described about seventy species, inventing new specific and generic names as seemed necessary. "The cold and ice of winter will unavoidably retard the progress of this interesting work," noted the editors of the *American Medical and Philosophical Register.* "But with the approach of warm weather, and the arrival of the migrating tribes, there will be the means of rapid acceleration."

In the same volume of the *Register* that noticed Mitchill's "Report" appeared his "Arrangement and Description of the Codfishes of New-York," in which eleven species and six varieties were described. Mitchill had become interested in codfish in 1803 in his capacity as chairman of the Committee of Commerce and Manufactures of the House of Representatives and was careful to point out the economic as well as the scien-

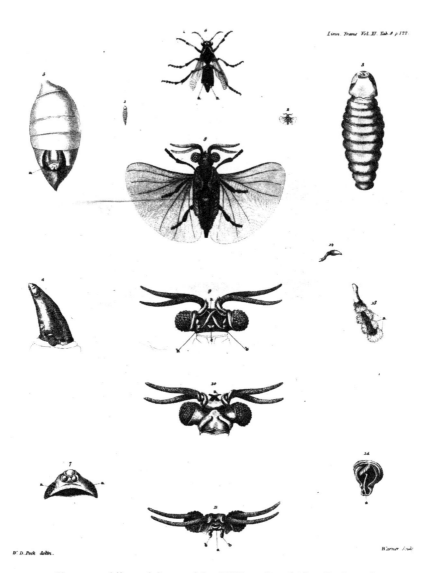

Linn. Trans. Vol. XI. Tab. 9 p 177.

W. D. Peck delin. Warner Sculp.

Xenos peckii *as delineated by William Dandridge Peck and published in the* Transactions *of the Linnean Society of London, vol. 11 (1815). This species belongs to an order (Strepsiptera) of minute parasitic insects known as "twisted-wing parasites" that feed within wasps. Courtesy of the Museum of Comparative Zoology, Harvard University* tific importance of that fish.[39] He continued to publish memoirs of the fishes of New York in the *Transactions* of the Literary and Philosophical Society of New York, Biglow's *American Monthly Magazine and Critical Review,* and Silliman's *American Journal of Science and Arts,* assigning new and different names to some of the species he had described in his "Report."

When Theodore Gill, professor of zoology at Columbia University, reprinted Mitchill's "Report" in 1898, he found it "remarkable for crudities and misconceptions." Mitchill, he said, "had an independent and positive mind, merging not infrequently into rashness, & was too much disposed to assume that what was not known to him was not known at all." Only twelve of Mitchill's thirty-eight supposed new species were really new, said Gill, and nearly all of these were "happy hits rather than the mature fruit of systematic research. . . . No diagnostic characters were given by Mitchill and he not only failed to appreciate their characteristics and relationships but associated with each, species which had no affinity to them." Nevertheless, Gill noted, Mitchill's "Report" was the starting point for part of the nomenclature of American fishes. "Three of the generic and twelve of the specific names of some of our most common and conspicuous fishes date from the publication of that little work."[40]

Rafinesque was as wide-ranging in his zoological interests as Mitchill and considerably more prolific as a writer. He flooded the *American Monthly Magazine* and the early issues of Silliman's journal with his contributions, including memoirs on North American fishes, crustacea, mammals, tortoises, and fossil invertebrates. On moving to Transylvania University in Kentucky in 1818, he continued to turn out publications at an astonishing rate. In the *Kentucky Reporter* for September 30, 1818, he announced that he had already discovered 22 new undescribed species of quadrupeds (mostly bats and rats), 3 new turtles, several new lizards and snakes, many new birds, 64 new fishes, and 85 species (mostly new) of shells and had sent specimens and drawings of these to the Lyceum of Natural History of New York and to the Museum of Natural History in Paris.[41]

Finding difficulty in publishing his memoirs in eastern journals (Silliman declared that he would have room for nothing else if he continued accepting Rafinesque's productions) Rafinesque was forced to send them abroad or publish them in western magazines or at his own expense. His *Monograph of the Fluviatile Bivalve Shells of the River Ohio, Containing Twelve Genera & Sixty-Eight Species* was first published in French in the *Annales Générales des Sciences Physiques* (1820) in Brussels. It was not translated into English until 1832, when Charles A. Poulson, who had befriended Rafinesque during his last years in Philadelphia, made a translation for the Philadelphia edition. In his letter introducing this edition

Poulson noted that only four of the many species discovered and described by Rafinesque were known by the names he had given them, "either in the works of American authors or in our collections." In 1864 all Rafinesque's writings on shelled animals, living and extinct, were brought together in one publication by William G. Binney and George W. Tryon, Jr. Brushing aside the controversies that had arisen regarding the propriety of adopting Rafinesque's generic and specific names, the editors paid tribute to his achievement. "The numerous valuable writings of our author on terrestrial and marine Mollusca (universally acknowledged as such) together with the great interest which has been awakened in his descriptions of our Naiades," said Binney and Tryon, "will doubtless render this volume an acceptable addition to conchological literature."[42]

Rafinesque's *Ichthyologica Ohiensis. Or Natural History of the Fishes Inhabiting the River Ohio and Its Tributary Streams* was first published in serial form in the *Western Review and Miscellaneous Magazine* (1819–1820) and subsequently issued as a separate publication at the author's expense. Dedicated to Mitchill and the French naturalist Charles A. Lesueur, it bore the motto: "The art of seeing well, or of noticing and distinguishing with accuracy the objects which we perceive, is a high faculty of the mind, unfolded in few individuals, and despised by those who can neither acquire it, nor appreciate its results."[43] To what extent Rafinesque possessed these qualities has been a matter of controversy to the present day. The British zoologist John Richardson had few good words to say for either Mitchill or Rafinesque in his "Report on North American Zoology" before the British Association for the Advancement of Science in 1836. Mitchill's descriptions, he declared, were "almost always imperfect, and often inaccurate, and he has arranged the species without judgment in Linnean genera, so that but for the accompanying figures it would be difficult to recognise the fish he mentions."[44] Rafinesque, he said, had proposed many new genera, "but characterising them with so little skill, that there is little chance of their being adopted by future naturalists."

A later evaluation of Rafinesque by the American ichthyologist David S. Jordan in 1877 was less severe. After three years of field work on the fishes of the Ohio valley, Jordan was in a good position to evaluate Rafinesque's work. It was evident, Jordan said, that Rafinesque's descriptions were drawn from memory and his measurements were made by the eye, not the tape measure. As a result his descriptions were often inexact. Nevertheless, said Jordan, "I have succeeded in identifying, more or less satisfactorily, nearly all of his species, and in restoring to a number of his names their rightful priority."[45] The unidentified species included some whose characters had not been adequately described by Rafinesque and others described by him on the testimony of Audubon, who had no compunction in assuring him of the existence of such mythical fish as the "Ohio Red-

Eye," described by Rafinesque (from a drawing by Audubon!) as a sort of sunfish with a dorsal fin "beginning behind the head with a single long, spiny ray, and ending close to the tail." Jordan concluded by endorsing the judgment of Rafinesque expressed by Louis Agassiz:

> I am satisfied that he was a better man than he appeared. His misfortune was his prurient desire for novelties and his rashness in publishing them, and yet both in Europe and America he has anticipated most of his contemporaries in the discovery of new genera and species in those departments of science which he has cultivated most perseveringly, and it is but justice to restore them to him, whenever it can be done.[46]

<h2 style="text-align:center">ZOOLOGY AT THE ACADEMY OF NATURAL SCIENCES OF PHILADELPHIA</h2>

Despite the foregoing developments in New England, New York, and Kentucky, Philadelphia continued to be the most active center of research and publication in natural history in the transitional years between the Jeffersonian and Jacksonian eras. Peale's Museum was still intact, and the Academy of Natural Sciences provided a meeting place, means of publication, and repository of collections for an avid group of naturalists. Among the zoologists were George Ord, Thomas Say, Charles A. Lesueur, Charles L. Bonaparte, John Godman, and Richard Harlan. Ord devoted himself wholeheartedly to advancing the ornithological work and reputation of his former colleague Alexander Wilson. In 1824–1825 he brought out a new edition of Wilson's *American Ornithology,* to which he added much material of his own. Meanwhile, Wilson's work had found a new admirer in Charles L. Bonaparte, Prince of Canino and nephew of Napoleon Bonaparte. Young Bonaparte had had excellent scientific training in France and was in touch with the best naturalists of Europe. He arrived in Philadelphia in 1822 and quickly assumed a prominent place in the scientific circles of that city. In March 1824 he presented to the Academy of Natural Sciences his "Observations on the Nomenclature of Wilson's Ornithology," subsequently published in the academy's *Journal* and issued as a separate publication. In this essay Bonaparte undertook to revise the nomenclature of Wilson's work in accordance with the best and latest systematic treatises in Europe, relating Wilson's names to those of Linnaeus, Gmelin, Latham, Temminck, Buffon, Vieillot, and others. Conceding that Wilson was "one of the most acute and accurate of Ornithologists," Bonaparte nevertheless found many errors in nomenclature arising from Wilson's lack of adequate library and museum facilities. To correct these errors, Bonaparte presented the genera in the order listed in

Wilson's *Ornithology,* correcting both generic and specific names as he went. Thus Wilson's *Sylvia pusilla* became *Sylvia americana,* since Latham had earlier designated it thus and had used the name *Sylvia pusilla* for an Australian warbler.[47]

Not content with correcting Wilson's errors in nomenclature, Bonaparte went on to produce his own *American Ornithology,* published in Philadelphia in 1825–1833. The illustrations for this work were drawn by Titian Peale and engraved by Alexander Lawson, Wilson's engraver, whom Bonaparte considered "the first ornithological engraver of our age." The plates were colored by one of Wilson's colorists, Alexander Rider, and Bonaparte's prose was edited for conformance to English idiom by Thomas Say and John Godman. Most of the new birds described in the first volume were discovered by Say and Peale on the Stephen H. Long expedition up the Missouri and Platte rivers to the Rocky Mountains and were drawn on the spot by Peale. In 1824 Bonaparte sent Peale to Florida to gather materials for the later volumes.[48]

Perhaps the most versatile of the zoologists at the Academy of Natural Sciences at this period was Thomas Say. The son of a Philadelphia physician-apothecary, Say married a granddaughter of John Bartram and tried his hand at the drug business for a while after a brief schooling under Quaker auspices, but his real love was collecting beetles and butterflies. In this enterprise he was encouraged by his great uncle William Bartram. When the business partnership Say had formed with John Speakman failed, he joined with Speakman, Philadelphia dentist Jacob Gilliams, and others in founding the Academy of Natural Sciences of Philadelphia and decided henceforth to devote his life to natural history. For several years he lived in the rooms of the academy on meager rations, devoting himself energetically to building up its collections and making contact with other devotees of his favorite subject.[49]

There were a good many of these scattered throughout the country, and some had already contributed to the literature of entomology, especially in regard to insects injurious to agriculture. As early as 1792 a paper on the Hessian fly appeared in the *Transactions* of the Society for the Promotion of Agriculture, Arts, and Manufactures. William D. Peck contributed a number of papers on injurious insects to the *Massachusetts Agricultural Repository and Journal,* and Benjamin S. Barton won the Magellanic Prize of the American Philosophical Society with his "Memoir on a Number of Pernicious Insects of the United States." During its brief existence Barton's *Philadelphia Medical and Physical Journal* carried a considerable number of entomological memoirs, including some observations of John Bartram.

Of systematic works, however, there were very few before Say. John Abbot's contribution to *The Natural History of the Rarer Lepidopterous*

Insects of Georgia, published in 1797 by James E. Smith, was a step in this direction, but the first American attempt at a systematic description of the insects of a region of the United States was Frederick V. Melsheimer's *Catalogue of Insects of Pennsylvania,* published at Hanover, Pa., in 1806. Melsheimer, a native of Regenborn in Brunswick, Germany, had served as chaplain in General Burgoyne's army during the revolutionary war. After the war he settled in Pennsylvania and began collecting insects during the hours he could spare from his parish duties, corresponding and exchanging insects with his boyhood friend A. W. Knoch in Brunswick. His *Catalogue of Insects,* based on the classificatory system of the German naturalist Johann Fabricius, was of limited scientific value since it listed names from unpublished manuscripts, but it proved an invaluable aid to Say in his early efforts, especially when reinforced by his correspondence and exchanges with Melsheimer's oldest son, John F. Melsheimer.

In April 1816 Say sent John Melsheimer a list of 231 species of insects included in the senior Melsheimer's *Catalogue,* requesting a specimen of each. Say also sought advice about ordering entomological books from Europe, remarking that the only general work in his possession at that time was William Turton's translation of Gmelin's last edition of Linnaeus's *Systema Naturae.*

The Melsheimers were by no means Say's only auxiliary resource. Peale's Museum contained a collection of several thousand insects, including species from India, China, South America, and Europe as well as North America. Say's Philadelphia friends, both at the Academy of Natural Sciences and elsewhere, contributed specimens for his collection. Besides Melsheimer, his correspondents included Thaddeus W. Harris and Charles Pickering in Massachusetts, Ezekiel Holmes in Maine, John P. Brace in Connecticut, Joseph Barabino in New Orleans, and Augustus Oemler in Savannah. Apparently Say was not in touch with Peck, many of whose specimens and descriptions were sent to William Kirby in England. Say's extensive foreign correspondence eventually put him in touch with William Kirby, Pierre F. Déjean, August Ahrens in Prussia, Jacob Sturm in Nürnberg, Carl J. Schoenherr in Sweden, Count Mannerheim in St. Petersburg, and many others.

Say's collecting trips took him far beyond the limits of Pennsylvania and New Jersey. In 1817–1818 he accompanied Maclure, Ord, and Titian Peale on a field trip to the Sea Islands off Georgia and to eastern Florida. The party intended to follow Bartram's earlier track up the St. John's River, but they turned back when they discovered that the Indians were up in arms "in consequence of this most cruel & inhuman war that our government is unrighteously & unconstitutionally waging against these poor wretches whom we call savages," as Say described the situation in a letter to John

Melsheimer. In 1819 Say was official zoologist on Long's expedition. He collected several thousand insects, hundreds of which were new to science, and described fossil shells, birds, snakes, shrews, squirrels, a deer, a lizard, a newt, and a sand rat, not to mention animals already familiar to science. In 1823 he was again with Long on an expedition to the upper Mississippi valley and the Red River, returning by way of the Great Lakes. The results of his researches appeared not only in the official reports of the Long expeditions but also in a series of articles in the *Transactions* of the American Philosophical Society, the *Journal* of the Academy of Natural Sciences, the *Annals* of the Lyceum of Natural History of New York, and other scientific journals.[50]

Meanwhile, Say was making progress on a synthetic work designed "to exemplify the genera and species of the insects of the United States, by means of coloured engravings." In 1817 he had published one brief part of a work entitled *American Entomology*. It was only ten pages, including six plates each with a descriptive text, but it set forth Say's larger purpose of producing a comprehensive American entomology on the model of Edward Donovan's *Natural History of British Insects* (1798–1801), adopting the corrections and improvements of the Linnaean system of classification suggested by the French entomologist Pierre A. Latreille.[51]

Say continued work on the *Entomology* as time permitted, and in 1824 the first volume, incorporating and adding greatly to the publication of 1817, was published. A second volume appeared in 1824 and a third in 1828 after Say had left Philadelphia with Maclure, Troost, Lesueur, and others for the socialistic experiment at New Harmony, Ind. No further volumes were ever published, probably because the market for them seemed too small to the publisher, Samuel A. Mitchell. The work was handsomely illustrated with fifty-four plates, twenty-eight of them done by Titian Peale and nine by Lesueur.[52] Ord considered the work "the most beautiful publication of the kind which has ever been issued from the American press," but Say did not have the literary talents or poetic vision of nature of an Alexander Wilson. Then, too, insects were less interesting to the public than birds.

Among entomologists, however, Say's three volumes and his numerous other publications on entomology were much appreciated. He was elected a foreign member of the Linnean Society of London. Latreille spoke highly of his talents, although he expressed a wish that Say's descriptions had more often been accompanied by illustrations to aid the reader in cases of doubt. Ord thought Say's descriptions much too brief and often unclear, but the entomologist Wilhelm Erichson praised Say for his brevity and clarity: "An abundance of words does not make descriptions more definite, and in brevity I see that no one excels the American Say, who

Cover page and illustration (Sphinx moth) from Thomas Say's American Entomology *(1817) and portrait of Say (by Joseph Wood) (below). Portrait by courtesy of the American Philosophical Society*

published descriptions so concise that they hardly go beyond the extent of a diagnosis, nevertheless so clear that you will hardly ever find doubtful a form exhibited by him."[53]

John G. Morris credited Say with having described 1,150 species of Coleoptera, 225 species of Diptera, 100 species of Hemiptera, and 100 species belonging to other groups. Another American entomologist, John L. Le Conte, called Say "the founder of that branch of science in this country" and declared that "the basis of all knowledge of our species rests upon a correct determination of those known to him." Accordingly, Le Conte assembled all Say's entomological writings and published them in 1859. "The entire destruction of [Say's] original specimens," Le Conte noted in the preface to the *Complete Writings,* "would be the subject of much greater regret, were it not for the fact that his descriptions are so clear as to leave scarcely a doubt regarding the object designated. I am thus enabled to assign nearly all of his Coleoptera their proper place in the modern system."[54]

Say's contributions to American zoology were not limited to his entomological writings. He published memoirs on freshwater, marine, and land mollusks, both living and extinct; on crustacea; and on reptiles, birds, and mammals. Of these nonentomological publications the most extensive and important were those on shelled animals, beginning with his account of them in Nicholson's *Encyclopaedia* in 1818 and culminating in his *American Conchology,* six numbers of which were published at New Harmony before his death in 1834, the seventh appearing four years later.[55] His article in the *Encyclopaedia,* described as "the foundation of conchology in America," was quoted by Lamarck and used by Jean B. Férusàc and other European writers. Nor did Say neglect fossil shells or fail to grasp their importance for stratigraphical geology. In a memoir, "Observations on Some Species of Zoophytes, Shells, etc., Principally Fossil," in the first volume of the *American Journal of Science and Arts* he declared his conviction that progress in geology "must be in part founded on a knowledge of the different genera and species of reliquiae, which the various accessible strata of the earth present." Both his conchological and paleontological writings were collected and republished after his death. His memoir on the crustacea of New Jersey, published in 1818, was described by a later expert in that field, Henry W. Fowler of the New Jersey State Museum, as "the very foundation of North American carcinology."[56]

Like Alexander Wilson, Say was a self-made naturalist, driven on by an enthusiastic love of nature and natural science, heedless of his physical limitations, content to live as best he could. Unlike Wilson, however, he had a patron (William Maclure), a wife, and the support of numerous colleagues at the Academy of Natural Sciences. Tall and spare, with black hair and dark complexion, he pleased everyone by his amiable disposition, modesty, honesty, and sincerity. An idealist like Maclure, he pledged himself to the socialist experiment at New Harmony even though his scientific work suffered. He was, one might say, the workhorse of descriptive natural history at a time when they were much needed in American science.

Scarcely less versatile as a zoologist than Say was the French naturalist Charles A. Lesueur, who came to Philadelphia in 1816 in company with William Maclure. Maclure had met Lesueur in Paris and admired his contribution to the official report of the voyage of the *Géographe* to Australia (1800–1804), an expedition that brought back 100,000 zoological specimens, of which 2,500 were new species. In August 1815 Lesueur signed a contract with Maclure pledging to accompany him on his travels for the purpose of making zoological drawings and observations and collecting specimens in the regions visited. After a tour of the West Indies, they landed in New York and proceeded westward to the Great Lakes, returning by way of New England.

For the next nine years in Philadelphia, Lesueur supported himself as an engraver and a teacher of drawing, painting, and the like. It was difficult and frustrating trying to do scientific research under these conditions. "My work forces me to be sedentary," he complained to his Parisian friend Anselme Desmarest. "I am not able to go about the country as I wish, to linger on shore and observe marine life, to add to my drawings and increase my collections. My means will not permit me. All that I have done thus far is due to the liberality of Mr. Maclure; few men would do what he has done and what he is still doing."[57]

Lesueur managed brief trips to Kentucky and to the Hudson River valley, however, and served for a time with the commissioner for the Canadian-American boundary. Meanwhile, he contributed regularly to the *Journal* of the Academy of Natural Sciences and to foreign journals as well. The first volume of the academy's *Journal* contained ten articles by Lesueur, and most of the plates for the volume were drawn and engraved by him. The next four volumes carried sixteen more of his memoirs and forty-one plates from his hand. He also served as one of the academy's three curators.

Like Rafinesque, Lesueur cast his zoological net widely. He published papers on reptiles, crustaceans, fossil shells, cephalopods, ascidians, and worms, but his main contribution was his work on fishes, notably his review of the family of suckers (Catostomidae) in the first volume of the academy's *Journal*. "Lesueur was the first to study the fishes of the Great Lakes of North America," wrote David S. Jordan. Lesueur's descriptions were detailed and accurate, his drawings well done. In the opinion of Louis Agassiz, as reported to Jordan by the English zoologist Richard Owen, Lesueur's contributions to American ichthyology were second only to those of Agassiz himself.[58]

During the years Lesueur spent in Philadelphia the first steps were taken toward producing a general account of American mammals. George Ord had led the way in 1815 with his anonymous contribution to the second American edition of William Guthrie's *New Geographical, Historical, and Commercial Grammar.* In a section entitled "Zoology of North America" Ord catalogued the mammals, birds, and reptiles known to him or described in Pennant's works or in William Turton's edition of Linnaeus's *Systema Naturae.* The catalogue of each group of animals was followed by extended descriptions of selected species, based in part on Ord's own observations but to a greater extent on the narratives of Lewis and Clark, Mackenzie, Pike, Brackenridge, and other travelers and also on books such as Pennant's *Arctic Zoology,* Thomas Bewick's *History of Quadrupeds,* and Felix d'Azara's *Natural History of the Quadrupeds of the Province of Paraguay.* From his reading in d'Azara, Ord was led into the mistake of listing monkeys and anteaters as inhabitants of North America. His account of mammals and birds was rather extensive, his description of reptiles very brief; fishes and insects were barely mentioned. In view of Ord's earlier connection with Alexander Wilson, it is not surprising that the ornithological section of his "Zoology" was best. His acquaintance with American mammals was chiefly from books and specimens in Peale's Museum, but he performed a valuable service in listing and describing the mammals discovered by Lewis and Clark. In many cases, however, he did not assign technical names to them, probably because he was planning to publish an illustrated work on the zoology of the Lewis and Clark expedition, which he never completed.

Modern zoologists recognize 860 species of North American mammals. According to Samuel Rhoads, the American zoologist who republished Ord's "Zoology" in 1894, Ord listed about 100 of these.

> Of those remaining in [Ord's] list, fifteen are undeterminable, twenty-four are Mexican and South American species, eighteen are synonyms of other names in the list and ten are old world forms having no close specific affinities with those of America. . . .
>
> That Ord did much toward reformation is well proven; in view of which we may charitably forget that he included in his 'Zoology of North America,' not only forms exclusively South American, but several species which neither the 'ingenious Mr. Pennant,' nor the versatile Dr. Turton had ever assigned to the fauna of the New World.[59]

The idea of producing a general account of American fauna was taken up again in the 1820s by Richard Harlan and John Godman, though neither succeeded in carrying the project beyond a description of American mammals. Harlan was at the beginning of his career when in 1825 he published his *Fauna Americana,* designed as the first part of a larger work that was

Four Philadelphians who contributed substantially to American zoology and paleontology: Caspar Wistar (by Thomas Sully after Bass Otis, 1817) (top left), Richard Harlan (bust in possession of his son, Dr. George Cuvier Harlan) (top right), George Ord (by J. Henry Smith, 1894, after John Neagle, 1829) (lower left), and John Godman (lithograph by Cephas Childs after a sketch by Henry Inman). Courtesy of the American Philosophical Society

never completed. Graduated from the medical school of the University of Pennsylvania only seven years earlier, Harlan, like his teacher Caspar Wistar, devoted himself to comparative anatomy and vertebrate paleontology. He immersed himself in the writings of Georges Cuvier, Anselme Desmarest, Étienne St. Hilaire, and other leading European zoologists and set out to apply their methods of naming, classifying, and describing animals to the fauna of America. His *Fauna Americana* was based in large part, as he himself conceded, on the *Mammalogie* (1820) of Desmarest, many of whose descriptions Harlan translated vebatim. But he was able to expand Desmarest's list of 100 species to 147 species of American mammals by drawing on the discoveries of the Long expedition and the explorations of William Parry and John Franklin in the arctic region. "Of these," said Harlan, "*Several* are entirely new, and not before described: *eleven* species are fossil . . . ; many were [formerly] imperfectly noticed, or erroneously described; others merely indicated. In several instances *species,* and in three or four cases, *genera,* have been confounded."[60]

Harlan's inclusion of fossil species in his book was especially important. He was the first American to apply Linnaean names to American fossil vertebrates. Wistar had laid the groundwork for an American, one might say a Philadelphian, school of vertebrate paleontology. Harlan built on that foundation by applying the methods of his master Cuvier to the reconstruction, naming, classifying, and describing of fossil animals. Before his death in 1843 he gave Linnaean names to seventeen fossil vertebrates, including two fishes, five reptiles, and ten mammals. "The list is not long, but it is very impressive when viewed in its historical position," wrote the twentieth-century American paleontologist George G. Simpson.

> No one else, either American or European, named nearly so many or nearly such a variety of American fossil vertebrates before [Joseph] Leidy. Besides the species named by Harlan, he described and discussed others named by his contemporaries or then still without distinctive names. Practically all . . . are valid taxonomic units, despite the fact that their names have mostly been changed by retroactive rules of nomenclature far in the future when Harlan wrote and by a narrowing of the scope of such units, also long subsequent to his period.
>
> Harlan's views as to the affinities of these fossil vertebrates were also generally correct, according to the best classification of recent vertebrates then available.[61]

Harlan's *Fauna Americana,* though highly useful, was largely a compilation, drawn not only from Desmarest's *Mammalogie* but also from Cuvier's *Animal Kingdom,* John Richardson's *Fauna Boreali-Americana,* Joseph Sabine's zoological appendix to the report of John Franklin's polar explorations (1819–1822), and Say's descriptions of mammals discovered

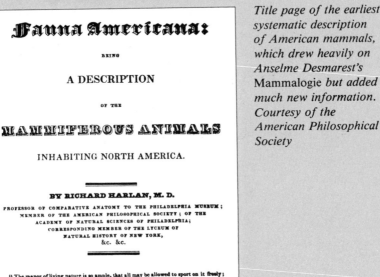

Fauna Americana:

BEING

A DESCRIPTION

OF THE

MAMMIFEROUS ANIMALS

INHABITING NORTH AMERICA.

BY RICHARD HARLAN, M. D.

PROFESSOR OF COMPARATIVE ANATOMY TO THE PHILADELPHIA MUSEUM;
MEMBER OF THE AMERICAN PHILOSOPHICAL SOCIETY; OF THE
ACADEMY OF NATURAL SCIENCES OF PHILADELPHIA;
CORRESPONDING MEMBER OF THE LYCEUM OF
NATURAL HISTORY OF NEW YORK,
&c. &c.

" The manor of living nature is so ample, that all may be allowed to sport on it freely; the most jealous proprietor cannot entertain any apprehension that the game will be exhausted, or even perceptibly thinned."

PHILADELPHIA:
PUBLISHED BY ANTHONY FINLEY.
J. HARDING, PRINTER.

1825

Title page of the earliest systematic description of American mammals, which drew heavily on Anselme Desmarest's Mammalogie *but added much new information. Courtesy of the American Philosophical Society*

on the Long expedition. Unfortunately for John Godman, it appeared just as he was preparing to publish his *American Natural History,* a work more original than Harlan's but covering precisely the same ground. Not content to let the public judge the two works comparatively, Godman published a vitriolic review of Harlan's book in the *Franklin Journal and American Mechanics' Magazine,* charging the author with plagiarism, inaccuracy in matters of nomenclature and classification, and extreme narrowness in his conception of natural history.[62] Harlan responded in kind, and a feud developed that spread to other matters and embroiled partisans on both sides.

That the two men should have quarreled is not surprising, for Godman was a physician-naturalist of a different kind from Harlan. Like Cuvier and his colleagues at the Museum of Natural History in Paris, Harlan was a museum naturalist interested chiefly in anatomy, nomenclature, and classification. Godman, on the contrary, was a naturalist in the tradition of William Bartram and Alexander Wilson, a lover of nature to whom natural history was a romantic, poetic, and semireligious calling. Orphaned at a tender age and apprenticed to a printer in Baltimore, he managed by dint of heroic exertions and precocious talent to earn a medical degree from the Medical College of Maryland. After practicing awhile in Anne Arundel County, Md., where he composed his *Rambles of a Naturalist,* he moved to Philadelphia, married Angelica Peale, daughter of Rembrandt Peale, and accepted an appointment at the Medical College of Ohio.

During his brief stay in Ohio, Godman launched the *Western Quarterly Reporter,* the first medical journal in the Ohio valley. When Daniel Drake was ousted from the Medical College, Godman returned to Philadelphia and set himself up as a private teacher of anatomy. In 1826, the year in which the first volume of his *American Natural History* appeared, he set forth his ideas about natural history in an article in the *Franklin Journal.* The study of natural history, he declared, was open to everyone who had eyes to see. To observe animals in the wild — "their peculiar construction and adaptation to the places they occupy, their modes of living, and the relations they bear to other animals, and to man himself" — this was natural history, the true natural history, as distinguished from "the fruitless and wearying discussion of technical phrases, or the propriety of various classico-barbarous appellations, with hair splitting distinctions of genera, sub-genera, species and varieties." Properly pursued, natural history has a tendency "to enlighten the mind, and warm the heart; to enlarge our feelings of respect for our own race, and enable us more correctly to appreciate the beneficent wisdom of the Creator."[63]

In keeping with these views, Godman tried to compose his *American Natural History* from living nature. He had, he said, frequently found it necessary to suspend work on this book "in order to procure certain animals, to observe their habits in captivity, or to make daily visits to the woods and fields for the sake of witnessing their actions in a state of nature." But he had also learned much from the specimens in Peale's Museum, from the library and collections of Charles Bonaparte, and from the notes and conversation of Thomas Say, George Ord, Titian Peale, and Robert Best, who had been curator of the Western Museum in Cincinnati before moving to Lexington. From the artist John Neagle, Godman obtained portraits of two Indian chiefs to be used as illustrations of the Indian race. Unlike Harlan's *Fauna Americana,* Godman's book was well illustrated with engravings, many of them done by Lesueur from specimens in Peale's Museum.[64]

In matters of nomenclature and classification Godman followed Linnaeus and Georges Cuvier, retaining the Linnaean names wherever possible. For the dentition of the mammals he drew on the work of Frédéric

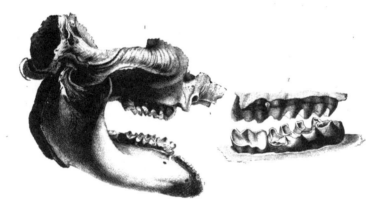

Two illustrations, bison and mastodon jaw and teeth, from John Godman's American Natural History *(1826–1828). Like Harlan, Godman included both living and fossil species in his account of American mammals. Courtesy of the American Philosophical Society*

Cuvier. At the end of the third volume was a "General Synopsis of Mammalia Inhabiting North America," prepared by Bonaparte. This was followed by an "Analytical Table of the North American Genera."

The relative merits of Godman's and Harlan's treatises on American mammals have been aptly summarized by Keir B. Sterling, a zoologist who paid tribute to these early American naturalists by reprinting their works with appropriate introductions. Godman's *American Natural History,* writes Sterling, was "the first American study of mammals which was both comprehensive and based upon the author's own research and observations."

> It is considered the best work of its type since Pennant's *Arctic Zoology,* published half a century before. While his work also reflected dependence upon other authorities, Godman used a greater variety of sources, his descriptions were fuller and his original observations much more conspicuous than Harlan's. . . . Harlan's book, on the other hand, emphasized anatomical detail and other 'essential characters' in more succinct form, with less detail given to life history material.[65]

Godman's work, like Wilson's, was completed in a race with death. In 1830, two years after the third volume of his magnum opus appeared, Godman died of tuberculosis.

CONCLUSION

In zoology and paleontology, by contrast to botany, Americans led the way in naming, classifying, and describing nature's productions, past and present. Through Jefferson's description of the megalonyx, Peale's exhumation of the mastodon, Wistar's account of two skulls from Big Bone Lick, and the subsequent work of Godman and Harlan, the groundwork was laid for Philadelphia's emergence as a major center of vertebrate paleontology in the nineteenth century. In zoology there were important contributions by naturalists from abroad like Rafinesque, Lesueur, and Bonaparte, but the main burden of investigation was carried by Americans. In Philadelphia, where the best work was done, the way was paved by the contributions of William Bartram, Benjamin S. Barton, and Charles W. Peale, whose museum played a role analogous to that played by Bartram's garden in the development of American botany. Alexander Wilson then produced his heroic *American Ornithology,* and this in turn inspired the subsequent efforts of George Ord, Thomas Say, and John Godman.

Throughout the period there was a noticeable tension between naturalists like Bartram, Peale, Wilson, and Godman for whom the study of nature was a semireligious calling and those like Barton, Say, and Harlan who were content to name, classify, and describe. The conflict between Godman and Harlan at the end of the Jeffersonian era was symptomatic of a broad change taking place in American natural history. The day of the poet-artist-naturalist was not completely over — Audubon would prove that — but the predominance of the systematist, the technical description in the learned journal, the comprehensive catalogue of orders, genera, and species was clearly at hand. The Linnaean spirit had long ago triumphed in Europe. It was now taking possession in America. Natural history, declared Georges Cuvier, was *nommer, classer, et décrire*. And so it became, as scientific botany, zoology, paleontology, and anthropology emerged from the mother matrix. A new name, biology, would eventually embrace all these sciences, leaving the old name, natural history, to describe the more popular, romantic, and human study of nature practiced by Peale, Bartram, Wilson, Godman and their spiritual descendants to the present day.

12

The Sciences of Man: Physical Anthropology

JEFFERSONIAN science did not limit itself to the study of the nonhuman world. Like eighteenth-century and early nineteenth-century science generally, it was intensely interested in the human race—its origins, racial varieties, "antiquities," history, customs, and languages. In America, moreover, there were special incentives to the study of topics like these. On the frontier were the Indian tribes, their hunting grounds strewn with impressive mounds and earthworks concerning whose origin they had only legendary knowledge. In the settled areas, particularly in the South, were the Negroes, a race transplanted from Africa to do hard labor for their white masters. The white population itself was a mixture of northern European peoples undergoing slow transformation by intermarriage and exposure to a new physical and social environment. Were all these human types distinct varieties of a single species? If so, how had they become so different from each other? Where had the Indians come from, and how were they related to the mound builders about whose constructions they knew so little?

If, as some writers suggested, the Negroes, Caucasians, Asians, Malays, and Indians were separately created species rather than varieties, how was this view of humanity to be reconciled with the biblical story of human history and what were its political and social implications for a nation whose Declaration of Independence proclaimed all men created free and equal, whose Constitution recognized and protected Negro slavery as an existing institution, and whose white inhabitants were bent on possessing the lands inhabited by Indian tribes? Given these circumstances and issues, it is not surprising that the current of scientific inquiry ran strongly in the direction of physical anthropology, archaeology, and comparative linguistics, each of which promised to throw light on the peculiarities, relationships, and history of the diverse human types inhabiting the American continent.

320

As in so many other fields of investigation, Thomas Jefferson's *Notes on the State of Virginia* served to bring these topics to public attention. As a defender of nature's productions in the New World, Jefferson was duty bound to refute Buffon's characterization of the American Indian as feeble, cowardly, sluggish in mind and body, destitute of body hair, and lacking in familial affection and sexual ardor. As a Virginia planter and author of the Declaration of Independence he felt equally obliged to discuss the issue of slavery and the related question of the Negro's place in nature. In Jefferson's time, as in our own, racial anthropology could not be separated from politics.[1]

Jefferson's knowledge of the black race was largely that of a slave-owning planter. It is not surprising, then, that he thought the Negro inferior to the white man in both body and mind. The blacks, he asserted, were deficient in foresight and imagination; they were equal to whites in memory but vastly inferior in powers of reason, more ardent but less delicate in sexual passion. "In general," he concluded, "their existence appears to participate more of sensation than of reflection."[2] He conceded that the life situation of the blacks in America was not favorable to the development of their faculties, but he pointed to the intellectual and artistic achievements of Roman slaves and the art and oratory of the Indians as proof that adverse conditions cannot completely suppress natural genius.

As to whether the Negro was a separate species of man, "distinct from all others since the day of creation," or merely a variety, "made distinct by time and circumstances," Jefferson was undecided. He was not concerned to corroborate the Bible, nor did he think that any moral issue hung upon the answer to this problem in natural history. His denunciation of slavery in another passage in the *Notes* and his reply in 1809 to the remonstrances of the Abbé Grégoire show that he viewed the question of the Negro's rights as one quite distinct from that of his rank in the scale of nature. Jefferson was certain, however, that steps should be taken to prevent miscegenation between blacks and whites: "Among the Romans emancipation required but one effort. The slave, when made free, might mix with, without staining the blood of his master. But with us a second is necessary, unknown to history. When freed, he is to be removed beyond the reach of mixture."[3] Unfortunately some of Jefferson's remarks were noticed and quoted by European writers bent on proving that the Negro was a separate and inferior species.

One turns with relief from Jefferson's prejudice-laden account of the blacks to his discussion of the American Indian. Jefferson, who had been familiar with Indians since boyhood, had read everything he could lay his hands on concerning them and had collected vocabularies and artifacts for many years. In opposition to Buffon he presented the Indian as active, brave, eloquent, indulgent toward children, ardent and fertile as the white man, and possessed of a high sense of honor and great vivacity of mind. "I do not mean to deny," Jefferson wrote, "that there are varieties in the race of man, distinguished by their powers both of body and mind. I believe there are, as I see to be the case in the races of other animals. I only mean to suggest a doubt, whether the bulk and faculties of animals depend on the side of the Atlantic on which their food happens to grow, or which furnishes the elements of which they are compounded." About the Indians of South America he professed to know nothing and to doubt most of what had been written about them.[4]

With respect to the hotly debated question of the origin of the Indian race Jefferson was tentative and open-minded. From the resemblance of most of the American aborigines to Asiatic tribes he concluded that they were derived from Asia, but both the appearance and the language of the Eskimos persuaded him that they had come from Greenland and, ultimately, from Europe. On the other hand, the Indian languages were so many as to suggest that they had undergone diversification during a very long period of time and that the inhabitants of America were in fact older than the Old World peoples, whose languages exhibited fewer differences. In this speculation Jefferson leaned toward the view that the Indians had been separately created, though he conceded that "the mind finds it difficult to conceive that so many tribes have inhabited it [America] from so remote an antiquity as would be necessary to have divided them into languages so radically different"[5] (see Chap. 14).

Jefferson's *Notes* was a miscellaneous collection of information and observations on a great variety of topics, not an anthropological treatise. The first such treatise published in the new American republic was Samuel S. Smith's *Essay on the Causes of the Variety of Complexion and Figure in the Human Species* (1787). Smith was professor of moral philosophy (later president) at the College of New Jersey in Princeton, where he had graduated in 1769. After graduation he was ordained to the Presbyterian ministry, preached in Virginia, served briefly as president of Hampden-Sydney College, and then returned to Princeton in 1779 when he was offered the chair in moral philosophy by President Witherspoon, whose daughter he had married.

Moral philosophy covered a wide range of topics in those days, and Smith chose as the subject of his researches the natural history of man, a topic which had been brought to public attention by Buffon and the great Swedish naturalist Linnaeus, who classified man as one of the primates and subdivided the human species into four races (European, American, African, and Asian) in his *Systema Naturae*. Soon afterward Johann F. Blumenbach, professor of medicine at the University of Göttingen, added a fifth race, the Malay, to Linnaeus's list and undertook to show, follow-

Samuel Stanhope Smith, by Charles B. Lawrence, president of the College of New Jersey (Princeton) and author of the first American treatise devoted to the causes of racial variation in the human species. Courtesy of The Art Museum, Princeton University

ing Buffon, that the differentiation of the various human races from a single original stock had been brought about by the influence of climate, diet, and other environmental factors. Meanwhile this monogenist view of human races, as it was called, had been challenged by polygenist writers like Voltaire and Lord Kames, a Scottish jurist, literary critic, and amateur anthropologist whose *Sketches of the Natural History of Man* appeared in 1774.[6] Smith had read some though not all of these writers, especially Kames, and he saw a golden opportunity to contribute simultaneously to the progress of science and the defense of the biblical revelation by showing that the differences among human races advanced by Kames as evidence of their separate origins could be explained by natural causes. "As the character, and manners, and state of society among the savages, would make a very important part of the history of human nature," he wrote to Charles Nisbet in 1784, "it appears to me to be an object that merits the attention of literary societies, not less than the discovery of new islands and seas." Hitherto the Indians had been observed chiefly by soldiers and traders dealing with tribes corrupted by contact with Europeans. In the future, Smith predicted, learned societies would send "good philosophers" to live among Indians untainted by contact with civilization, "to reside among them on a familiar footing; to dress and live as they do; and to observe them when they should be under no bias or constraint."[7] Inspired by visions of anthropological research like this, Smith threw himself into his studies and in February 1787 journeyed to Philadelphia to present the results of his researches before the American Philosophical Society, to whose membership he had been elected two years earlier.

In his discourse, Smith made no attempt to classify the races of man, observing that various efforts to draw sharp lines between them had been unsuccessful and that it was "a useless labor" to pursue the investigation further. The important thing, he insisted, was to account for the differences among human races and to show that they could be explained by natural causes acting on a single original stock. To accomplish this objective, Smith relied on two principal causes of variation: climate and the state of society. On the subject of climate he followed Buffon, arguing the slow and imperceptible modification of the human constitution by heat and cold, the transmissibility of modifications produced by natural causes, and the general correlation of latitude and skin color. Unaware that Johann F. Blumenbach had advanced a similar theory in 1775, he speculated that "the complexion in any climate will be changed toward black, in proportion to the degree of heat in the atmosphere, and to the quantity of bile in the skin."[8]

In support of his arguments, Smith pointed to the dark complexion of the western pioneers and the laboring poor in the Carolinas and Georgia. The hair of the Anglo-American was losing its curl and the skin of the American Negro its peculiar odor, he asserted. In short, the recent inhabitants of North America had begun to resemble the native Indians. The resemblance would never be complete, he added, since the changes in appearance would take place on facial structures formed in different climates, and the effects of climate would always be modified by the influence of civilization. Eventually, however, the temperate zone of America, like that of the Old World, would produce a type not very different from the perfection of form displayed in man's first parents. "When time shall have accommodated the constitution to its new state, and cultivation shall have meliorated the climate, the beauties of Greece and Circassia may be renewed in America; as there are not a few already who rival those of any other quarter of the globe." A happy prospect for American manhood and womanhood!

In discussing the influence of society on the human constitution, Smith went well beyond anything Buffon had suggested and anticipated many of the ideas James C. Prichard was to advance a quarter of a century later. By state of society, Smith explained, he meant "diet, clothing, lodging, manners, habits, face of the country, objects of science, religion, interests, passions, and ideas of all kinds, infinite in number and variety." The human form could not retain or achieve its original perfection apart from civilization, he declared. Savages living exposed to the influence of raw nature and habituated to uncleanliness, neglect, and the practice of painting and anointing their bodies could never be fair in complexion. Animals were most beautiful in their wild condition, but man, having been made for society and civilization, attained his proper form and aspect only in a civilized state. Thus a wise Providence had made the perfection of human nature to depend partly on man's own efforts.

In Smith's view, the social state did not merely qualify the influence of climate; it was a causal influence in its own right. In a society divided into classes the complexion, gait, figure, and aspect of the patrician differed markedly from that of the plebeian. In America, said Smith, the general equality of condition among the whites had prevented their differentiation into distinct types. In the case of the Negro, however, the domestic servants were far in advance of the field slaves in acquiring the regular features and expressive face of civilized peoples. If they were given freedom and permitted to enjoy property and social respectability, the change in their physical appearance would proceed even more rapidly.

The standard of beauty in a society was another powerful influence toward uniformity of appearance, Smith went on. By education and example, by establishing certain preferences in the choice of marriage partners, and by sanctioning deliberate alterations in face and figure, society molded the appearance of individuals to the prevailing standard. Thus the Germans, Swedes, and French in America, living by themselves and continuing the habits and practices of their native countries, had retained their original features despite their new physical environment.

Here, as elsewhere in the *Essay,* Smith failed to distinguish the influence of the physical and social environment from the effect of isolation and interbreeding on the hereditary constitution. He thought that the coarse features of laboring people were produced by their hard mode of life and transmitted to their posterity by inheritance and conjectured that deliberate modifications of face and figure could likewise be inherited. "We continually see the effect of this principle on the inferior animals," he wrote. "The figure, the colour and the properties of the horse are easily changed according to the reigning taste." But he added in a footnote: "By choosing horses of the requisite qualities, to supply the studs." At times, however, he seemed to realize that the inheritance of acquired characters could not be taken for granted. Any alteration must become a "habit of the body" before it became hereditary, he suggested. "The effect proceeds[,] increasing from one generation to another, till it arrives at that point where the constitution can yield no farther to the power of the operating cause. Here it assumes a permanent form and becomes the character of the climate or the nations."[9] So Buffon had argued.

Having vindicated the inspiration of the Bible by explaining the variety of mankind in terms of the operation of natural causes, Smith carried the attack to the enemy in a series of "Strictures on Lord Kaims's Discourse on the Original Diversity of Mankind," appended to the *Essay*. In these remarks Smith took exception both to Kames's reasoning and his statements of fact. Smith denied that the Indians were naturally beardless and scoffed at the assertion that the South Carolina climate tended to exterminate Europeans before they had time to degenerate to the American model. He did not question, however, that Europeans tended to degenerate in hot climates. To Kames's argument that chance causes could not produce uniform varieties he answered that a uniform climate could be expected to produce uniform effects by the operation of uniform laws of nature. Nor was it unbecoming of the Creator to have made mankind subject to changes of this kind. "The goodness of the Creator appears in forming the whole world for man, and not confining *him,* like the inferior animals, in a bounded range, beyond which he cannot pass either for the acquisition of science, or for the enlargement of his habitation."

Finally, said Smith, there was no difficulty in reconciling the historical account in Genesis with the widespread existence of savage peoples in ancient and modern times. After the Deluge, civilization was preserved among some of Noah's descendants, but others, being too lazy or stupid to pursue agriculture profitably, took to hunting and fishing and wandered gradually into the wilderness created by the Deluge. In time these wandering bands became separated and developed their own languages and constitutional peculiarities. Thus the savagery of many human tribes and the diversity of their languages could be explained without assuming divine interposition in human affairs. To rest the whole explanation on the story of the Tower of Babel, as Kames did, was to discredit the Bible by invoking divine aid unnecessarily, a favorite device of infidels.

Smith's reply to Kames reveals how essentially traditional both men were in their conception of nature. In Kames's discourse the sentiment of the fixity of nature and the perfect adaptation of every created thing to its environment was particularly strong. Smith's *Essay* stressed the mutability of the human constitution, but this exception to nature's rule was justified by the highly traditional notion that the earth and its productions were made for man, who in turn was made for all climates and seasons. Thus Smith was able to stress the pliancy of human nature without extending the idea of indefinite plasticity to animal nature. In any case, Smith's time perspective was too limited to make room for the possibility that degeneration might eventually affect the essential nature of man. Kames leaned toward the view that man had progressed to a civilized state from rude beginnings instead of declining from a state of original perfection, but he did not explicitly reject the contrary account in Genesis. Sensing Kames's doubts on this point, Smith dismissed as a mere subterfuge his Lordship's resort to the story of the Tower of Babel to explain human diversity. In the second edition of his *Essay* in 1810 Smith would return to the question of man's original condition with a much keener sense of its importance for the philosophy of human nature.

Smith's *Essay* created little stir in the United States, but it was reviewed in London — favorably in the *Monthly Review,* unfavorably in the *Critical Review* — and was republished in Edinburgh through the efforts of "a Gentleman of the University of Edinburgh." The gentleman in question was probably Benjamin S. Barton, who was then studying medicine at Edinburgh and who had himself submitted to the Royal Medical Society of Edinburgh "An Essay towards a Natural History of the North American Indians: Being an Attempt to Describe and Investigate the Causes of Some of the Varieties in Figure, in Complexion, &c. among Mankind," in which he sided with the monogenists, derived all the American aborigines except the Eskimos from Asia, and relied on "the operation of Cold, the Nature and Quantity of his food, together with the State of Society" to explain the peculiarities of the Eskimo's figure and complexion. The Edinburgh edition, the preface stated, had been "carefully corrected from a copy containing a great number of alterations, &c. in the author's own hand-writing, with which the Editor was fortunately furnished." It seems likely that Smith had asked Barton to arrange an Edinburgh edition of his *Essay* or that Barton had requested permission to do so and had received Smith's approval and suggested corrections to the text.[10]

In 1809 a belated attack on Smith's *Essay* appeared in the *New York Medical and Philosophical Journal and Review.* The author, a Virginian by birth, was John A. Smith, professor of anatomy and surgery at the College of Physicians and Surgeons in New York. During his medical studies in London he had been much impressed by a polygenist treatise, *An Account of the Regular Gradation in Man,* written by Charles White, an eminent Manchester surgeon. White had compared several anatomical traits of Europeans, Negroes, and apes and had concluded that in every case the Negro was intermediate between the European and the ape, the other human races occupying intermediate positions between the European and the Negro. Extending his comparisons to living persons, he attempted to show that there were constant constitutional differences between the Negro and the European in cartilages, muscles, tendons, skin, hair, sweat, odor, size of brain, reason, speech, and language, as well as in the skeleton. These differences, White asserted, could not be the result of climate and mode of life. Instead, the races of man must be separate species created in a graded series and adapted each to its own environment by Divine Wisdom.[11] The young American medical student adopted these views. On returning to the United States and assuming his teaching duties at the College of Physicians and Surgeons, he devoted most of his inaugural lecture in anatomy to the subject of the anatomical differences between the races of man, relying heavily on White's arguments but drawing also on the concept of the facial angle popularized by the Dutch anatomist Petrus Camper and on Georges Cuvier's investigation of the relative proportions of face and cranium in man, orangutan, and monkey. Upon examining a collection of human skulls in his own possession by these craniometrical criteria, John A. Smith said, he had discovered that they formed a graded series rising from the Negro to the European. He lamented with Jefferson the black man's inability to blush and recapitulated White's table of anatomical differences between the European and the Negro.[12]

Not until he turned to the question of the causes of these differences among human races did he allude to Samuel S. Smith's *Essay* of 1787. The Bible, he conceded, stated plainly that all mankind were offshoots of one stock, but those writers who had attempted to corroborate revelation by attributing the differences among men to the action of climate and the state of society had injured both science and religion. Samuel Smith, for example, had displayed an appalling ignorance of anatomy in his discussion of pigmentation and had grossly exaggerated the correlation between complexion and climate. Even supposing that climate could permanently alter complexion, how could it affect the structure of the bones? The supposition that cold weather had enlarged the heads of the Laplanders by contracting their faces and limbs was ludicrous. The causes of most of the variations in the human frame were unknown, and it was better to confess one's ignorance than to invent preposterous explanations. On this note of humility Smith closed his lecture, leaving his students to draw their own conclusions. "Differing minds are satisfied with different degrees of evidence," he philosophized; "and far be it from me to fix the bounds of your faith."

If the New York Smith expected the Princeton Smith to bow humbly before the onslaught of superior anatomical wisdom, he was badly mistaken. The latter had been gathering materials for a second edition of his *Essay;* it appeared in 1810, enlarged to nearly twice the original size. By this time, however, the argument had broadened beyond the monogenist-polygenist issue to include the question of whether the original condition of mankind had been a bestial one. On that point, Samuel Smith pointed out, the traditions of all nations agreed. Man had originated in central Asia in a state of civilization. And how could he have survived if he had not been instructed and provided for by his Creator? Without such aid, mankind would soon have perished. Nomads like the Indians, far from being the prototypes of civilized man, were the descendants of restless spirits who threw off the restraints of civilization and wandered into the wilderness, there to forget the arts of their fathers, although they preserved faint memories of the events and divinely revealed truths of man's early existence.

Smith then confronted those writers who argued that man had not been divinely created but had emerged as a dumb brute from the primeval slime. If so, said Smith, why had this process of spontaneous generation never been repeated? Had nature lost her ancient powers? The speculative mind must stop somewhere in its search for human beginnings. "If so, why not stop with religion at the beginning of this world, where we may behold man coming from the hand of his Creator, not like a casual clod of the valley, nor thrown from Him like a wretched and abandoned orphan, but so instructed and assisted by Him who deigned to form him, and endow him with reason, that he should be worthy to be the parent of his numerous posterity, and lord of the new creation?"[13] For all his open-mindedness, Smith was still confined within the limits of a biblical view of history.

Turning to the main subject of his *Essay,* Smith welcomed to the side of the monogenists the celebrated Professor Blumenbach, whose work had by this time come to Smith's attention. Blumenbach's view of the physical agency of climate was much like his own, even in respect to the physiological mechanisms involved, Smith pointed out. Adopting Blumenbach's definition of a species as a group of animals possessing distinctive characters that could not be explained by known processes of degeneration, Smith set out to prove again that all humans were of one species. His arguments to this end were much the same as in the first edition, though more profusely illustrated with examples. From the works of Lavater, Gall, and Spurzheim on physiognomy Smith drew support for his contention that skull form is determined partly by intellectual and emotional activity. Conversely, the original form of the head conditioned the operations of the mind and thus favored or retarded the development of particular mental or moral abilities. Mind and matter affected each other reciprocally. On the question of whether modifications in head form produced by mental activity could be inherited, Smith remained undecided.

The addenda to the second edition of the *Essay* contained, besides the "Strictures" on Lord Kames's discourse, replies to Charles White and John A. Smith and a discussion of Jefferson's remarks on the mental endowment of the Negro. Samuel Smith conceded that White's description of the Negro was fairly accurate. He argued, however, that climate and mode of life had made the Negro what he was and that changes in these causes would eventually produce changes in the Negro's appearance and behavior. In proof of this, Smith offered the example of a Maryland Negro named Henry Moss who had turned from black to white in the course of a lifetime. This case, Smith argued, disposed of the notion that the Negro was a species immutably different from the European.

In reply to the skeletal measurements cited by White, Smith reported that he had made some measurements of whites and blacks in Princeton and had discovered that the anatomical differences within the white and the black races were more striking than those between them. The blacks of Princeton, Smith reported, exhibited a range of variation from seventy to seventy-eight degrees in their facial angles. Many of them had high and prominent foreheads. Who, then, was the smatterer in science, the Princeton Smith or the New York Smith?

Toward Jefferson, Samuel Smith was more lenient. He pointed out that the arguments by which Jefferson had exonerated the Indian and the Anglo-American from the charge of mental and physical debility applied with equal force to the Negro — with greater force, in fact, since the Negro had the added disadvantage of being a slave. No wonder he had not produced flights of oratory! The Indians themselves, said Smith, tended to degenerate into lazy, wretched, demoralized creatures when subjected to the influences of civilization. "They afford a proof of the deterioration of the mental faculties which may be produced by certain states of society, which ought to make a philosopher cautious of proscribing any race of men from the class of human beings, merely because their unfortunate condition has presented to them no incentives to awaken genius, or afforded no opportunities to display its powers."[14]

The second edition of Smith's *Essay* drew fire from several quarters. In England, James C. Prichard acknowledged his debt to Smith for the idea that varying standards of beauty in different societies might have the effect of creating or accentuating racial differences, but he rejected completely the notion that characteristics acquired during an individual's lifetime were transmitted to offspring. In the United States the *Essay* was reviewed in the *Medical Repository,* presumably by its chief editor, Samuel L. Mitchill, who lectured regularly on natural history, including that of man, to his class at the College of Physicians and Surgeons. The review began with a long summary of the biblical account of man's origin and went on to state that Smith was a firmer believer in the inspiration of the Bible, bent on corroborating Genesis by showing that all mankind were descended from a single stock. Mitchill agreed that this hypothesis was more defensible than the gratuitous assumption of several species of human beings, but he objected to limiting the causes of human diversity to the agency of climate and mode of life. The most important single source of variety in successive generations was the mechanism of heredity, said Mitchill. "Enough for our present purposes is the statement of the fact, established on broad induction, 'that a man and woman may beget a child of a different complexion from either of the parents, and the complexion of the offspring may be perpetuated in his or her descendants.' "[15] Secondly, it was possible there were differences in the "primitive family" of mankind that were propagated to the offspring. Finally, it should be recognized that social imitation operated powerfully to recreate family, national, and other group resemblances in each generation. Thus the agency of climate was merely one, and by no means the most important, source of variation in man.

Mitchill's own views on the classification and origin of human races had been outlined in the *Medical Repository* during the previous year in an abstract of his lectures at the College of Physicians and Surgeons. Mitchill divided the human species into six races: Caucasian, Hyperborean or Laplander, Tartar, South American, Malay, and Negro. The early inhabitants of North America were said to be derived from the Tartars by way of Alaska and, in the case of the Eskimos, from the Laplanders by way of Greenland.[16] By 1816, however, Mitchill's views had been altered and expanded. He now recognized a tawny race, including Tartars, Malays, Chinese, and American Indians, from which he derived two other races, the white and the black. In his lectures at the College of Physicians and Surgeons that year he suggested that North America had been the scene of a series of conflicts between a southern people derived from the Malays and some northern tribes akin to the Tartars, the "Tartarians" emerging victorious. The Indian mounds and other earthworks of the Ohio and Mississippi valleys he attributed to the defeated Malays, several of whose bodies he believed to have been found in caves in Tennessee and Kentucky. One of these mummies, found in a cave in the vicinity of Glasgow, Ky., had been sent to Mitchilll in 1814.[17] The wrappings in which the mummy was encased seemed to Mitchill to resemble fabrics brought to New York from the Fiji Islands by Americans who had lived there.

Mitchill derived the aborigines of northeastern America from Scandinavia, Lapland, and Wales. These peoples, he conjectured, had filtered as far south as the Hudson River, only to be driven back to Labrador by the victorious Tartarians. "Think," Mitchill told his students, "what a memorable spot is our Onondaga, where men of the Malay race from the southwest, and of the Tartar blood from the northwest, and of the Gothick stock from the northeast have successively contended for supremacy and rule; and which may be considered as having been possessed by each before the French, the Dutch, and the English visited the tract, or indeed knew anything about it."[18]

By this time Mitchill had come to favor Jefferson's suggestion in the *Notes* that Asia might have been peopled from America rather than the other way around. He was hesitant to oppose the prevailing ideas about man's creation and dispersion, he wrote to De Witt Clinton, but could see many advantages in Jefferson's hypothesis and much evidence to support it.[19]

Mitchill was no comparative anatomist. His views about human races were based largely on ocular inspection of living exemplars of the different races of man. As a member of a New York commission to negotiate a purchase of Indian lands in 1791 he had become acquainted with the Iroquois. Later, as a United States senator, he chaired a committee on Indian affairs and had extensive contacts with representatives of various tribes. "I soon became convinced, that the opinions of the European historians and naturalists were so full of hypothesis and errour that they ought to be discarded," he wrote to the secretary of the American Antiquarian Society in 1817. In the previous year he had carefully inspected several Chinese sailors who had assisted in navigating a ship from Macon to New York. "The thinness of their beards, the bay complexion, the black lank hair, the aspect of the eyes, the contour of the face, and in short the general external character," he reported to De Witt Clinton, "induced every person who observed them to remark, how nearly they resembled the Mohegans and Oneidas of Newyork."[20] Mitchill added that the American painter John Smibert, in executing paintings of Tartars for the Grand Duke of Tuscany, had been struck with the similarity of their features to those of the Narragansett Indians.

As a physical anthropologist, Mitchill's main contribution was his insistence that racial differences were to be explained by "generative influence" as well as by environmental circumstances. "If by the act of modelling the constitution in the embryo and foetus, a predisposition to gout, madness, scrofula and consumption, may be engendered," he wrote, "we may rationally conclude, with the sagacious [Felix] d'Azara, that the procreative power may also shape the features, tinge the skin, and give other peculiarities to man."[21] This emphasis on the role of heredity in supplying the materials of racial variation was a needed corrective in an age dominated by the environmentalism of Buffon and Blumenbach.

The sharpest attack on the 1810 edition of Samuel S. Smith's *Essay* was that of the Philadelphia physician Charles Caldwell, published in *The American Review of History and Politics* in 1811 and subseqently republished and greatly expanded in the *Port Folio*. Caldwell objected strenuously to Smith's dragging the issue of the inspiration of the Bible into the discussion of a scientific problem and branding as infidels all those who questioned the monogenist position with respect to the races of man. He himself had no intention of questioning the authority of Scripture "under a liberal interpretation," he declared, but he felt certain that the Bible was never meant to instruct mankind in scientific matters. "Its province is morality, not physics, — its end[,] practice, not theory. It makes plain to us the whole path of our duty, but is silent on the subject of natural science."[22] If the Bible must be invoked, Caldwell added, the relevant passage in connection with Smith's *Essay* was that in which the prophet asked: "Can the Ethiopian change his skin or the leopard his spots?" Apparently the scriptural question was a touchy one, for the editors of the *Port Folio* twice went out of their way to deny any intention of undermining the authority of the "sacred oracles." They also disavowed any purpose of defending slavery.

Caldwell also affirmed his belief in the unity of the human species, but he proposed to show that the environmental causes Smith had employed to explain the origin and modifications of the various human races were totally incapable of producing the effects attributed to them. Unlike Prichard, Caldwell did not deny the transmissibility of physical traits acquired during an individual's lifetime through environmental influences. Instead, he made a distinction between organic and inorganic matter. The latter, he said, underwent steady changes as a result of the uniform operation of natural laws. In living matter, however, there was an "ever active principle . . . which offers resistance to the impressions of new and unfriendly causes, and so completely accommodates itself to the existing state of things, as finally to paralyze and even completely destroy all susceptibility to their actions." "This principle," he added, "absolutely prohibits the continued operation of causes of the same power from being uniformly progressive in its effect." A hot climate might produce some change in the color of the skin, but it could never transform a white man into a Negro.[23]

On the whole, however, Caldwell spent little time expounding his own views about the causes of racial differentiation. His lengthy review was devoted almost entirely to questioning Smith's statements of supposed facts and the efficacy of the environmental causes invoked to explain them. Whether it was a question of the general correlation of climate and skin color, the differences between domestic and field slaves in America, the characteristics of the Jews in different environments, the supposed metamorphosis of a colony of Hungarians into Laplanders and a colony of Portuguese into Hottentots, or the physical transformations undergone by northern European stocks in America, Caldwell advanced arguments to show that the alleged facts were incorrect, that they were due to miscegenation or a diseased condition of the body, or could not in any case have been produced by environmental causes. Not until 1830 did he present his own ideas about the origin of races in a book entitled *Thoughts on the Original Unity of the Human Race,* which turned out to be a polygenist argument for the original diversity of mankind, based on the old idea that God had created separate and unequal races adapted to the environments and ways of life which He had ordained for them. Caldwell's antimonogenist arguments were now aimed at Prichard's *Researches on the Physical History of Man* rather than Smith's *Essay,* but they were in many ways the same as those he had used against Smith.[24]

It seems odd that Caldwell made no mention in his long review of

Smith's *Essay* of Hugh Williamson's *Observations on the Climate in Different Parts of America* (1811). Williamson, like Samuel S. Smith, was a devout believer in the efficacy of environmental influences in shaping the features and physique of the various human races. His attention had been drawn to the problem of the origin of these differences by Johann D. Hahn and Pieter Luchtmans while studying medicine at the University of Utrecht. After completing his studies, Williamson settled in Philadelphia, where he was active in the American Philosophical Society. He then moved to North Carolina, returning to Philadelphia as a delegate to the Continental Congress and, later, as a delegate to the Federal Convention of 1787. Doubtless he was among the audience who heard Smith's "ingenious and learned oration" on the varieties of mankind in Philosophical Hall on the evening of February 28, 1787.

Williamson's own notes on anthropological subjects had been destroyed during the American Revolution; but years later, after he had left North Carolina for New York City and retired to devote himself to literary and scientific pursuits, he decided to publish his thoughts on the climate and geography of the globe in relation to the peoples inhabiting it as an introduction to his *History of North Carolina*. "The publication of this discourse and of the history to which it refers," he informed his readers, "has been suspended many years, with the hope of getting from the Western Country, according to the promise of sundry correspondents, some farther account of Indian antiquities, and obtaining conclusive materials towards the history of the original inhabitants. But people who live near the ancient monuments, and see them every day, are less disposed than we could wish to dig in quest of antiquities."

The argument of Williamson's *Observations* was similar to the one already advanced by Smith in his *Essay*. Like Smith, Williamson undertook to refute Charles White's invidious comparison of the Negro and the European, pointing out that many of the peculiarities White attributed to the Negro — for example, having six lumbar vertebrae instead of five — were found in some whites as well. As for the facial angle:

> If the greater facial angle may be supposed to mark a superior order of beings, it will follow, that the ancient Grecians were of an order superior to the present Europeans; for their skulls measure an angle of 90 degrees. Is that race of men extinct? Or is it not more probable that the shape of the head, by different modes of dressing, and other causes, is altered?[25]

Williamson saw clearly that skin color in men or fur color in animals was related to the regulation of body heat in different climates, but it did not occur to him that this adaptation was the result of natural selection. "We see the changes that are produced by warm climates, and we discover in some cases, the reason why such changes are necessary or useful," he wrote; "but the method by which they are effected is beyond our reach." Whatever the cause, he was convinced that the peoples of every nation and race underwent changes in pigmentation and physique "according to the soil, situation and climate, by the most regular and insensible deviations and shades" and hence that the differences among the human races were no proof of their separate origins. The American climate and topography, he declared, "seems to produce animals of equal strength and firmness, but less ferocity of disposition, than are produced on the other continent," contrary to the assertions of Buffon, the Abbé Raynal, and others. It was probable, therefore, that the American continent would produce a race of men "less prone to destroy one another, and more desirous to improve the understanding and cultivate social virtues." Unlike Jefferson, Williamson and Smith were convinced that Nature had enlisted herself on the side of the *cis*atlantic partisans.

Whatever the deficiencies of Williamson's treatise from a modern point of view, it was apparently well received by his contemporaries, for it won him election to the Holland Society of Science and the Society of Arts and Sciences of Utrecht and an honorary degree from the University of Leyden.

Of all the speculative theories to explain the origin of human races proposed in the Jeffersonian period, the one that came closest to anticipating modern views was proposed by William Wells, a native of Charleston who sided with the Tories during the American Revolution and subsequently achieved a successful medical career in London. In 1813 Wells's attention was drawn to the question of racial differences by the case of a white woman with patches of black skin on her body. From his investigation of this case he became convinced that difference of skin color was no sure indication of a difference of species and that the dark color of the Negro was not caused by long exposure to the tropical sun. White-skinned individuals might become deeply tanned, but they never transmitted this characteristic to their offspring.

At the same time, Wells was impressed with the susceptibility of both Negroes and whites to disease when they were transplanted to unaccustomed climates. It then occurred to him that the prevalence of dark-skinned people in Africa might be explained by assuming an unknown cause correlating darkness of skin color with resistance to the diseases of the African continent. If such were the case, tribes with a hereditary tendency to dark skin would eventually drive out those tending toward a fair complexion.

> Of the accidental varieties of men, which would occur among the first few scattered inhabitants of the middle region of Africa, some one would be better fitted than others to bear the diseases of the country. This race would consequently multiply, while the others would de-

crease, not only from their inability to sustain the attacks of disease, but from their incapacity of contending with their more vigorous neighbors. The colour of this race I take for granted, from what has been already said, would be dark. But the same disposition to form varieties still existing, a darker and a darker race would in the course of time occur, and as the darkest would be the best fitted for the climate, this would at length become the most prevalent, if not the only race, in the particular country in which it had originated.[26]

Wells offered no explanation of the "disposition to form varieties," but he noted that variants of one sort or another occurred constantly throughout the animal kingdom. In freely interbreeding populations these variants would be lost by intermixture, he said, but in regions isolated by geographic or other barriers they might become established and persist over many generations. Among domestic animals, breeds were established by artificial selection; but might not nature practice its own selection on the human species, "chiefly during its infancy, when a few wandering savages, from ignorance and improvidence, must have found it difficult to subsist throughout the various seasons of the year, even in countries the most favourable to their health?"

These speculations, presented to the Royal Society of London in 1813 and published five years later as an addendum to Wells's *Two Essays* foreshadowed the theory of natural selection later developed by Charles Darwin and Alfred R. Wallace in their researches. But Wells himself did nothing further with his brilliant speculation, and there is no evidence that either Darwin or Wallace heard of it until after the publication of Darwin's *Origin of Species*. Doubtful whether the theory of natural selection could explain the origin of human races — racial characteristics seemed to have no clear survival value — Darwin elaborated a theory of sexual selection to account for racial differences. But modern biologists, profiting by the rise and development of genetics and correlative advances in human physiology, believe that natural selection may have had more to do with producing racial differences than Darwin suspected.[27]

Theoretical speculation was one thing; solid empirical investigation was quite another. The best example of the latter among the Americans who investigated the origin of human races was provided by John C. Warren, professor of anatomy and surgery at the Harvard medical school. Warren had studied anatomy, including some comparative anatomy, with the best teachers in Paris. On his return to Boston he became deeply interested in the claims made for phrenological science by Franz Gall, whose novel theories linking psychology with the anatomy of the brain had been the subject of a special report by a commission of the Institute of France, written by Georges Cuvier. About 1808 Warren secured a copy of this

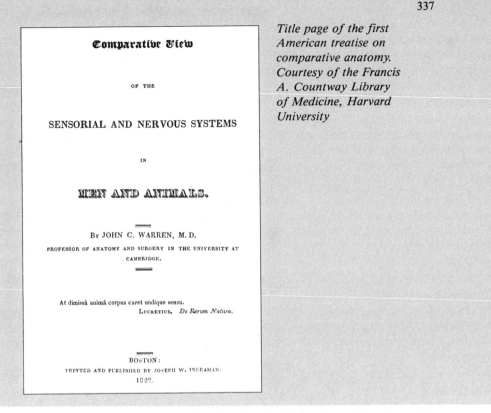

Comparative View

OF THE

SENSORIAL AND NERVOUS SYSTEMS

IN

MEN AND ANIMALS.

By JOHN C. WARREN, M. D.
PROFESSOR OF ANATOMY AND SURGERY IN THE UNIVERSITY AT
CAMBRIDGE.

At dimissâ animâ corpus caret undique sensu.
LUCRETIUS. *De Rerum Natura.*

BOSTON:
PRINTED AND PUBLISHED BY JOSEPH W. INGRAHAM.
1822.

Title page of the first American treatise on comparative anatomy. Courtesy of the Francis A. Countway Library of Medicine, Harvard University

report and found it so interesting that he decided to investigate the subject himself.

> I . . . went to work with Cuvier's report in my hand, and examined the structure of the brain as laid down by Gall, with the corrections of Cuvier. First I obtained a great number of human brains; then those of various animals, — as the sheep, hog, cat; various birds; among amphibia, those of turtles, frogs, and various fishes, also the brain and nervous system in the invertebral animals, — the lobster, sepia, cuttlefish, oyster, scorpion, and medusa.
>
> This pursuit I continued for many years with unabating interest; especially after the publication, by Gall and Spurzheim, of their 'Anatomy and Physiology of the Brain,' in 1810.[28]

In 1820 Warren presented the results of his researches to the Massachusetts Medical Society at their annual meeting. His discourse, published two years later under the title *Comparative View of the Sensorial and Nervous Systems in Men and Animals,* was the first American treatise on comparative anatomy.

After comparing the brain and nervous system in the various classes of animals, Warren took up the question of whether the anatomical differences between man and other animals were a sufficient explanation of the difference in their intellectual powers. He was especially concerned to refute Charles White's assertion of a correlation between anatomical structure and the degrees of intelligence in men and animals. Warren could find no evidence of any such correlation. Not only was the structure of the brain essentially the same in man, apes, and monkeys, but Cuvier's tables showed that there was no correlation between brain size and intelligence in the animal kingdom. Nor was there any such correlation among the various types of human beings. Warren had had occasion to examine the brains of a considerable number of persons of outstanding ability and social position during the course of his medical career and had found that some such persons had large brains and others small ones. One especially outstanding person turned out to have had an uncommonly small brain.[29]

Gall and Spurzheim had attempted to link various intellectual powers and psychological propensities with definite regions of the brain, Warren noted, but had failed to do so. Comparative anatomy, he concluded, lent no support to the supposed gradation of intellectual powers matching a similar gradation in anatomical form nor to the idea that intellectual powers were dependent solely on brain configuration.

On the question of the relative intellectual capacities of the various races of man Warren maintained an open, if somewhat ethnocentric, mind. The Chinese and Japanese, he noted, had ancient cultures and it seemed likely that, "if their prejudices were broken down by intercourse with foreign nations, and their minds enlarged by the introduction of Christianity," they might rival the Western nations in literary and scientific achievement. The North American Indians had been slow to adopt the white man's ways, but in places where they had been treated well and taught by Christian missionaries, as among the Cherokees and Choctaws, they had made rapid progress toward civilization. The question of their innate abilities, Warren declared, could not be said to have been fairly tried "until they have received the education of whites, have come forward to perform professional and political duties, and have themselves had an opportunity of answering the question, whether nature has made them inferiour to the Caucasian race." As for the blacks, their climate and mode of life in Africa had not been favorable to physical, intellectual, and moral development, but Warren was convinced from what he had seen and read of New England blacks that they tended to improve both physically and mentally in their new environment. Those who had been brought up from infancy with white children, he testified, were "as intelligent, as gay, as ready to imbibe the rudiments of learnings as the whites; and if their education had kept pace with that of the latter, they might generally have continued on the same level through life." Perhaps the black experiment in republican government then being tried in Haiti would settle once and for all the question of the black man's ability to take his place in the onward march of civilization.

Warren's *Comparative View* was especially notable for its detailed technical descriptions of the skulls of various races, illustrated by engravings. These plates were added to an appendix entitled "Account of the Crania of Some of the Aborigines of the United States." Plates 6 and 7 depicted Indian skulls sent to Boston from the mouth of the Columbia River by a Mr. Keith, who lived in a settlement there. One of the skulls had been artificially flattened, and Warren was pleased to note that the donor had stated that the practice of flattening the head "appears no wise to affect their intellectual or corporeal faculties." This statement was corroborated, said Warren, by Lewis and Clark's description of the Flathead Indians as "inquisitive and loquacious, with understandings by no means deficient in acuteness, and with very retentive memories."[30] It was further corroborated by Warren's own experience with a Flathead Indian who had been brought to Boston at the age of ten from Port Bucarely in one of the Queen Charlotte Islands by a Bostonian named Dorr, "who purchased him for a trifle to prevent his being put to death as a prisoner, taken in war with a neighbouring nation." The boy's head had been flattened in keeping with the custom of his tribe. "He lived here many years as a servant; was found to be in no respect wanting in understanding; his apprehension was quick, and his capacity for retaining what he learnt, quite equal to that of other boys of his age." From these examples it was evident, said Warren, that the brain could undergo substantial alterations in shape without any corresponding change in intellectual ability.

In his eighth plate Warren depicted the skull of a native of the Marquesas Islands in the Pacific. The skull had been purchased by the master of an American ship from one of the islanders, who wore it around his waist as a trophy of war. This head, said Warren, resembled that of a Negro in the shape of the face but seemed Caucasian in its cranial capacity. This was the last of the plates, but Warren also described several skulls of Indians found in the Boston area and another skull found near Marietta, Ohio, and procured for the collections of the Linnaean Society of New England, from whom Warren borrowed it. The skull was presumably that of one of the mound builders of the Ohio country, for it had been found in a cavity fifty feet above river level along with the rest of a perfectly preserved skeleton and a bowl similar to others found in the mounds at Marietta.[31]

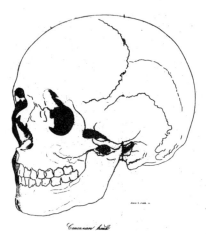

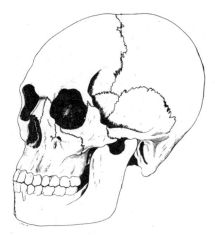

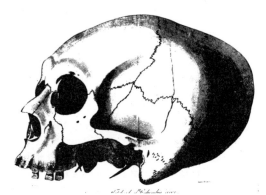

Four skulls from the collection of Dr. John Collins Warren, as shown in the "Account of the Crania of some of the Aborigines of the United States" appended to his Comparative View of the Sensorial and Nervous Systems in Men and Animals. *Courtesy of the National Museum of American History, Smithsonian Institution*

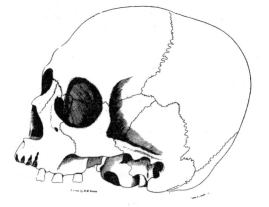

On examining this specimen, Warren was at a loss to classify it. It was unlike any he had seen heretofore. "The forehead," he explained, "is remarkable for being more nearly vertical, from the superciliary edge of the orbit to the frontal protuberances, than in any head of Caucasian origin, out of at least *forty* in my possession; yet as a contrast to this, the upper jaw has a decided projection like that of the aboriginal American, so that a line in the direction of the forehead would cut off the face." From these anatomical peculiarities and from the situation in which the skeleton had been found Warren concluded that the skull had belonged to an individual of an extinct race of mound builders.

> Whether to refer it to the Chichimecs, or Aztecs who established themselves in Mexico in the twelfth century; to the polished Toltecas, who, coming from the north, appeared in Mexico twelve hundred years ago; or to go farther back to the Hiongnous from Asia, or to people who, in the earlier ages of the world, inhabited this quarter of the globe, are curious and interesting questions, but not to be answered at present.[32]

With the publication of Warren's *Comparative View,* comparative anatomy and physical anthropology had attained a solid footing in the United States. From his "Account of the Crania of Some of the Aborigines of the United States," based on an extensive collection of skulls, it was but a step to Samuel G. Morton's *Crania Americana* (1839), which drew on a skull collection rivaling that of Blumenbach himself. The purely speculative stage of American physical anthropology was over.

Warren realized that the problem of Indian origins could not be solved by physical comparison alone. In support of his own ideas on this subject he cited the journals of Lewis and Clark, a description of the mounds and other ancient constructions of the Ohio valley published in 1820 by Caleb Atwater of Ohio in the *Transactions and Collections* of the American Antiquarian Society, and John Heckewelder's account of the North American Indians in the report of the Historical and Literary Committee of the American Philosophical Society (1818). Heckewelder, he noted, had lived with the Lenape, or Delaware, Indians and had recorded a tradition that their ancestors had migrated from the western regions of America, driving the inhabitants of the eastern parts, called Alligewi, southward and occupying their lands. In the light of this tradition it seemed probable that the original inhabitants of eastern America had migrated from the civilized parts of Asia at an early period in world history and that the ancestors of the Delawares had followed later from the more barbarous, northern parts of Asia, settling along the western coast of America and then moving eastward in search of better lands. "We may expect from these facts," Warren concluded, "to find a difference in the osseous fabrick of the ancient inhabitants of this country and the Indians still existing; but in what particulars they differ must be determined by future observations." Clearly, the task of tracing Indian origins would require the combined resources of physical anthropology, ethnology, archaeology, and comparative linguistics. To the first of these Warren had made a substantial contribution with a craniological essay that was, in the words of the twentieth-century physical anthropologist Aleš Hrdlička, "remarkable for the systematic, technical descriptions of the specimens."[33]

CONCLUSION

In both Europe and America physical anthropology was in its infancy, characterized more by speculative virtuosity than by empirical research. At the speculative level most American writers followed the lead of the Count de Buffon, Johann F. Blumenbach, and other Europeans who explained the origin of human races by the cumulative inherited effects of climate, diet, and other environmental influences. The most important American contribution to this school of thought was Samuel S. Smith's discussion of the influence of social and cultural factors on the human constitution. A few Americans, however, joined the ranks of those who questioned the heritability of traits acquired during the individual's lifetime and looked to congenital variations as the source of racial diversity. The most notable of these was the Anglo-American William Wells, whose speculation on the origin of the black race in Africa anticipated many of the features of Darwin's theory of natural selection. At the level of empirical research the best American work was done by John C. Warren, author of the first solid American contribution to craniometry and comparative anatomy generally. Throughout the Jeffersonian era, as indeed for many decades thereafter, investigation of the origins and relative capacities of the various human races was intricately interwoven with religious and political debates concerning the inspiration of the Bible and the morality or immorality of slavery. Science, religion, and politics could not easily be separated.

13

The Sciences of Man: Archaeology

SPEAKING before the American Academy of Arts and Sciences on the occasion of his inauguration as its president in November 1780, James Bowdoin called attention to the academy's stated purpose "to promote and encourage the knowledge of the antiquities of America." "A knowledge in the antiquities of a country," he told his auditors, "implies a knowledge of its antient history, and the researches into them lead directly to the source and original of things." In the case of America, he added, there were both European and aboriginal antiquities to be studied, but the latter would probably prove to be very few in number. The Indians were uncivilized, unprogressive, and without a written language, hence it seemed likely that their history would show little novelty or development and that consequently "it would be in vain to search among them for antiquities."[1]

THE MYSTERY OF THE MOUND BUILDERS

Events were soon to disprove Bowdoin's estimate of the prospects for archaeological research in America. In the same year that he voiced his opinion in the matter the Abbé Francisco Clavigero, for thirty-six years a Jesuit missionary in Mexico, published his *Storia antica del Messico,* in which he described the historical traditions and paintings of the natives of Mexico and their impressive temples, or *teocalli,* resembling in some respects the monuments of the Near East. Four years later John Filson, in his *Discovery, Settlement, and Present State of Kentucke,* described "several ancient remains . . . which seem to prove, that this country was formerly inhabited by a nation farther advanced in the arts of life than the Indians." In the neighborhood of Lexington, Filson reported, there were the remains of two ancient fortifications, one occupying about six acres of land and the other, three. The trees growing in their midst were at least 160 years old, and there were fragments of pottery different from any produced by present-day Indians. There were also burying grounds similar to those described by Scottish antiquaries and exhibiting a mode of burial very different from any practiced by the Indians. The bodies were laid on large, broad stones and separated from each other by other stones; there were several such layers of bodies rising in a narrowing pile to the height of a man. Filson was greatly mystified by these "curious sepulchres, full of human skeletons."[2]

With the organization of the Ohio Company by Manasseh Cutler, Rufus Putnam, Samuel H. Parsons, James Varnum, Winthrop Sargent, and other New Englanders in 1787 and the founding of the company's settlement at Marietta, Ohio, in the following year, still further news of western antiquities reached the eastern seaboard, for Marietta was located on an ancient mound builder site overlooking the Muskingum River near its junction with the Ohio. A full description of these earthworks by Jonathan Heart, with an accompanying map, appeared in the *Columbian Magazine* for May 1787. Soon afterward the nature and origin of these works was made the subject of an extended debate between Noah Webster and Ezra Stiles, president of Yale, in Webster's *American Magazine.*

Webster opened the discussion in December 1787, adopting the hypothesis, which he attributed to Benjamin Franklin, that the mounds at Marietta and Lexington were built by Ferdinand De Soto and his men. In Franklin's library, Webster declared, there was an account of De Soto's expedition that seemed to show he had penetrated as far north as the Muskingum River. In a later issue of the magazine Webster concluded that only the military earthworks had been constructed by De Soto; the ceremonial and burial mounds were probably the work of the present Indian tribes. Stiles replied that the monuments of the western country were too numerous, too large, and too far north to be the work of Spaniards. The Spaniards had constructed no massive forts in Mexico or Peru and would certainly have needed none in their conflicts with the Indians of the Ohio and Mississippi valleys. Stiles gave short shrift to those "flighty geniuses" who imagined that the world had existed for three or four hundred thousand ages and that the inhabitants of America had passed through all the stages of society and subsequently fallen into decline. The similarity between the religious practices of the Indians and those of the peoples of Asia shortly after Noah's flood was proof, he said, that the Indians came to America shortly after that great catastrophe. Perhaps the study of Indian and Asiatic languages would place these conjectures on a firmer footing before long.[3]

To Thomas Jefferson, who was taking Franklin's place as minister to France at this time, Stiles sent a drawing of the Muskingum earthworks done by Parsons, adding: "This, with Bricks, and even peices of Earthen

Ware dug up in the Kentucky Country, shew that there have been European or Asiatic Inhabitants there in antient ages, altho' long extirpated." Jefferson, who was convinced that the Muskingum earthworks were constructed by Indians, was skeptical about Parsons's reference to brick work. "The art of making it [brick] may have preceded the use of iron," he wrote to Stiles. "But it would suppose a greater degree of industry than men in the hunter state usually possess. I should like to know whether General Parsons himself saw actual bricks among the remains of fortification."[4]

As usual, Jefferson had little use for speculation uninformed by careful research. In thanking his Philadelphia friend Charles Thomson for the extract of a letter on western antiquities written by John Symmes of Ohio, Jefferson urged that exact descriptions of the western mounds be made without proposing any theory concerning them. It was too early to begin forming theories, he declared. "We must wait with patience till more facts are collected." In particular he hoped that the American Philosophical Society would make it their business to procure detailed descriptions of the mounds and publish them "naked" in their *Transactions*. "Patience and observation may enable us in time to solve the problem whether those who formed the scattering monuments in our Western country, were colonies sent off from Mexico, or the founders of Mexico itself," Jefferson concluded.[5]

True to his own precepts, Jefferson included in his *Notes on the State of Virginia* a detailed account of his excavation of an Indian mound located on the low ground near the Rivanna River not far from Monticello. The mound, he wrote, was spheroidal in form. Once about twelve feet high, it had been reduced by cultivation to seven and a half feet. "Before this it was covered with trees of twelve inches diameter, and round the base was an excavation of five feet depth and width, from whence the earth had been taken of which the hillock was formed."[6] Digging at random, Jefferson came upon several lots of human bones scattered haphazardly, as if dumped from a basket. He then made a perpendicular cut through the entire barrow and found the bones lying in strata separated by layers of earth and stones. Examining the bones carefully, he concluded that they were not those of warriors but rather the remains of successive village burials accumulated over the course of many years. He noted the existence of similar mounds elsewhere in the Appalachian country but offered no hypothesis concerning the people who had built them.[7]

THE CONTRIBUTIONS OF BARTON AND BARTRAM

The decade of the 1780s was a banner one for American publications on Indian mounds and related topics. Besides the publications already mentioned, Clavigero's *History of Mexico* appeared in English translation in London, and Benjamin S. Barton, then a medical student at Edinburgh, published his *Observations on Some Parts of Natural History*. This work, "written as a recreation from the laborious studies of medicine, in a bad state of health," began with a summary of previous accounts of mounds, barrows, and fortifications in the region west of the Appalachians. Barton then described some of the Muskingum earthworks from information sent to him by William Tilton of Philadelphia, derived apparently from the same sources Franklin and others had seen. Barton's description was accompanied by a map "accurately and elegantly engraved from the original plan, which was done from actual survey." These works, he declared, could never have been built by Indians. The Indians had no traditions concerning them; they never buried their dead in mounds, nor did they build earthen fortifications. The discovery of pottery in some of these works proved beyond doubt, said Barton, that they were erected by a people much more advanced in civilization than the Indians.

Who, then, were the mound builders? The answer, Barton thought, was to be found by comparing their constructions with those found elsewhere. Thus the mounds of the Mississippi valley were remarkably similar to the pyramidal constructions described by Clavigero. These in turn resembled the tumuli opened in Ireland by British antiquaries and attributed by them to the Danes. Was it not possible, then, that the same race of people had erected all these structures? Might not necessity, love of fame and fortune, or some nautical accident have brought the venturesome Scandinavians to America? Who could doubt that America was peopled from some part of the Old World? "Even the warmest favourer of the doctrine of separate creations," wrote Barton, "cannot but view the posterity of the *Greenlanders* in the wretched inhabitants of LABRADOR; he cannot but confess the amazing similitude of the *Iroquois* to some of the nations inhabiting the northeast parts of Asia."[8] Why not suppose, then, that the Danish mound builders of Ireland had pressed westward to the Ohio River and thence onward to Mexico? Barton submitted this conjecture to the learned world with great diffidence, begging its indulgence for "the first effort of a very young man."

Barton's interest in western antiquities continued after his return to the United States in 1789 and his appointment as professor of botany and natural history at the University of Pennsylvania. From Winthrop Sargent, who had been appointed secretary of the Northwest Territory when Congress established a territorial government in 1787, he secured a description and drawings of some ornaments and utensils found in an ancient grave near Cincinnati. In 1799 he published this communication in the *Transactions* of the American Philosophical Society with some comments

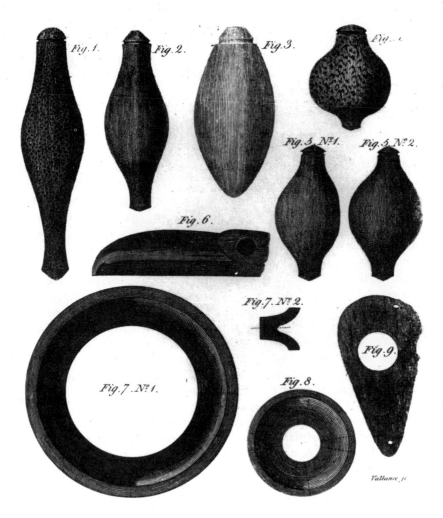

Artifacts from an "ancient grave" near Cincinnati. Courtesy of the American Philosophical Society

of his own. According to Sargent these artifacts, made variously of stone, copper, or mica, had been found buried with a human body "lying in nearly a horizontal position, about five feet from the surface of the earth, with the head towards the setting sun, and at the South West side of, or about fifteen feet from, an extensive artificial mound of earth, raised probably for the purpose of a burial grounds upon the margin of the second bank of the Ohio River."[9]

Barton's comments on this discovery took the form of a letter to Joseph Priestley, who was then living in Northumberland, Pa. Dividing the objects described by Sargent into two groups, ornamental and superstitious, Barton compared them with Aztec artifacts described by Clavigero and other Spanish writers and concluded that the American artifacts greatly resembled the Mexican, indicating a historical connection between the two cultures. But which of the two was derived from the other? In his *Observations* Barton had pictured a race of mound builders, different from the Indians, migrating to America from northwestern Europe and spreading slowly southwestward to the Mississippi valley and onward to Mexico. Now, however, he suggested that the Mexicans, descended from some Asiatic people, were responsible for the earthworks of the Mississippi valley and beyond. All the Indian nations of eastern North America, he declared, believed that their ancestors had come from the west, the northwest, or the southwest. And since all these tribes preserved the memory of a more polished state of society, they were probably the degenerate descendants of more highly civilized peoples. "This opinion," he wrote to Priestley, "is rendered more probable, when it is remembered that one of the most polished nations of America, I mean the Mexicans, migrated from certain countries situated north of the Vermillion-Sea; and that in the progress of their migrations these tribes moved far towards the east. The Mexicans, a number of circumstances have induced me to believe, were the ancestors of the nations known by the name of Choktah, Chikkasah, &c."[10]

The unacknowledged source of Barton's belief that the southern Indian tribes in particular were descendants of the Mexicans may well have been William Bartram, who had described the pyramidal mounds, tetragonal terraces, wooden obelisks, and "chunky-yards" of the Creek and Cherokee country in his famous *Travels*. In 1789, two years before the *Travels* appeared, Barton sent Bartram an extensive questionnaire about the southern Indians, asking about their traditions; paintings and hieroglyphics; religious beliefs and practices; government, manners, and languages; physical characteristics; diseases and medicinal remedies; ideas about property; food and methods of cultivation; and earthworks and chunky-yards.

In his written responses to these inquiries Bartram included drawings of pyramidal mounds, chunky-yards, and other constructions, showing the plan of the terraces, public squares, mounds, obelisks, and the like. The chunky-yard of the Creek Indians, he explained, was a cubiform area, usually in the center of the town, with the rotunda and public square at one corner and the winter council house diagonally across in another. The chunky-yard was sunk two or three feet below the surrounding banks or terraces built from the earth excavated to form the yard. On a low circular mound in the center of the yard stood the chunky-pole, an obelisk thirty or forty feet high formed from four square wooden pillars inclining to a point and capped by a flag or other object serving as a target for marksmen armed with rifles or bows and arrows. At one end of the yard were two smaller pillars, about twelve feet high, to which captives condemned to be burned were bound. These "slave-posts" were usually decorated with the scalps of slain enemies, and each was crowned with a dry white skull.

> I have counted six or eight scalps fluttering on one pole in these yards. Thus it appears evidently enough that this area is designed for a public place of exhibition of shows and games [the spectators sitting on the terraces surrounding the yard], and formerly some of the scenes were of the most tragical and barbarous nature, as torturing the miserable captives with fire in various ways, and causing or forcing them to run the gauntlet naked, chunked and beat almost to death with burning chunks and fire-brands, and at last burnt to ashes.[11]

Captives were no longer tortured in this manner by these tribes, Bartram reported, but the council houses, rotundas, and other public buildings were still erected on the ancient earthworks in the older Creek towns, and the chunky-yards were swept clean every day and kept in good repair.

Bartram was convinced that the chunky-yards were very old—"not the formation of the present Indians." He was skeptical of Barton's suggestion that the southern tribes might be descended from Mexicans. The traditions of all these tribes, he said, indicated that they had migrated from lands west of the Mississippi River, probably after the Spanish conquest of Mexico. The Cherokees were the oldest existing tribes in the southeastern United States, and their empire had once extended over much of the region in which the mounds, terraces, and chunky-yards were found. Yet the Cherokees admitted that they had not built these structures and denied that they had been built by the people who inhabited the country before them. These constructions, Bartram concluded, were a difficult puzzle. "But as Time changes the face of things, I wish they could be searched out and faithfully recorded, before the devastations of artificial refinements, ambition, and avarice, totally deface these simple and most ancient remains of the American aborigines." His own account of them, he testified, was "to the best of my remembrance, as near the truth as I could express."[12]

Unfortunately Bartram's responses, like much other information Barton collected, remained unpublished during Barton's lifetime. It was not until 1853 that they were published in the *Transactions* of the American Ethnological Society, and not until 1927 did John R. Swanton use Bartram's descriptions to argue what Bartram had denied—the historical continuity between the mound builders and the Indian tribes Bartram had visited. "We may conclude," Swanton wrote in the *Annual Report* of the Smithsonian Institution for 1927, "that the historic ceremonies and ceremonial mounds of our southeastern Indians, or, for that matter, of the Creeks alone, suggest physical and technical forces sufficient to account for all of the mounds of the Mississippi Valley and the districts north of the Gulf of Mexico."[13]

BRACKENRIDGE, HUMBOLDT, AND CLINTON

To return to 1799 and the American Philosophical Society, in that year the society prefaced its *Transactions* with a circular letter signed by Jefferson, inviting communications on a variety of topics, including "ancient Fortifications, Tumuli, and other Indian works of art." On behalf of the committee that issued the letter, Jefferson urged that tumuli and other earthworks be investigated systematically by cutting trenches through them, that their age be estimated by counting the rings of trees growing on them, and that masonry found in connection with them be carefully measured and specimens of the stones and cement be sent to the committee. There was no immediate response to this invitation, but eventually Bishop James Madison, president of William and Mary College, sent the society a description of a large earthwork on the Kanawha River in Western Virginia, Charles W. Short described an ancient fortification near Lexington, Ky., and in 1813 Henry M. Brackenridge sent Jefferson an extensive account of the aboriginal monuments he had encountered in his travels up and down the Mississippi valley.

Brackenridge's communication was the most interesting and useful. Born and reared in Pittsburgh, Brackenridge had acquired an early interest in Indian mounds from reading Jefferson's *Notes* and from visiting a large mound near Pittsburgh. Later in life he practiced law in St. Louis and New Orleans, devoting his spare time to researches on the geography, ethnology, and archaeology of the Mississippi valley and publishing his findings in the *Missouri Gazette* and other newspapers. He was especially interested in mounds and other earthworks. "Since the year 1810," he wrote to Jefferson, "I have visited almost everything of this kind, worthy of note on the Ohio and Mississippi; and from examination and reflection, something like hypothesis, has taken the place of the vague wanderings of fancy. The following is a sketch of the result of those observations."[14] It was 1818 before Brackenridge's letter to Jefferson appeared in the *Transactions* of the American Philosophical Society. In 1814, however, he included substantially the same material in his *Views of Louisiana.*

In both his memoir and book Brackenridge distinguished between two kinds of western antiquities, which he attributed to two different races of people. To the first class he assigned the barrows, small mounds, and remnants of "pallisadoed" towns and villages found throughout the western country by the thousands. These he attributed to the Indians of the post-Columbian and late pre-Columbian periods, having himself seen palisaded villages among the Mandan Indians on his trip up the Missouri River with fur traders. In the second class he placed the more substantial mounds and earthworks, numbering perhaps 3,000 or more, "the smallest not less than twenty feet in height, and one hundred in diameter at the base," resembling in many respects the *teocalli* of Mexico. These included the Grave Creek Mound below Wheeling, the mound near Pittsburgh, and the impressive earthworks at Marietta, Cincinnati, and New Madrid; at St. Genevieve and fifteen miles below it; at St. Louis and the mouth of the Missouri River; along the Cahokia River in Illinois Territory; near Washington in the Missouri Territory; at Baton Rouge on the Bayou Manchac; and on the Black River in Louisiana.

Of all these constructions the most impressive were the earthworks on the banks of the Cahokia across from St. Louis. Brackenridge visited these in 1811 and described them in some detail in the *Missouri Gazette* and his *Views of Louisiana*. On an open plain about a half a mile from the Cahokia River he found a group of mounds, mostly circular in shape, "resembling enormous haystacks scattered through a meadow." One of the largest was 200 paces in circumference, with a level space at the top sufficient to contain several hundred persons. In a semicircle around this mound he counted forty-five others of considerable size and many smaller ones. Walking for several miles along the river, he came to an even larger assemblage of mounds. The largest, standing immediately on the bank of the Cahokia and covered with lofty trees, had a circumference of 800 yards and a height of ninety feet. It formed a parallelogram, and on the south side was a broad apron or step about halfway down, from which jutted another projection into the plain about fifteen feet wide. To the west were forty other mounds scattered about. Viewing the largest of these, Brackenridge felt an astonishment "not unlike that which is experienced in contemplating the Egyptian pyramids. . . . What a stupendous pile of earth! To heap up such a mass must have required years, and the labors of thousands."[15]

From the location of this and similar groups of mounds on the banks of major rivers in the midst of fertile bottomlands, and from the size and number of these constructions, Brackenridge inferred that these river valleys must once have sustained a large population numbering hundreds of thousands. The center of population was undoubtedly between the Ohio,

Mississippi, Missouri, and Illinois rivers, and the teeming towns there must have resembled those encountered by the Spanish conquistadors in the valley of Mexico. Like the Mexican *teocalli,* the mounds of the Mississippi valley were oriented to the cardinal points of the compass. The larger mounds had several stages, and in every group there were two mounds much larger than the others, the smaller being placed around these symmetrically. These resemblances, said Brackenridge, suggested the existence of similar arts and customs and the likelihood of intercourse between the Mexicans and the inhabitants of the Mississippi valley. "The distance from the large mound on Red River, to the nearest in New Spain," he wrote, "is not so great but that they might be considered as existing in the same country."[16] It was even possible that the mounds of the Mississippi region were more ancient than the Mexican *teocalli,* which Mexican traditions attributed to the Toltecs or to the Olmecs, who probably migrated from the Mississippi valley to Mexico. In any case, Brackenridge concluded, these ancient monuments, like the mountains, volcanoes, and waterfalls of America and the diversity of its aboriginal languages, proved that the New World was as ancient as the Old and, like it, had undergone innumerable revolutions in which countless races of men had flourished and disappeared.

If Brackenridge's conclusions seem highly speculative to the modern reader, they were no more so than those of most writers of his day, including the great German traveler and scientist Alexander von Humboldt, whose *Political Essay on the Kingdom of New Spain* was quoted at length in Brackenridge's appendix and whose *Researches Concerning the Institutions and Monuments of America* appeared in English translation in the same year that Brackenridge's *Views of Louisiana* was published. In his *Political Essay,* Humboldt discussed the origin of the Indians largely in terms of their physical appearance. Conceding their general resemblance to the Mongols and Malays, he drew attention to several striking peculiarities in the cranial anatomy of the Americans — the facial line, the lower jaw, and the protuberances corresponding to the cerebellum — and conjectured that the Indians might be the product of a mixture of Asiatics with aboriginal American races now extinct. The solution of the problem of origins, he suggested, would require new researches in comparative anatomy, linguistics, and plant geography.[17]

In his *Researches* Humboldt attacked the problem of the peopling of America with evidence drawn from the architecture, sculpture, historical paintings, hieroglyphics, and languages of the early Central and South American civilizations. While recognizing east Asian influences in these civilizations, Humboldt believed that they had developed independently for a series of ages before becoming subject to Asian influence. He attributed the slow progress of the American aborigines in intellectual and

*Alexander von Humboldt (by Rembrandt Peale), scientific explorer of South and Central America, painted during Humboldt's visit to Philadelphia in 1804 on his way back to Europe.
Courtesy of the Independence National Historical Park Collection*

artistic development to this long period of isolation in a wild and thinly populated continent.[18] Thus Humboldt, like Brackenridge, gave free reign to speculative imagination in interpreting the results of his researches.

About the same time that Brackenridge informed the American public about the antiquities of the Mississippi valley, De Witt Clinton called attention to those of western New York. In December 1811, in an address before the New-York Historical Society, he observed that the Iroquois country was studded with ancient fortifications that the Iroquois themselves could never have constructed. The antiquity of these earthworks, he argued, was attested not only by the great trees growing on them but also by their geographical location. All were south of a great ridge that ran parallel to Lake Ontario at a distance of six to ten miles and extended from the Genesee River to Lewiston on the Niagara River. From geological evidence it was clear that this ridge was once the shore of the lake, perhaps one or two thousand years ago. "Considering the distance to be, say seventy miles in length, and eight in breadth, and that the border of the lake is the very place that would be selected for habitation, and consequently for works of defense, on account of the facilities it would afford for subsistence, for safety, for all domestic accommodations and military purposes; and that on the south shore of Lake Erie these ancient fortresses exist in great number," said Clinton, "there can be no doubt that these works were erected, when this ridge was the southern boundary of Lake Ontario, and, consequently, that their origin must be sought in a very remote age."[19]

Who were these early Americans? Perhaps, suggested Clinton, they were the first of several waves of people driven into North America by wars and migrations in the heart of Asia. The first immigrants may have had time to settle down and develop the arts of peace and war before they were overrun by new hordes. History was full of cycles of this kind.

THE AMERICAN ANTIQUARIAN SOCIETY AND
Archaeologia Americana

New England had nothing to match the antiquities of the western country, whether in New York or in the Ohio and Mississippi valleys, but its learned men were in close touch with the settlers in southeastern Ohio, many of whom were from Massachusetts and Connecticut. In March 1789 Manasseh Cutler, recently returned from the settlement at Marietta, informed his ministerial colleague Jeremy Belknap that he had become convinced from personal inspection of the works there that they were built long before the discovery of America, the mounds being overgrown by large trees. It was ironical, he added, that the military men in Ohio had given up their former idea that the earthworks were intended as fortifications in favor of Barton's hypothesis that they had a religious use, whereas Cutler, although a clergyman, insisted that they had been built for defense. He promised a full account of the works for the *Memoirs* of the American Academy of Arts and Sciences or the *Transactions* of the American Philosophical Society as soon as Rufus Putnam forwarded the measurements he had agreed to make. As to who the mound builders were, Cutler was in doubt and in search of further light. Perhaps Belknap could suggest some accounts of the mounds in Denmark, Norway, and eastern Asia for him to read.[20]

Putnam did make a map of the earthworks at Marietta, a map described by the twentieth-century archaeologist Henry C. Shetrone as "the genesis of the science of archaeology in the United States," but Cutler never produced the memoir he had promised. He did, however, give some account of the mounds at Marietta in a note appended to his sermon at the ordination of Daniel Story on August 15, 1798, in Marietta. The largest square, he noted, contained forty acres. There were three openings, or gateways, on each side. At the angles of the squares were openings similar to those at the sides. The walls of the squares, made of earth brought in from a distance, were 4 to 8 feet in height and from 25 to 26 feet in breadth at the base. Two still larger walls, measuring 24 feet in height and 42 feet

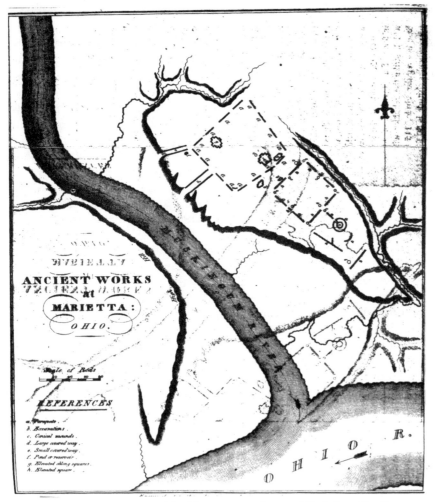

Map of the mound-builder constructions at Marietta, site of the first settlement in the Northwest Territory (1787). From Caleb Atwater's memoir in the Transactions and Collections of the American Philosophical Society, *vol. 1 (1820) by courtesy of the society*

at the base, ran parallel to each other from a corner of the largest square down toward the Muskingum River. Within the various squares were many conical mounds of different sizes and three huge mounds in the form of a parallelogram. The largest of these was 1,888 feet in length, 1,000 feet wide, and 9 feet high. Near the smallest square was a conical mound 115 feet in diameter and 30 feet high, surrounded by a ditch 15 feet wide and 4 feet deep, with a gateway leading toward the square.[21]

The note went on to describe the felling of several of the largest trees growing on top of the mounds and the counting of the tree rings in the presence of Governor St. Clair and other dignitaries of the Northwest Territory. The largest trees were hollow. Of the sound trees, the largest had from 300 to 400 rings, and there was plenty of evidence of a previous growth of trees fully as large as the present ones. It was apparent, said Cutler, that the earthworks had been deserted at least 900 years. "If they were occupied one hundred years, they were erected more than a thousand years ago."[22] Upon opening the largest of the conical mounds, the excavators found the bones of an adult stretched out in a horizontal position and covered with a flat stone. But the digging was not continued. Respect for the dead and a desire to preserve the monument intact led to an order to fill in the excavation.

Cutler was not present at the opening of this mound, but he was convinced from Winthrop Sargent's account that the Marietta mounds were built for religious purposes, probably by the same people who built the pyramids of Mexico. "If the Mexican tribes, agreeably to their historic paintings and traditions, came from the northward, and some of them in their migrations went far to the eastward," Cutler wrote, "it is not improbable that either some of those tribes, or others similar to them in their customs and manners, and who practiced the same religious rites, were the constructors of those works. The present natives bear a general resemblance in their complexion, form and size to the ancient Mexicans. Though their rites and ceremonies differ, they profess the general principles of the Mexican religion."[23]

In 1812 a group of New Englanders formed a society for the exclusive purpose of investigating and collecting American antiquities. In that year the highly successful printer and publisher Isaiah Thomas organized the American Antiquarian Society, dedicated to "the collection and preservation of the Antiquities of our country, and of curious and valuable productions in Art and Nature." In 1815 the society moved its headquarters from Boston to Worcester, Mass., and construction was soon begun on a building to house its collections. Members were elected from all the states of the Union and exhorted to send the society the antiquities—natural, artificial, and literary—of their regions.

On May 30, 1818, the recording secretary of the society, Rejoice Newton, received a letter from Caleb Atwater, a lawyer residing in Circleville, Ohio, announcing that he had been collecting information about the "aboriginal antiquities" of the western country for some years and would be glad to assist the society in its endeavors. Atwater was a a native of Massachusetts, born at North Adams on Christmas Day 1778.

Isaiah Thomas (by Ethan Allen Greenwood), founder and president of the American Antiquarian Society (upper left), and Caleb Atwater (by Philip Skardon), pioneer archaeologist and contributor to the society's Archaeologia Americana *(upper right), with a view of the society's first building. Courtesy of the American Antiquarian Society (Thomas) and the Ohio Historical Society (Atwater)*

After graduating from Williams College in 1804, he moved to New York state, where he practiced law from 1809 to 1814, when financial reverses led him to migrate to Circleville. There his attention was immediately attracted by the aboriginal earthworks from which the town had taken its name and by similar constructions he encountered as he traveled about southern Ohio on legal business. He communicated his interest in these matters to De Witt Clinton, Samuel L. Mitchill, and other New York literati and was encouraged by them to prepare an extended account of the Ohio earthworks. In 1817 President James Monroe requested him to undertake a description of the so-called forts in Circleville and appointed him postmaster of that town, a position that enabled him to carry on an extensive scientific correspondence free of charge.[24]

Atwater seems to have included a brief essay on western antiquities in his initial communication to the American Antiquarian Society, for his letter acknowledging his election to membership in the society expressed satisfaction that his memoir on the aboriginal antiquities of the west had been well received in Worcester. He indicated that Governor Lewis Cass had proposed to him the formation of a society dedicated to collecting information about the mounds of the western country before they were despoiled by brickmakers and curiosity seekers. He had, he said, spent $200 during the past year to procure reliable information about the mounds and would be glad to prepare a full account of them if the American Antiquarian Society would give him financial support. "I have a number of very interesting articles of antiquity, which I wish you had in your museum," he added by way of inducement.[25]

Apparently Isaiah Thomas's curiosity was whetted by Atwater's communications, for he now wrote requesting him to obtain surveys and detailed accounts of the Ohio mounds. In February 1819 Atwater replied that he would gladly work at this for a year or two if the society would advance him $150 or $200 to pay for surveys already made and would agree to bear all the expenses of printing and binding the publication and engraving the plates. The society would receive half the profits from its sale and Atwater the other half. He would get 3,000 copies of each engraving and 3,000 copies of an engraved map of Ohio to be included in the memoir. The rest of his share would be paid in cash. "By the 1st of Septr next 100 pages octavo with plates representing at least 30 different works, such as forts, temples, altars, cemeteries, places of amusement, towns, villages, habitations of chieftains, wells &c. &c. shall be sent by mail, free of postage (being myself a postmaster) either to Worcester or Boston as you may direct, unless sickness or death prevent it," Atwater promised.[26] Besides financial assistance Atwater asked only that Thomas assist him in obtaining needed reference works.

As it is I have consulted Clark's travels in Tartary, &c., the Bible, Lewis & Clark, Dr. Robertson, Abbe Clavigero, Humboldt, &c. &c., the writings of father Henipen, Creuxieux, Jontel &c French travellers who accompanied La Salle in his expeditions from Canada to the Mississippi. By the aid of these *great lamps,* assisted by my own dim taper, I have ventured to enter the labyrinth, where some . . . unfortunate travellers, have heretofore lost their . . . [*torn*] and bewildered those, who have undertaken to follow such blind guides. If my propositions are acceded to, you may *instantly announce* the publication of my work.[27]

Without waiting for a reply, Atwater plunged enthusiastically into the task he had set for himself. On March 25 he reported to Thomas that he had made arrangements for surveying the works not yet surveyed and that he would try to borrow money at a bank to enable him to travel at least as far as the lower part of Kentucky to inspect mounds and earthworks. "I have now about 30 drawings ready to be forwarded by mail," he reported. "Those of Marietta are surveyed but I have not recd the sketches of them."[28]

Atwater was at great pains to obtain accurate surveys of the mounds and other earthworks he described, employing county surveyors whenever possible. He wrote several years later,

When the works were surveyed, it was done, almost always, in the presence, and with the aid of the most intelligent and respectable men in the several counties where the works are situated. And I was either present myself and assisted in the surveys, or carefully examined everything described — sometimes before the survey, sometimes afterwards; but I always assured myself of the entire correctness of every thing stated as a fact.[29]

In pursuing his researches, Atwater sought and received assistance from many people. For the Cincinnati area he drew on Daniel Drake's excellent chapter on antiquities in his *Natural and Statistical View.* Drake, a Cincinnati physician, had examined the mounds and other earthworks in that vicinity with great care, describing their appearance, location, dimensions, manner of construction, and contents (so far as they were known) in some detail. He compared his observations with those of Anthony Wayne and William H. Harrison, who had reconnoitered the earthworks in and around Cincinnati in the summer of 1793. (General Wayne ordered eight feet cut off the largest of the Cincinnati mounds to make a sentry post for the watch against Indians.) Like Wayne and Harrison, Drake was convinced that the embankments following the contours of high ground above river valleys were defensive fortifications. He was in doubt concerning the embankments situated on lower ground. The conical mounds, he thought, were burying places of "various grades and kinds of distinction." As to the mound builders themselves, Drake agreed with Benjamin S. Barton, Hugh Williamson, and Henry M. Brackenridge that they were probably the ancestors of the Indians currently living in the northeastern United States. He had collected some human bones and a single skull from some of the tumuli and compared them with those of modern Indians, using Johann F. Blumenbach's method of comparison. There was, he reported, no great difference between the ancient and the modern crania. Some of the earthenware vessels found in the mounds, he noted, "have in their composition a perfect identity with that fabricated since the discovery of America, even up to the present time, by many of the tribes low on the Mississippi."[30]

Drake kept Atwater informed of the latest discoveries in the Cincinnati area, Samuel P. Hildreth performed a similar service in Marietta, and a Dr. Rhodes did likewise in Zanesville, Ohio. In Lexington, Ky., Atwater found a valuable collaborator in John D. Clifford, a merchant who had assembled "the best collection of aboriginal antiquities in the Ohio Valley" and who acted as an intermediary between Atwater and collectors in Tennessee and Mississippi. Clifford's sister assisted by making drawings of artifacts found in the Kentucky and Tennessee mounds. Constantine Rafinesque also corresponded with Atwater and sent him descriptions and drawings, but he and Atwater soon became bitter rivals in western archaeology.[31] Even the postmaster-general of the United States, R. J. Meigs, Jr., was pressed into service in Atwater's project.

Atwater was careful to check the accuracy of his informants in every way he could. "I have traversed country of greater extent than all New England at no little expense both of time & money, to collect this information," he wrote to Thomas in September 1819. He was ready and eager to extend his researches all the way to the Gulf of Mexico, he told Thomas, if only he could be supported financially in his efforts.

> With my slender means, poor, embarrassed and afflicted with several years sickness, I have done more than was ever done by any other person in North America and have never recd one cent. . . . If the sale of the book furnishes me with funds, my labors will be redoubled and I may be able to produce a work, which may be placed on the shelf with Pennant, Humboldt & Denon and the gap be thus filled up, which now exists in the antiquities of the world. From my humble labors in the West, the Mosaic account of the creation and dispersion of man — the doctrine of the Trinity and many other things, will receive additional support.[32]

While Atwater was struggling to find time and money to complete his researches, Thomas and the Committee on Publication of the American

Antiquarian Society were wrestling with the problem of piecing together a sustained and coherent memoir from Atwater's letters and pencil drawings. Thomas found a competent engraver in East Windsor, Conn., but since Atwater had sent his drawings to Worcester in advance of the letters describing them, the committee had difficulty in associating the engravings with the text. Fearing that some of his letters had failed to reach Worcester, Atwater sent a list of the forty-four letters he had dispatched as of September 24, 1819, with the dates they had been sent. "I generally write in great haste, when the ideas flow with ease, and I have not always the leisure to write my compositions over again," he wrote to Thomas, adding that he had no copies in case the letters went astray.

By November 1819 Atwater was beside himself with frustration at Thomas's repeated queries about the arrangement of the materials.

> By reading *all* I have written, your other questions you will find answered. . . . If my papers are shifted about from place to place and from hand to hand, they had better be returned to me, to page, correct and replace anew. . . . I am entirely discouraged. My mortification and regret are greater than I can express.[33]

To Thomas's warning that he must check his facts carefully and avoid flights of speculation, Atwater replied that he had exercised the greatest caution. The mounds at Marietta had been opened at his request, and great care had been taken to record the exact situation of the bones and articles found.

> The roots of the largest trees passed through & quite below the bones, which had been burnt in a hot fire, on a funeral pile. One of the 'bosses' was quite destroyed by rusting—the sword blade was oxidized completely—nor could any water have reached these articles which lay in a high, dry, sandy soil. These articles had lain there at least . . . 1000 years.[34]

As to theories, said Atwater, he had indulged none but such as were fully warranted by his discoveries. His main object had been to describe these antiquities before they were destroyed by the westward march of civilization, "to snatch from the destroying, ruthless hand of man, as many articles as my feeble arm could, and bear them to the place of deposit, which your enlightened liberality has erected, there to remain for the inspection of future generations."

The final pages of Atwater's manuscript, devoted to the question of what had become of the mound builders, would be written, he said, as soon as he could procure a copy of Humboldt's work on Mexican antiquities. Meanwhile, his financial situation was desperate. "Unless I can get funds from some source before many months, I shall be ruined, as to property. I have given up an extensive practice at the bar, in all the counties around me, except my own, so as to enable me to attend to the antiquities, and it will be no easy task to regain it," he wrote to Thomas. Nevertheless, he added, he would be willing to do much more for the society if he could obtain the necessary funds.

In June 1820 Atwater's perseverance and Thomas's patience and liberality finally bore fruit. In that year was published *Archaeologia Americana,* volume 1, devoted chiefly though not entirely to Atwater's *Description of the Antiquities Discovered in the State of Ohio.* In presenting his material, Atwater began by distinguishing three kinds of American antiquities: those left by early European explorers, those deriving from the tribes and nations of Indians with whom these explorers had come into contact, and those attributable to a race of mound builders who had arrived in the Ohio and Mississippi valleys after the ancestors of the present Indians had occupied the eastern coast of America but before the advent of the white man. Like most of his contemporaries, Atwater found it hard to believe that the people who built the mounds and other earthworks of the Ohio valley were of the same stock as the Indians whom the Europeans had encountered in the forests of North America. These Indians, he said, had never been known to construct earthworks like those in southern Ohio; work silver, iron, and copper; burn the bodies of their chiefs on piles and raise lofty tumuli over the cremated remains; erect walled towns; dig deep wells; manufacture handsome pottery; or worship idols reminiscent of those of ancient India. Moreover, the Indians were generally tall, slender, and straight limbed, whereas the skeletons found in the mounds were short and stocky. "The limbs of our fossils," Atwater wrote, "are short and very thick, and resemble the Germans, more than any European with whom I am acquainted." The mound builders were "rarely over five feet high, and few indeed were six. Their foreheads were low, cheek bones rather high; their faces were very short and broad; their eyes were very large; and they had broad chins. . . . I have examined more than fifty skulls found in the tumuli," Atwater added, supplying a drawing of one of these skulls.[35]

From the geographical distribution of the mounds Atwater drew a similar conclusion. None, or very few, were found east of the Alleghenies. Beginning with small and rather rude forts in western New York and northern Ohio, the constructions increased in number and size as one proceeded southward along the Ohio and Mississippi valleys to the Gulf of Mexico and thence southwestward into Texas and Mexico. And, whereas the skeletons and artifacts of present-day Indians were always found on or near the surface of the earth unless buried in a grave, the articles associated with the mound builders were often found many feet below the surface,

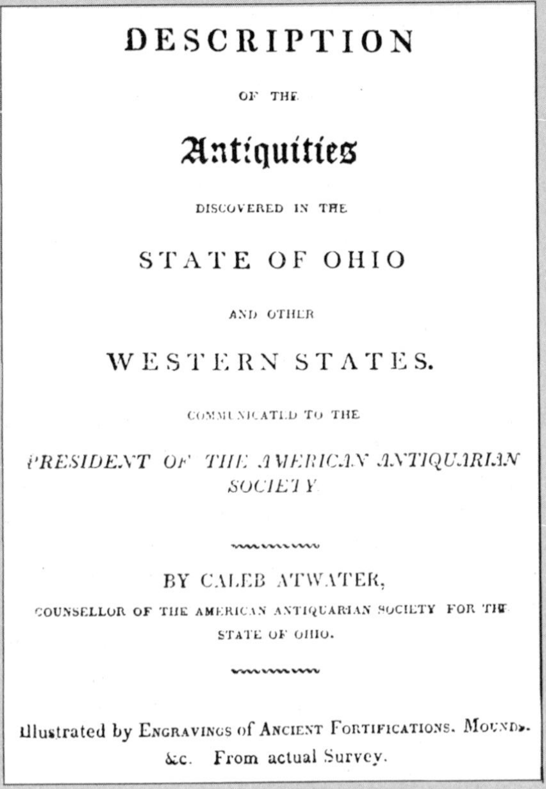

DESCRIPTION

OF THE

𝕬𝖓𝖙𝖎𝖖𝖚𝖎𝖙𝖎𝖊𝖘

DISCOVERED IN THE

STATE OF OHIO

AND OTHER

WESTERN STATES.

COMMUNICATED TO THE

*PRESIDENT OF THE AMERICAN ANTIQUARIAN
SOCIETY*

BY CALEB ATWATER,

COUNSELLOR OF THE AMERICAN ANTIQUARIAN SOCIETY FOR THE
STATE OF OHIO.

Illustrated by ENGRAVINGS of ANCIENT FORTIFICATIONS. MOUNDS.
&c. From actual Survey.

Title page of the first substantial archaeological monograph published in the United States. From Transactions and Collections of the American Antiquarian Society *vol. 1 (1820). Courtesy of the American Philosophical Society*

especially in river-bottom sites. The skeletons found in Indian burials were usually in a sitting or upright position. Those of the mound builders were invariably horizontal.[36]

After distinguishing the mound builders from the Indians by these cultural and osteological traits, Atwater proceeded to describe their various constructions in detail, dividing them into "ancient works" and "ancient tumuli." Proceeding southward along the tributaries of the Ohio River, he described, with the aid of engraved surveys, the constructions of earth and stone at Newark, Circleville, Marietta, Paint Creek, Portsmouth, Cincinnati, and other sites and speculated as to the uses for which they had been intended. The larger works at Newark, situated on a high fertile plain forty feet above the surrounding country, had probably been built to defend the population, whether at work in the fields or passing from one part of the town to another between parallel walls. The works at Circleville must also have been erected for purposes of defense. One of the fortifications was circular, the other square, and both had been laid out with a precision that astonished Atwater.

> As the square fort is a *perfect* square, so the gateways or openings
> are at equal distance from each other, and on a right line parallel with
> the wall. The walls of this work vary a few degrees from north and
> south, east and west, but not more than the needle varies. . . . Let
> those consider this circumstance, who affect to believe these antiquities
> were raised by the ancestors of the present race of Indians.[37]

Having described these constructions to the best of his ability, Atwater turned next to the conical tumuli of earth and stones found in or near them, describing their appearance and the contents of those that had been excavated. Atwater himself was present when a tumulus nearly ten feet high and several rods in diameter near the center of the round fort at Circleville was removed. In it he found two human skeletons, a large number of arrowheads, a sword or knife handle made of elk horn, the remains of what appeared to be a brick crematory, a large mirror made of isinglass, and the oxidized remains of an iron plate. Forty rods distant on a large hill was another mound more than ninety feet in height containing an immense number of human skeletons of various ages laid horizontally with the heads toward the center of the tumulus. From the large number of these tumuli and the skeletons found in them, as well as from the extent and number of the fortifications and other works of earth and stone, Atwater concluded that a huge population had once inhabited the region where these remains were found.

Atwater's descriptions of the artifacts found in the mounds and earthworks were supplemented with twenty-three engravings made from his pencil sketches. These included silver, copper, and stone ornaments; stone tools; pottery; and clay and stone idols. In a section on aboriginal arts Atwater described these objects in some detail, noting that they ranged from crude implements to highly wrought works of art. The pottery found at or near the surface of the earth was rather crude, made of sandstone and clay in the Lake Erie region, of clay along the northern reaches of the Scioto River, and of clay and shells in the mounds along the Ohio and Mississippi rivers. None of this ware was glazed. Near the bottom of the mounds and in connection with the burial of distinguished persons, however, much finer pottery was found.

> Two covers of vessels were found in a stone mound in Ross county,
> very ingeniously wrought by the artist, and highly polished. These
> were made of a calcareous breccia; fragments of which were examined
> by Professor Silliman, of Yale College, Connecticut. These covers
> resembled almost exactly . . . vessels of that material manufactured
> in Italy at the present time.[38]

Copper had been found in more than twenty mounds, Atwater said, "but not very well wrought. . . . Pipe bowls of copper, hammered out, and not welded together, but lapped over, have been found in many tumuli." No brass or cast iron had been discovered, nor any glass. Silver, well plated, had been found in several mounds. Gold ornaments were said to have been found, but Atwater had never seen any. There were a few traces of oxidized iron, apparently from sword or knife blades. At the bank of the Muskingum River near its mouth, the river had exposed fireplaces and remains of chimneys, formerly from two to six feet under the surface of the earth.

> Around them are deposited immense quantities of muscle shells, bones of animals, &c. From the depth of many of these remains of chimnies, below the present surface of the earth, on which, at the settlement of this country by its present inhabitants, grew as large trees as any in the surrounding forest, the conclusion is, that a long period, perhaps a thousand years, has elapsed since these hearths were deserted.[39]

In the nitrous caves of Kentucky, Atwater noted, mummies had been discovered enveloped in three kinds of covering, the innermost of coarse linen cloth, the next of bird feathers woven into a mesh of coarse threads, and the outer envelope like the inner one or sometimes made of leather. There were also mats made of a substance resembling hemp, "or possibly the bark of some kind of vegetable." No kind of writing had been found. Several reports of discoveries of this kind had reached Atwater, but he had not been able to confirm any of them. Smoking pipes made of stone and clay were abundant, Atwater said, "and the teeth of many of the fossil skulls, show that their owners were in the constant habit of using them.[40]

His descriptive account completed, Atwater turned to the vexing question of who the mound builders were, where they had come from, and what had become of them. Twentieth-century archaeologists have attributed the works Atwater described to peoples of a number of different cultures, but especially two: the Adena culture, typified by the Miamisburg Mound in Montgomery County, Ohio, and the Grave Creek Mound in Marshall County, W.Va.; and the Hopewell culture, characterized by highly developed artistic productions and geometric earthworks such as those at Marietta, Newark, Chillicothe, Portsmouth, and Circleville. Radiocarbon dates for the Adena culture range between 800 B.C. and A.D. 900; for the Hopewell they range from 600 B.C. to A.D. 1500. Atwater himself was uncertain whether all the antiquities he described had been produced by one people, but he was inclined to think that they had. Like many of his contemporaries, he accepted the biblical account of early human history as "the most authentick, the most ancient history of man" and sought to connect it with developments on the American continent. Absolute certainty was unattainable, he conceded, but it should be possible to arrive at probable conclusions "by obtaining a thorough knowledge of the geology and botany of the country where these works are found; by a careful examination of the [*torn*] of the people themselves; their dress; their ornaments . . . ; their places of amusement, burial and worship; their buildings, and the materials used in their structure; their wells; domestic utensils; weapons of offence and defence; their medals and monuments . . .; their idols; their modes of burial and of worship; their fortifications, and the form, size, situation, and materials with which they were constructed."[41] True to his own precepts, Atwater had studied the mineralogy, geology, and natural history of the Ohio valley. Much of the information he gathered appeared later in his *History of Ohio, Natural and Civil* (1838).

Looking about in the Old World for a nation whose productions resembled those of the mound builders, Atwater selected the Hindus as the most likely progenitors of the American aborigines. In particular, he was impressed by a clay vessel found in a tumulus near Nashville, Tenn., and placed in John Clifford's museum in Lexington, Ky., where it was called "the Triune vessel." Around the bottom of the neck were three hollow heads facing outward from the center and forming the supports of the container.

> These heads are all of the same dimensions, being about four inches from the top to the chin. The face at the eyes is three inches broad, decreasing in breadth all the way to the chin. All the strong marks of the Tartar countenance are distinctly preserved, and expressed with so much skill, that even a modern artist might be proud of the performance. The countenances are all different each from the other, and denote an old person and two younger ones.[42]

Comparing these faces with those described or illustrated in William Jones's *Asiatic Researches* and in Claudius Buchanan's *Star in the East,* Atwater concluded that they were representations of Brahma, Vishnu, and Siva. "What tends to strengthen this inference," he declared, "is, that nine murex shells, the same as described by Sir William Jones . . . and by Symmes [Michael Symes] in his 'Essay to Ava,' have been found within twenty miles of Lexington, Kentucky, in an ancient work. . . . These shells, so rare in India, are highly esteemed and consecrated to their God Mahadeva."

From this and other bits of evidence Atwater inferred that the mound builders were descended from the peoples of India. The progenitors of the American Indian, on the contrary, were probably Tartars, wild hunters who found their way across the Bering Strait at an early epoch and moved eastward along the Great Lakes to the eastern shore of North America. Years later they were followed by more civilized peoples from southern

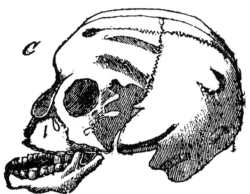

Clay vessel (the "Triune vessel") and human skull from the Indian mounds of Ohio, as depicted in Caleb Atwater's Description of the Antiquities Discovered in the State of Ohio. . . . *Courtesy of the American Philosophical Society*

Tartary and India, who settled along the tributaries of the Mississippi and the Ohio, erecting their temples, altars, burial mounds, and other sacred places on the banks of these rivers.

> In this country, their sacred places were uniformly on the bank of some river; and who knows but that the Muskingum, the Scioto, the Miami, the Ohio, the Cumberland, and the Mississippi, were once deemed as sacred, and their banks as thickly settled, and as well cultivated, as are now the Indus, the Ganges, and the Burrampooter?
>
> Ablution, from the situation of all the works which appear to have been devoted to sacred uses, was a rite as religiously observed by the authors of our idols, as it was neglected by our North American Indians. If the coincidences between the worship of our people, and that of the Hindoos and southern Tartars, furnish no evidence of a common origin, then I am no judge of the nature and weight of testimony.[43]

These migrations, Atwater continued, must have taken place at an early age of the world, as mankind was passing from a hunting to a pastoral state of society. The fortifications in western New York described by Clinton, he noted, were located on a ridge that must once have been the southern shore of Lake Erie. Likewise, many of the works on the Scioto and the Miami had gateways and walls leading down to creeks that had once washed the foot of the hills but had since formed extensive new deposits and worn down their channels ten or fifteen feet.

What, then, had become of the mound builders? In Atwater's opinion, they had moved gradually southward, increasing in numbers and in the variety and sophistication of their edifices, arts, and sciences until they arrived in Central and South America, where they created the Aztec and Incan civilizations discovered by Cortez and Pizzaro. A stone had been found in the Grave Creek Mound near Wheeling, Va., having on it a brand exactly like the one commonly used by the Mexicans in marking their cattle and horses; in a saltpeter cave in Kentucky the head of a Mexican hog, perfectly preserved, had been discovered. To clinch his argument, Atwater included an appendix containing extracts from Humboldt's *Views of the Cordilleras,* "to shew the correspondence which exists between the Teocalli of the Mexicans, and the tumuli of the North Americans."

However inconclusive Atwater's argument may seem to modern readers, it should be realized that as late as 1936 Henry C. Shetrone, an Ohio archaeologist whose treatise on the mound builders was long the leading work on that subject, stressed the similarities between their culture and that of the Aztecs. Shetrone supposed that immigrants from Asia, after crossing the Bering Strait, moved southward along the Pacific Coast into Mexico and Central America, where they developed a settled agricultural life, and that some of them subsequently migrated northward into New Mexico and the Mississippi and Ohio river valleys, carrying their newly developed culture with them. Atwater's migrants moved in the opposite direction, from the Great Lakes southward to Mexico, but both hypotheses were based on cultural and osteological similarities between the peoples of Central and North America.[44]

The osteological similarity reported by Atwater caught the attention of Humboldt. Writing to the American historian George Bancroft in 1821 after having received a copy of Atwater's memoir, he said, "I have been particularly struck with what is said of the short & thick skeletons of the mounds, . . . compared with the taller form of the present race of Indians. The ancient nations must then have had a greater resemblance to the Mexicans & Peruvians of the present day. I have for a long time insisted on these osteological differences."[45] Humboldt then urged that skulls of the cave mummies in Kentucky be collected and compared with those of the mound builders and those of present-day Indians and that specimens of all these types be sent to Professor Blumenbach at the University of Göttingen or to Georges Cuvier at the National Museum in Paris for comparison with American skulls in their collections.

The most critical American review of Atwater's monograph was that of Rafinesque, published in the *Western Review* for September 1820. He complimented Atwater for his industry but criticized him for overlooking many mounds, describing others inaccurately, and making unacknowledged use of information derived from Rafinesque himself and from John Clifford. Atwater's style, said Rafinesque, was "animated" but "diffuse, and not always correct." His monograph was lacking in originality and badly arranged. Nevertheless, it was a useful supplement to the earlier contributions of others, including Rafinesque.[46]

Rafinesque was scarcely an impartial critic. Ever since his arrival in Lexington in the summer of 1818 he had been exploring Indian mounds in the vicinity, making sketches and publishing accounts of them in the *Kentucky Reporter* and the *Western Review*. In September 1820 the month in which his review of *Archaeologia Americana* appeared, he wrote to Isaiah Thomas offering his services as a student of western antiquities. He claimed to have been the source of much of Atwater's information and accused Atwater of failing to forward to Thomas the maps of western earthworks Rafinesque had sent to him for this purpose. "I have already surveyed about fifteen groups and 50 monuments, besides common mounds and graves," Rafinesque added. "I send you now as a specimen a Plan of a singular monument surrounded by water . . . 9 miles N. W. from this town. . . . I offer you the whole for publication in your next volume."[47] Rafinesque enclosed his "Map of the Monument at Boardman's Mill on South Elkhorn Creek" along with sketches of three artifacts found

Constantine Samuel Rafinesque's sketch of Indian mounds on
Elkhorn Creek in Kentucky. Courtesy of Special Collections,
King Library, University of Kentucky, Lexington.

therein and offered to send further memoirs and drawings for the society's next volume, including a description of all the Alleghawee earthworks
of Kentucky with fifteen maps, an essay on the affinities between American
Indian languages and those of Asia, a memoir on the arts and sciences
of the Alleghawee, and an account of the physical revolutions of North
America, its original population, subsequent development, and the like,
with six maps. These were all projects then burgeoning in Rafinesque's
fertile mind.

Thomas was sufficiently impressed by Rafinesque's letters and contributions to have him elected a member of the American Antiquarian
Society. He directed the corresponding secretary, Samuel Burnside, to
solicit further contributions from Rafinesque and to seek his help in acquiring the antiquities in the museum of John D. Clifford, who had died
suddenly in 1820. In January 1821 Rafinesque reported that the executors
of Clifford's estate had decided against selling the antiquities separately
from the rest of the collection. He then went on to describe the
"monuments" of Fayette County, Ky., adding that he had "already during last year examined thoroughly and delineated those existing as yet in
the counties of Fayette, Woodford, Jessamine, Montgomery, Bourbon,
Harrison, &c. and hope to complete this year the survey of those which
I have heard of in the counties of Garrard, Madison, Clarke, Bath, Warren, Jefferson, &c. . . . I have already sent you the drafts of my surveys
of . . . the two groups on North Elkhorn . . . through the care of Mr.
Atwater," Rafinesque concluded. "I now send you the plans of the Dromus,
the group on Town Fork, the group on South Elkhorn and several views
of the graves."[48]

A copy of this letter was reprinted in Rafinesque's ill-fated periodical
Western Minerva in its first and only issue (1821), along with a not very
flattering description of Atwater as "an able man . . . but a diffuse writer,
his style being deprived of order, perspicuity and elegance." Atwater and
Daniel Drake, Rafinesque observed, were "writers of some talent," but
afflicted with "the moral diseases called selfishness and conceit, adding
thereto a proportionate share of jealousy of each other, and every body
else who may attempt to be on a par with them."[49] Rafinesque was quick
both to take and to give offense and always more inclined to see the mote
in someone else's eye than in his own. Atwater was less paranoid than
his Kentucky rival, but he was nonetheless infuriated by Rafinesque's aspersions on his character and abilities. "I am informed from sources entitled
to my full confidence," he wrote to Parker Cleaveland at Bowdoin College, "that any statement of facts cannot be depended on [as] such, if derived from the pretended 'Professor' Rafinesque. . . . I regret extremely,
that I had not sooner known the real character of that Italian [Rafinesque
was of French and German ancestry]. I am also informed that he is not

an officer of the University in which he pretends to be a Professor—that he merely teaches such as please to attend upon his lectures."[50] So distrustful of Rafinesque had Atwater become that he declared his intention to rewrite his memoir on western antiquities, striking out everything derived from Rafinesque, "who ought to be ranked among the worst of impostors, in literature and science, now living in the world."

Despite Atwater's warning, Thomas continued to press Rafinesque for further information about western antiquities. But Rafinesque was reluctant to comply with these requests without solemn assurances as to the ultimate fate of his communications.

> I have collected descriptions &c of nearly 2000 ancient monuments of North America, compared them and detected to whom each kind belongs. I have written the history of the revolutions of nature and mankind in this Continent, ascertained the primitive origins and geneaology, of the greatest part of American tribes, divided into 7 or 8 stocks, and 25 languages and 2000 dialects! traced their migrations, wars, manners, religions, and affinities with eastern nations.[51]

All this information and more, Rafinesque assured Thomas, would be made available to the American Antiquarian Society if only pledges were given that it would be published.

In 1824 Rafinesque's views on all these subjects were set forth in his *Ancient History, or Annals of Kentucky,* which served as a preface to Humphry Marshall's *History of Kentucky.* The *Ancient History* was subsequently printed separately, with a "Philological and Ethnological Table" added. It was dedicated, appropriately enough, to Alexander von Humboldt, whose researches on America surpassed those of Rafinesque in their ambitious scale. The main text of the work said nothing about Indian mounds, but there was an appendix containing a list of sites of "ancient towns and monuments of Kentucky," with a very brief description of each. Rafinesque also listed similar monuments that had been reported in other states, concluding:

> The actual number of ancient seats of population or sites already ascertained by me, in North America, amount therefore to 541, of which 393 out of Kentucky, and 148 in Kentucky, while the ancient monuments found in those sites amount already to 1830, of which 505 in Kentucky and 1325 out of it.
>
> If by my researches during 4 years, I have been able thus to increase the knowledge of the number of ancient sites and monuments in the single State of Kentucky, from 25 sites to 148, and from 100 monuments to 505: it is very probable that when equal industry will be exercised in the other States, that number will be more than

> doubled; since I entertain no doubt that 1000 sites and 4000
> monuments exist still in the United States, exclusive of Mexico, besides
> the small burrows, and those that have been destroyed.[52]

Unfortunately none of Rafinesque's sketches and extended descriptions
of the Kentucky mounds was ever published, though several have survived
in manuscript.[53]

As for Atwater, his fervent hope that the sale of *Archaeologia
Americana* and reproductions of the engravings in it, especially the map
of Ohio, would enable him to pursue his researches all the way to the Gulf
of Mexico and beyond was never realized. Most of the 2,000 copies of
the volume, offered to the public at three dollars a copy, remained unsold
(the author of this book bought one in the 1960s!), and the map of Ohio
was copied and sold by the engraver. "I have resigned the post office here
and of course, cannot correspond much, even if my professional
engagements permitted it," Atwater wrote to Benjamin Silliman. "I have
stopped all newspapers, journals &c but yours, and shall never probably
write another line for publication except it be in my professional business.
My age, numerous and increasing family, added to my poverty, must be
my excuse for this course."[54]

But Atwater's disaffection with science and scholarship was only temporary. "He is a great antiquarian, and exists more in the antiquities of
Ohio, than in the present world," wrote the Duke of Saxe-Weimar after
visiting Atwater in 1826. "I spent the evening with this interesting man,
and was very agreeably entertained; he possesses a collection of objects
which were found in different mounds; it contains fragments of urns,
arrow-heads of a large size, battle axes made of flintstone, and several
human bones. Mr. Atwater likewise possesses a very handsome collection
of minerals."[55] Henry Howe, who visited Atwater twenty years later, found
him "a queer talker" and a disappointed, unhappy man.[56]

In 1835 Atwater wrote once more to the American Antiquarian Society requesting financial support for new surveys of Indian mounds and the
publication of a revised edition of his monograph of 1820. "I have matter
enough now, for two good sized 8vos, and I wish to add several plates
to the former ones," he wrote.[57] Nothing ever came of this proposal, but
Atwater had the satisfaction of knowing that his *Description of the Antiquities Discovered in the State of Ohio* had been well received in Europe
by Humboldt and others. The next important monograph on the mound
builders was produced, not by Atwater, but by two fellow Ohioans,
Ephraim G. Squier and Edwin H. Davis, whose classic *Ancient Monuments
of the Mississippi Valley* was published by the Smithsonian Institution in
1848 as the first of the *Smithsonian Contributions to Knowledge*. Squier
and Davis acknowledged Atwater's pioneering effort in their study, and

Samuel F. Haven, secretary of the American Antiquarian Society and himself an able archaeologist, paid a final tribute for the society in his obituary notice of Atwater in the *Proceedings* for April 1867:

> Although [Atwater's] work met with a most flattering reception abroad, its author did not obtain at home the full credit to which he was entitled. . . . It was not until Messrs. Squier and Davis had gone over the same ground for the same purpose, with the advantage of having before them a country cleared and populated, . . . that the real merits of Mr. Atwater's production under unfavorable conditions were tested and established. The Memoir of those gentlemen . . . is a most satisfactory vindication of the ability and general accuracy of Mr. Atwater's papers as published by this Society.[58]

With the publication of Atwater's *Description,* archaeology had gained a secure foothold in the United States. For several decades afterward the American Antiquarian Society continued to promote archaeological research, especially during the secretaryship of Samuel F. Haven. Not until the end of the nineteenth century did the society relinquish this role and turn its museum collections over to the Smithsonian Institution and the Peabody Museum of Archaeology and Ethnology.[59] Among the items donated to the Smithsonian Institution was the famous "Mammoth Cave Mummy," the remains of an Indian woman discovered in Mammoth Cave, Ky., in 1816, pirated and exhibited far and wide by an entrepreneur from Marietta, Ohio, named Nahum Ward, recovered by the American Antiquarian Society and given a place of honor in its museum, and subsequently exhibited in the world fairs of 1876 and 1893 before reaching a final resting place at the Smithsonian. In more ways than one, Indian antiquities had come a long way since James Bowdoin predicted a bleak future for them in 1780.

CONCLUSION

In America, as in Europe, the wide-ranging scientific curiosity of the Age of Enlightenment extended to every aspect of human nature and history and led gradually to the exploration of the burial mounds and other constructions of earth and stone that were scattered across the face of the globe from the steppes of Asia to the Mississippi valley and the plateaus of Mexico and Peru. Jefferson set an example for Americans with his excavation of a burial mound on the Rivanna river. Benjamin S. Barton, drawing on William Bartram's extensive knowledge of the southern Indians and the earthworks in their territories and on correspondents in the western country, followed hard in Jefferson's footsteps, and the American Philosophical Society began actively to solicit accounts of archaeological sites. Henry M. Brackenridge produced a general account of the mounds of the Mississippi valley, and De Witt Clinton did likewise for those of northern New York, but neither found time for detailed investigations. Meanwhile the New Englanders, stimulated by archaeological discoveries at their settlement in Marietta, Ohio, formed the American Antiquarian Society under the leadership of Isaiah Thomas, who found a willing amateur archaeologist in Caleb Atwater of Circleville. Sustained by Thomas's support and encouragement, Atwater completed and published in the *Transactions* of the society an extended description of the Indian mounds of Ohio based on actual surveys, with some account of the artifacts found in them. Like many of his contemporaries, Atwater held the erroneous belief that the mound builders were not ancestral to the Indians inhabiting eastern North America in his own day. Nevertheless, his careful investigation of their constructions laid a foundation for the further development of archaeological science in the United States.

14

The Sciences of Man:
Comparative Linguistics and the
Problem of Indian Origins

TO the modern mind comparative linguistics seems far removed from the sciences of nature, but for Jefferson and his contemporaries it took its place along with physical anthropology, archaeology, and ethnology as an indispensable tool for solving the problem of Indian origins, which had fascinated scholars and scientists ever since the discovery of America. The problem had both a scientific and a religious aspect. Scientifically it bore on the hotly debated question of whether all the races of mankind were varieties of a single species, as the monogenists asserted, or whether they were separately created species, as the polygenists claimed. And this issue was indissolubly connected with the question of the credibility of the biblical account of early human history and hence with the debate between Christians and deists concerning the inspiration of Scripture.

For Jefferson himself the problem of Indian origins was primarily a scientific one, and he regarded Indian languages as the key to its solution. "A knowledge of their several languages would be the most certain evidence of their derivation which could be produced," he wrote in his *Notes on the State of Virginia.*

> How many ages have elapsed since the English, the Dutch, the Germans, the Swiss, the Norwegians, Danes and Swedes have separated from their common stock? Yet how many more must elapse before the proofs of their common origin, which exist in their several languages, will disappear? It is to be lamented then . . . that we have suffered so many of the Indian tribes already to extinguish, without our having previously collected and deposited in the records of literature, the general rudiments at least of the languages they spoke. Were vocabularies formed of all the languages spoken in North and South

America, preserving their appellations of the most common objects in nature, of those which must be present to every nation barbarous or civilised, with the inflections of their nouns and verbs, their principles of regimen and concord, and these deposited in all the public libraries, it would furnish opportunities to those skilled in the languages of the old world to compare them with these, now or at a future time, and hence to construct the best evidence of the derivation of this part of the human race.[1]

Apparently Jefferson practiced what he preached, for he had already collected enough information about Indian languages to convince himself that they were much more diverse than those of the Old World. The New World, he ventured to predict, would be found to possess twenty times the number of radical languages — "so-called because, if they were ever the same, they have lost all resemblance to one another" — as the "red men of Asia" possessed. If this were so, the American aborigines must be of greater antiquity than those in Asia. A difference of dialect might develop in "a few ages only," but a separation into radically different languages would require a much greater period of time, "perhaps not less than many people give to the age of the earth."[2]

In the years following the publication of the *Notes,* Jefferson continued to collect Indian vocabularies by every means at his disposal. From Paris he thanked the Indian agent Benjamin Hawkins for responding to his queries. "I have sent many copies to other correspondents," Jefferson added, "but as yet [August 13, 1786] have heard nothing from them. I shall proceed however, in my endeavors particularly with respect to their language and shall take care so to dispose of what I collect thereon as that it shall not be lost."[3] A year late Jefferson acknowledged receipt of several Indian vocabularies from Hawkins, adding: "This is an object I mean to pursue, as I am persuaded that the only method of investigating the filiation of the Indian nations is by that of their languages."[4] To aid his informants, Jefferson supplied them with a printed list of words for which he wanted Indian equivalents.

Jefferson's main competitor in the collection of Indian vocabularies was Benjamin S. Barton, whose interest in everything connected with the Indians rivaled Jefferson's. The range of Barton's researches in this field is indicated by the queries he submitted to his friend William Bartram and to John Heckewelder. Barton asked Heckewelder about the languages of the Indians, their traditions, relations with each other, manner of living, graphic arts, diseases, medical remedies, religion, government, customs, physical characteristics, treatment of women, property systems, agriculture, food, names for plants and animals, stature, complexion, color at birth, child-rearing practices, longevity, mental health, and general vigor.[5]

Like Jefferson, Barton was especially diligent in collecting Indian vocabularies. He ranged far and wide among printed works, combing the writings of missionaries, travelers, traders, soldiers, and scholars from the days of Samuel de Champlain and Roger Williams to those of Francisco Clavigero, Bernard Romans, and James Adair. But he did not confine himself to printed sources. From Heckewelder he solicited and obtained valuable manuscript materials compiled by him, David Zeisberger, and other Moravian missionaries. He interviewed traders, travelers, and soldiers and took advantage of every opportunity to question the Indians themselves about their languages and customs, making a trip through the Iroquois country in 1797 for this purpose. Meanwhile, he maintained a wide correspondence through which he received useful information from every part of the United States. In short, he spared no labor or expense to make his collection of Indian vocabularies as extensive and accurate as possible.

These researches soon placed Barton in a position to compare Indian vocabularies with each other, but if he was to succeed in tracing Indian origins to the Old World he must have a comparable collection of vocabularies from that hemisphere. Fortunately for him such a collection was available. Since the time of Peter the Great the Russian monarchs had sponsored expeditions into their vast territories for economic, geographical, astronomical, zoological, botanical, and linguistic research, directing the explorers "to describe the peoples of Siberia and to study their languages philologically."[6] This activity, which reached a climax in the second Kamchatkan expedition (1733–1743), was revived under Catherine the Great. In 1773 she directed G. L. C. Bacmeister, a German scholar in her employ, to circulate an appeal to the scholars of the world requesting their assistance in collecting linguistic materials.

In 1784 the appeal was renewed, and copies were sent to Washington and Franklin. The Czarina took up the work herself and made up a list of 286 words, the equivalents for which in two hundred languages were to be published. Finding the task too time consuming for an empress, she turned it over to the German scholar and explorer Peter S. Pallas, whom she had brought to Russia in 1768 to organize scientific research. Under Pallas's direction the work was completed and published in Russian and Latin.[7] First issued in 1787–1789, it indicated in Russian the pronunciation of the names for a long list of objects in each of 200 European and Asiatic languages, beginning with the words for *God, heaven, father,* and *mother* and going on to the names of the parts of the body, the elements, and miscellaneous common things. Barton was very lucky to gain access to a copy of this work; only a few found their way abroad. One such copy came into the hands of Joseph Priestley, who lent it to Barton and told him to keep it as long as he had any use for it.

With the *Comparative Lexicons* at hand, Barton proceeded to compile his own contribution to comparative linguistics and to the problem of Indian origins. It was published in 1797 and again in 1798, "corrected and greatly enlarged," under the title *New Views of the Origin of the Tribes and Nations of America.* "All the Indian, Asiatic, and European words which are compared or otherwise mentioned . . . ," Barton explained in the preface, "are printed into different kinds of letter, vis. the Italic and the Roman. The former . . . are taken from printed books, or have been communicated to me by friends, in different parts of North America. . . . All the words printed in the Roman letter were collected by myself: the greater part of them as they were pronounced by the Indian interpreters, traders or gentlemen who have been among the Indians."[8]

Unfortunately Barton had no systematic way of indicating the sounds he wished to convey to the reader. This was a difficult problem, which was to baffle linguists for many years to come, hence it is not surprising that Barton failed to solve it. He did what he could, however. In the case of words taken from travelers, traders, missionaries, and the like, he was careful to indicate the nationality of the original collector. Whenever this collector had laid down rules of pronunciation, as Zeisberger had done, Barton took pains to reproduce these for the reader. He gave general instructions for pronouncing words he had collected himself, but his directions were not very clear.[9] For words taken from Pallas, transcribed from Russian into Roman letters, still another set of phonetic equivalents was indicated. Thus Barton's phonetic method, like so many others in his day, was extremely rough and ready, so much so that modern linguists find it all but impossible to determine exactly what sounds were intended to be represented by the letters used.

The word lists in Barton's *New Views* followed the order adopted in the first edition of Pallas's *Lexicons,* beginning with the words for *God.* In all, Barton was able to give Indian equivalents for seventy of Pallas's words. In each case Barton placed the Indian words first, beginning with the Delaware tribes and proceeding thence to other North American tribes and on to those of Central and South America. Then came a brief selection of European and Asiatic equivalents, taken mostly from Pallas. These were selected on the basis of their similarity to words in the Indian lists. Thus the purpose of the comparison was to establish the existence of a connection between the languages of the New World and those of the Old World by pointing out similarities in the words used to designate particular objects. In Barton's opinion his vocabularies rendered it very probable that the nations of America and those of Asia had a common origin. "I flatter myself," he wrote, "that this point is now established with some degree of certainty, though I by no means suppose that what I have said should preclude the necessity of farther inquiries on the subject."[10]

Title page of Benjamin
Smith Barton's pioneer
work on Indian
linguistics, with sample
vocabularies from the
second edition.
Courtesy of the
American Philosophical
Society

N E W   V I E W S

OF THE

O R I G I N

OF THE

T R I B E S   A N D   N A T I O N S

OF

A M E R I C A.

By BENJAMIN SMITH BARTON, M. D.

CORRESPONDENT-MEMBER OF THE SOCIETY OF THE ANT QUARIES
OF SCOTLAND; MEMBER OF THE AMERICAN PHILOSOPHICAL
SOCIETY; FELLOW OF THE AMERICAN ACADEMY OF
ARTS AND SCIENCES OF BOSTON; CORRESPONDING
MEMBER OF THE MASSACHUSETTS HIS-
TORICAL SOCIETY,

AND PROFESSOR OF MATERIA MEDICA, NATURAL HISTORY
AND BOTANY,

IN THE

UNIVERSITY OF PENNSYLVANIA.

PHILADELPHIA:

PRINTED, FOR THE AUTHOR,
BY JOHN BIOREN.
1798.

Since Barton based his conclusions concerning the origin of nations on the supposed affinities among their languages, it is important to understand what he considered satisfactory proof of linguistic affinity. On this crucial point Barton seems to have been rather naive, even for his day. He apparently believed that the existence of similar sounding words for several objects or ideas in two different languages created a strong presumption of linguistic affinity. He even argued that languages displaying very few word similarities might nevertheless be regarded as related if each contained several words similar to those found in a third language.

We find the Toungusian word for star in the dialects of the Mohawks, Onondagos, and other nations of the [Iroquois] confederacy. There are many words of this Asiatic nation in the languages of the Delaware

126	N I G H T.		N I G H T.	127

N I G H T.

Lenni-Lennápe.			Peefkéoh. *Piefkeu,* dark.
Minfi.	-	-	*T'pochcu.*
Mahicanni.	-	-	Tepockq, Neepauweh.
Miamis.	-	-ˎ	Pachkoantekeeh.
Pottawatameh.	-		Pecuneah.
Mohawks.	-	-	*Aghfuntbea.*
Onondagos.	-	-	*Achfuntha.*
Oneidas.	-	-	Kawoffondeak.
Tufcaroras.	-	-	Oofottoo, Autfonneah, Aucht-tfeeneeah, Yootfautheh.
Naudoweffies.	-	-	*Bafatfche, Bafatche.* ⎫
Iffati.	-	-	*Bafatfche, Bafatche.** ⎭
Cheerake.	-	-	*Tfennée* (n).
Mufkohge.	-	-	*Neethleeh,* Neethleeh.
Chikkafah.	-	-	*Neennak.*
Choktah.	-	-	*Neennak,* Neenak.
Woccons.	-	-	*Yantoba.*
Galibis.	-	-	*Cooquo.*
Brafilians.	-	-	*Putuna.*
Peruvians.	-	-	*Tuta.*
Chilefe.	-	-	*Pun, Paun.*

✦

Oftiaks,	-	-	75.	*Pect.*
Semoyads, 121, 122, 123, 124.			*Pee.*
Ofetti,	-	-	79.	*Achfaf, Achfew.*
Dugorri,	-	-	82.	*Achfawa.*

---

* On the authority of Father Hennepin. This author fometimes fpeaks of the Naudoweffies and Iffati as one tribe, and at other times, as two tribes.

Koriaki,	-	-	153.	*Neeg'inok, Neekeeneek.*
	-	-	154.	*Neketta.*
Semoyads,			126.	*Peen, Pete.*
	-	-	128.	*Peen.*
	-	-	127.	*Peetn.*

M O R N I N G.

Lenni-Lennápe.	-		Aullapaueh.
Natichs.	-	-	*Mahtompan.*
Miamis.	-	-	Chaieepauweh.
Mohawks.	-	-	*Yorheanfke.*
Choktah.	-	-	Oonnahheleh.

E V E N I N G.

Lenni-Lennápe.	-	-	*Wulacu,* Woolaukoo.
Miamis.	-	-	Allaqueekee.
Mohawks.	-	-	*Diyogarafkwe.*
Choktah.	-	-	Háfhe-cunne-é-chóme.

S U M M E R.

Lenni-Lennápe.	-	-	Neepun.
Minfi.	-	-	*Nichpen.*
Miamis.	-	-	Neepeenweeh, *Nipinwai.*
Indians of Virginia.	-		*Cohattayough* (CAPT. SMITH).
Senecas.	-	-	Kan-guit-tik-neh.
Oneidas.	-	-	Kau-wau-kun-hcak-kee.
Mufkohge.	-	-	Mifke, *Mifsa.*
Choktah.	-	-	Lufhpah, fummer, and warm.
Chilefe.	-	-	*Uan.*

tribes. . . . By a careful inspection of the vocabularies, the reader will find no difficulty in discovering that in Asia the languages of the confederates and the languages of the tribes of the Delaware-stock may be all traced to ONE COMMON SOURCE. Nor do I limit this observation to the languages of the American tribes just mentioned. It will be easy to trace the languages of the Cheerake, Muskohge, Chikkasah, Choktah, and even those of the Mexicans, the Peruvians, the Chilese, and many other nations . . . to the same sources from whence have sprung the languages of the confederates and Delawares. The inference . . . is obvious and interesting: THAT HITHERTO, WE HAVE NOT DISCOVERED IN AMERICA . . . ANY TWO, OR MORE LANGUAGES BETWEEN WHICH WE ARE INCAPABLE OF DETECTING AFFINITIES (AND THOSE OFTEN VERY STRIKING) EITHER IN AMERICA, OR IN THE OLD WORLD.[11]

Thus Barton took as his working hypothesis the idea that all the languages of the earth were related to each other and that their affinities might be discovered by compiling ever larger and more accurate vocabularies of all of them. He agreed with Jefferson, to whom he dedicated his *New Views,* that the apparent diversity of languages in the Americas had been produced by the modifying influence of time and circumstance, but he rejected Jeffer-

son's corollary that the inhabitants of the New World must be more ancient than those of the Old World since their languages were more diverse.
The time required to obliterate the original resemblance of languages would
depend on the conditions of life, said Barton. In the New World these must
have been much more conducive to diversification than in the Old. Even
so, the evolution of so many apparently dissimilar languages in America
must have required a long time, "many hundred, perhaps three or four
thousand years." But whether the American Indians were derived from the
Old World or the inhabitants of the Old World from America was a question that linguistic science in itself could not determine. The solution of
this problem, said Barton, would require supplementary investigations into
tribal myths and traditions, social customs and physical characteristics,
and the geographical distribution of aboriginal earthworks. When these
various types of evidence had been duly weighed, it would become apparent, Barton predicted, that central Asia was the "principal foundery"
of mankind, as Moses had declared long ago.[12]

As a treatise on comparative linguistics Barton's *New Views* left much
to be desired. It should be remembered, however, that the comparative
study of languages was in its infancy. It was not until 1786, eleven years
before Barton's work appeared, that William Jones startled the scholars
of Europe with his paper before the Royal Asiatic Society in Calcutta
demonstrating the kinship of Sanskrit, Greek, Latin, and the Germanic
languages and calling attention to the incredibly rich tradition of linguistic
scholarship in India. The way was now open for the development of a science of comparative and historical philology, but even in Germany, where
this science was to flourish most vigorously, the outburst of linguistic activity did not take place immediately. Friedrich Schlegel's *On the Language
and the Learning of the Indians* was not published until 1808. Franz Bopp's
*On the Conjugation System of Sanskrit, in Comparison with That of
Greek, Latin, Persian, and German* appeared in 1816, by which time comparative philology had become firmly seated in the German universities.
In the interim, German linguistic scholarship was typified by the four-
volume *Mithridates: Oder Allgemeine Sprachenkunde* (1806–1817), begun
by Johann C. Adelung and carried to completion by Johann S. Vater, professor of theology at the University of Königsberg. Peter S. Du Ponceau,
the Philadelphia lawyer-linguist who succeeded Barton in the study of Indian languages, considered the *Mithridates* "the most astonishing
philological collection that the world has ever seen."

> It contains an epitome of all the existing knowledge of the ancient and
> modern languages of the whole earth. It exhibits specimens of the
> words of each language, by means of which their affinities can be traced
> as far as etymology may help to discover them, with a delineation of

their forms, syntax, construction, and general grammatical character, exemplified in the greatest number of cases by the Lord's prayer in each language and dialect, with a literal German translation interlined, and followed by a commentary in which every sentence is parsed and the meaning of each word given, with an explanation of the grammatical sense and form in which it is employed.[13]

Vater reviewed Barton's *New Views* for the *Goettingische Anzeigen von den Gelehrten Sachen*. He praised the work as containing many new and interesting observations and a "treasure" of comparative Indian vocabularies but criticized Barton for failing to make systematic comparisons among them. Barton, he complained, asserted an affinity between the Delaware and Iroquois languages but left it to his readers to discover the evidence for this assertion in the relevant vocabularies. In like manner Barton declared that the languages of northeast Asia were reducible to a single stock and those of America to two stocks, but he failed to substantiate these assertions by detailed comparisons. More generally, Vater continued, Barton placed too much reliance on occasional similarities between words in different languages.[14]

Whatever the merits or shortcomings of Barton's *New Views,* neither Barton nor Jefferson regarded the work as having closed the door to further inquiries of the same kind. In March 1800 Jefferson wrote Benjamin Hawkins that he was preparing to publish his collection of vocabularies, "lest by some accident it might be lost," and asked Hawkins to procure equivalents in the Creek, Choctaw, and Chickasaw languages for the words on his enclosed list. "You need not take the trouble of having any others taken," wrote Jefferson, "because all my other vocabularies are confined to these words, and my object is only a comparative view. . . . All the despatch which can be conveniently used is desirable to me, because this summer I propose to arrange all my vocabularies for the press, and I wish to place every tongue in the column adjacent to its kindred tongues."[15]

In January of the following year Jefferson informed William Dunbar that his word lists were now fairly complete for about thirty Indian languages, "among which the number radically different, is truly wonderful. . . . I at first thought of reducing them all to one orthography," Jefferson added, "but I soon became sensible that this would occasion two sources of error instead of one. I therefore think it best to keep them in the form of orthography in which they were taken, only noting whether that were English, French, German, or what."[16] A year later Dunbar sent Jefferson vocabularies of the Atacapa and Chetimacha tribes obtained for him by Martin Duralde of Opelousas. About the same time, Chief Little Turtle and the interpreter William Wells visited the capital, and Jefferson obtained from them a vocabulary of the Miami language to supplement the one the Count de Volney had presented to him in 1798 after interviewing Little Turtle and his companion in Philadelphia.

In 1803 Jefferson seized the opportunity of purchasing the entire Louisiana Territory from France and with it the means of acquiring vocabularies of still other Indian tribes. In May 1805 he sent copies of his word list to John Sibley and asked him to fill them out for the numerous tribes encountered on his recent expedition into the Arkansas country.

> I have taken measures for obtaining those north of the Arcansa, and already possess most of the languages on this side of the Mississippi. A similar work, but on a much greater scale, has been executed under the auspices of the late empress of Russia, as to the red nations of Asia, which, however, I have never seen. A comparison of our collection with that will probably decide the question of the sameness or difference of origin, although it will not decide which is the mother country, and which the colony.[17]

Strange to say, Jefferson had never seen a copy of Pallas's *Lexicons,* although his friend Barton had possessed a copy when he wrote his *New Views.*[18] Apparently Jefferson and Barton continued on good terms, for Jefferson notified him in January 1806 that several Indian chiefs from the trans-Mississippi West were about to leave Washington for Philadelphia and New York, "and consequently will give you an opportunity at Philadelphia of making all the enquiries you desire." In September 1809 Jefferson wrote Barton that he would be glad to let him see any or all of his vocabularies if he were able to do so.[19]

But Jefferson was now unable to oblige Barton or anyone else with Indian vocabularies. The accident he had dreaded in his letter to Hawkins in 1800 had happened. He had put off arranging his vocabularies for publication until he could incorporate the word lists brought back by Lewis and Clark. His turbulent last term as president of the United States having expired, he had packed the vocabularies and related materials in a trunk and sent them to Monticello with the rest of his things, looking forward eagerly to completing work on them when he arrived at his beloved plantation. Alas, the trunk never arrived. It was stolen from the ship that carried it up the James River, and the thief, disappointed to find nothing but "worthless" papers in the trunk, had emptied its contents into the river. Only a few of the precious documents floated ashore and were rescued from the mud. Among these was Lewis's vocabulary of the Pani language, which Jefferson sent to Barton along with a fragment of another vocabulary in Lewis's hand.

> It is a specimen of the condition of the little which was recovered. I am the more concerned at this accident, as of the two hundred and fifty words of my vocabularies, and the one hundred and thirty words

of the great Russian vocabularies of the languages of the other quarters
of the globe, seventy-three were common to both, and would have furnished materials for a comparison from which something might have
resulted. . . . Perhaps I may make another attempt to collect, although
I am too old to expect to make much progress in it.[20]

What Barton's feelings were on learning of the shipwreck of Jefferson's linguistic project is not recorded. Since the publication of his *New
Views* Barton had continued to collect vocabularies and other materials
relating to the Indians, announcing from time to time his intention of
publishing a revised edition of *New Views* and a general work on the
American Indians. In his "Hints on the Etymology of Certain English
Words," read before the American Philosophical Society in 1803 and
published in their *Transactions* in 1809, Barton stretched his elastic concept of linguistic affinity to the limit by attempting to show that, for each
of about fifty English words, similar sounding words for the same object
or idea could be found in other languages in different parts of the world,
including the Indian languages.[21] For further proof of this proposition Barton referred his readers to Part 2 of his *New Views,* "which is preparing
for the press."[22]

The second part of Barton's *New Views* never appeared, but the original work seems to have proved useful to European scholars interested in
the Americas. When Humboldt came to consider the linguistic evidence
for the Asiatic origin of the Toltecs and Aztecs in his *Political Essay on
the Kingdom of New Spain* (1811), he had little except Barton's
vocabularies to guide him. These, he noted, showed only a few word
similarities between the languages of Tartary and those of the New World.
Three years later, when Humboldt published his *Researches Concerning
the Institutions and Monuments of the Ancient Inhabitants of America,*
he cited not only Barton but also Vater. These two scholars, he noted, had
demonstrated the existence of some common words in the languages of
Old and New World tribes.

> In eighty-three American languages, examined by Messrs. Barton and
> Vater, one hundred and seventy words have been found, the roots of
> which appear to be the same. . . . Of these . . . which have this con-
> nexion with each other, three fifths resemble the Mantchou, the
> Tongouse, the Mongul, and the Samoyede; and two fifths the Celtic
> and Tschoud, the Biscayan, the Coptic, and the Congo languages.
> These words have been found by comparing the whole of the American
> languages with the whole of those of the Old World; for hitherto we
> are acquainted with no American idiom, which seems to have an ex-
> clusive correspondence with any of the Asiatic, African, or European
> tongues.[23]

Although Humboldt seems to have valued the comparative researches of Barton and Vater, he found them inconclusive on the main point at issue. For his own part, he proposed to show that whatever the deficiencies in linguistic evidence there was plenty of evidence in the monuments, hieroglyphics, cosmogonies, and institutions of the peoples of America and Asia to establish the probability of an ancient communication between them.

The work by Vater to which Humboldt alluded bore the title *Inquiry into the Peopling of America Out of the Old World.* Published in German in 1810 and dedicated to Humboldt, this work was primarily a study in comparative linguistics. In its opening pages Vater paid his respects to that "worthy scholar in Philadelphia," Herr Smith-Barton, and proceeded to reproduce all fifty-four of the word similarities Barton had discovered in his comparison of the languages of the New and Old World. Vater took exception to some of the items in Barton's list, arguing that Barton had erred in transcribing from the Russian or, in other cases, had been misled by affixes and prefixes. Quite apart from these doubtful cases, Vater thought that Barton had jumped to unwarranted conclusions from his comparative survey. The word similarities were too few in number and too widely scattered geographically to provide a sure footing for generalization. The comparison, he added, should not be limited to words but should extend to linguistic structure. Structural similarity more than the resemblance of particular words proved the common derivation of the American languages. In all of them, Vater declared, a single grammatical tendency was evident: the tendency in conjugating verbs "to describe the relationship of action to the object and thus to differentiate from cases in which there is no object." This universal grammatical tendency, said Vater, was unmistakable evidence of a common origin.[24]

The extent of Vater's debt to Barton and other American writers for information concerning Indian tribes and languages may be judged from Part 3 of the third volume of the monumental *Mithridates* (1816). For information concerning the tribes west of the Mississippi River, Vater relied on the accounts of Pike, Mackenzie, Umfreville, La Pérouse, Krusenstern, Du Pratz, Humboldt, and other travelers; on the Spanish scholar Hervas y Panduro; and on the spurious *Travels of Capts. Lewis & Clarke* published in London in 1809 and plagiarized from the writings of Jonathan Carver and Alexander Mackenzie and from Jefferson's message to Congress in 1806. Vater translated this book into German, unaware that it was spurious.[25]

East of the Mississippi, Vater made good use of Barton's *New Views* and the sources he had used. In treating the languages of the Iroquois, for example, Vater listed words collected by Barton and his correspondents;

words derived from Zeisberger manuscripts in Barton's possession; those recorded by Father Gabriel Sagard, Baron La Hontan, and the traveler John Long; words from George Henry Loskiel's history of the Moravian missions in North America; and those contained in *The Order of Prayer, Translated into the Mohawk Language* by three Anglican missionaries. The extent and variety of the printed materials available to Vater in his researches in Königsberg is quite astonishing. From the religious texts translated into Indian languages by missionaries, notably the Lord's Prayer, he was able to present extensive commentaries on the grammatical structure of these languages. For word lists he drew heavily on Barton and also on missionaries, travelers, and Indian agents. One can only regret that Jefferson's vocabularies, including those collected by Lewis, were not available to him.

Vater's conclusions, though stated more tentatively than Barton's, were similar to them in many respects. He found that the languages of the New World exhibited traces of a common origin despite their superficial diversity and that there was a clear linguistic connection between some of the American tribes, especially those of the far North, and those of northeast Asia. "Linguistic relationships unite vast parts of North America, bringing many peoples into one race. . . . These languages lead us in the far north back to northeast Asia. . . . In Greenland and on the coast of Labrador, as on its western coast near Asia, there lives a people which is one and the same with the inhabitants of the north coast of Asia and the islands which lie between the two parts of the world. What a huge expansion of one race!"[26] Vater did not attempt to reduce all American languages to one stock deriving from Asia or to link particular languages in the eastern and western hemispheres, however. His caution in this respect was well considered. Twentieth-century linguists are impressed with the difficulty of both enterprises.[27]

To modern scholars the *Mithridates* seems little more than "a philological curiosity" or at best the most spectacular example of a tradition of linguistic research that was being rendered obsolete at the time it was published. The compilers of the *Mithridates,* said the Danish philologist Wilhelm Thomsen, had too little firsthand acquaintance with the languages they compared; they were too little concerned with the structural analysis of language and were handicapped by their reliance on the Lord's Prayer as the main comparative text.[28] To the generation of American linguists who succeeded Barton, and especially to his fellow-Philadelphian Peter S. Du Ponceau, however, the *Mithridates* was an inspiration and a revelation. Du Ponceau marveled at "the astonishing penetration of the great Vater" and warned that those who attempted research into Indian languages without first studying the *Mithridates* would find their "discoveries" anticipated in that work. "Strange!" exclaimed Du Ponceau, "that we should have to go to the German universities to become acquainted with our own country."[29]

THE DU PONCEAU-HECKEWELDER CORRESPONDENCE

Peter S. Du Ponceau was born on the Isle of Ré off the western coast of France in 1760. A precocious child, he showed an interest in languages at an early age, acquiring a knowledge of English and Italian from the foreign troops quartered in the vicinity of his home. After eighteen months of study at a Benedictine college he was sent to teach Latin at a school in Poitou. It was an unhappy experience for the fifteen-year-old boy, and he soon rebelled.

> On the 25th of December, (being Christmas day), leaving all my baggage behind, I sallied out at day break, with the *Paradise Lost* in one pocket, and a clean shirt in the other, and bravely took my way on foot towards the great Capital, where I arrived in the beginning of January following, with the firm resolution of depending, from that moment, on my own exertions alone for subsistence, and for whatever fortune might await me.[30]

Aided by his linguistic talents and his father's friends in Paris, he eventually secured a job as secretary to the celebrated Antoine Court de Gébelin, who was then deeply engrossed in a search for the primitive language of mankind. Somewhat later Du Ponceau met Frederick W. von Steuben, who was getting ready to depart for America to join the American patriots in their struggle for independence and was looking for an English-speaking secretary to accompany him. Eager to see the world, Du Ponceau accepted von Steuben's invitation to serve in this capacity, and in December 1777 he found himself on American soil in Portsmouth, N.H., mildly astonished to find that even the milkmaids spoke English fluently!

Four years later Du Ponceau became an American citizen and joined the new government as assistant to the Secretary for Foreign Affairs Robert R. Livingston, having been highly recommended by Richard Peters of Philadelphia. "French is his native tongue," wrote Peters. "English he has acquired perfectly, and he understands German, Italian, and Spanish. He can translate Danish and Low Dutch with the help of a dictionary, but a little application will make him master of these. He is also a good Latin scholar."[31] A year and a half later Du Ponceau entered on the study of law and was soon engaged in legal practice in his adopted city of Philadelphia.

For the next thirty years Du Ponceau's linguistic talents lay largely unutilized while he made a place for himself in his chosen profession. Then in March 1815, in the surge of patriotic fervor that followed the successful conclusion of the War of 1812, the American Philosophical Society decided to add a Committee of History, Moral Science, and General Literature to the six committees it had established earlier, and Du Ponceau was elected corresponding secretary. His duties were to solicit "Original Documents, such as Official and Private Letters, Indian Treaties, Ancient Records, Ancient Maps, and such other Papers as may be calculated to throw light on the History of the United States, but more particularly of this State, to be preserved among the archives of this Society for the public benefit." The committee was likewise to seek "correct information on matters connected with the History, Geography, Topography, Antiquities, and Statistics of this Country" and was authorized to publish such materials as it was able to collect.[32] William Tilghman was elected chairman, John Vaughan recording secretary. Among the other members were the Abbé Correa da Serra, Caspar Wistar, Thomas Jefferson, and John Heckewelder. In August 1815 the committee published a circular soliciting information on the subjects committed to its charge, and Du Ponceau was soon immersed in a heavy correspondence.

Acting on a suggestion from Thomas Jefferson, Wistar wrote to Heckewelder requesting information about "everything which throws light upon the nature of the Indians, their manners and customs; their opinions upon all interesting subjects, especially religion and government; their agriculture and modes of procuring subsistence; their treatment of their wives and children; their social intercourse with each other; and in short, every thing relating to them which is interesting to you, will be instructing to the Society."[33] Heckewelder had been engaged in missionary work for the Society of the United Brethren since 1762. From 1771 to 1781 he served as assistant to David Zeisberger in the Indian mission at Friedenstadt on the Beaver River in northwest Pennsylvania. During this period Heckewelder lived within two miles of the seat of government of the Delaware tribes and served for three years as a sort of clerk to the Great Council of that nation, taking down the speeches of the chiefs. In the 1780s the Moravians established missions in southeastern Ohio. When Indian uprisings forced their abandonment, Heckewelder and his associates spent four difficult years among the Chippewa Indians near Lake St. Clair, cut off from the fort at Detroit by snow in winter, plagued by mosquitoes and rattlesnakes in summer. Then came a series of hazardous missions for the American government in its dealings with the Indian tribes of Ohio, after which Heckewelder settled with his family at the Gnadenhütten mission on the Muskingum River. Finally in 1810, after nearly half a century of perilous travels, grueling hardships, and unswerving devotion to duty, this indomitable servant of Christ returned to Bethlehem, Pa., in order, as he said, "to spend the remainder of my days in rest and quietude."[34]

In reply to Wistar's request for information, Heckewelder wrote that he was completing a narrative of the mission of the United Brethren and would be glad to send a copy of the same to Wistar along with other materials relating to the Indians. He added that he had sent much information of this kind to Barton over the years in response to his assurances that all or most of it would eventually be published. "Had he not told me this repeatedly," wrote Heckewelder, "I should long since have tried to correct many gross errors, written and published, respecting the character and customs of the Indians." However, he would try again now that Barton was dead. As a first installment Heckewelder promised to send Zeisberger's manuscript grammar of the language of the Lenni Lenape, or Delawares.[35] Shortly thereafter the bulky manuscript was delivered to Wistar in Philadelphia.

When Du Ponceau examined the Zeisberger grammar, he was astonished to find that the Delaware language, far from being the primitive affair European writers had taught him to expect from savages, was a highly complex and elegant tongue, rich in grammatical forms and containing a rigorous inner logic. He resolved to translate the grammar from German into English, seeking guidance from Heckewelder as the work progressed.

> How delighted I am with the grammatical forms of the Indian languages, particularly of the Delaware, as explained by Mr. Zeisberger. I am inclined to believe that those forms are peculiar to this part of the world, and that they do not exist in the languages of the old hemisphere. At least, I am confident that their development will contribute much to the improvement of the science of universal grammar.[36]

As this letter indicates, Du Ponceau was studying Indian languages generally, avidly seeking every scrap of information he could obtain about them. As corresponding secretary of the Historical and Literary Committee he cast his net far and wide. From Thomas Jefferson the committee received the journals of the Lewis and Clark expedition, Benjamin Hawkins's account of the Creek country, a dozen Indian vocabularies, and such fragments of Jefferson's own comparative vocabulary as had survived its immersion in the waters of the James River. Through Heckewelder, Du Ponceau received not only detailed answers to his queries about Indian languages but vocabularies, manuscript dictionaries and grammars, and religious and educational tracts in Indian tongues from the Moravian library in Bethlehem. Through Heckewelder he was put in touch with two missionaries in the Cherokee country, John Gambold and Daniel Buttrick.

From Theodore Schultz of Schoeneck, Pa., formerly a missionary in Dutch Guiana, he obtained a grammar and dictionary of the Aruwak language. The bishop of Quebec enabled him to open a correspondence with missionaries to the Canadian Indians. He obtained information about the languages of the Chickasaws and Hurons from interpreters.

Meanwhile, Du Ponceau was spending a small fortune building up a library of books on the Indians of North and South America and their languages. The libraries of Philadelphia, notably that of the American Philosophical Society, served him well, but there were many books that had to be procured from Europe. Through the German geographer Christoph D. Ebeling he obtained a copy of the volumes of the *Mithridates* dealing with American languages, but he could find no copy of Pallas's *Lexicons*. "I have received precious books from Washington [D.C.]," he wrote to John Vaughan, librarian of the American Philosophical Society.

> I did not know the subject was so extensive, but zeal conquers every thing. I am getting a copy made of Sagard's Huron Dictionary, printed at Paris in 1632, a very scarce book — I shall present it to the Society's Library. I wish I could do the same with the Caribbee Grammar, another very scarce book, but I must be content with Extracts. . . .
>
> I am exerting myself for presenting a tolerable general view of the Indian languages, of their construction & grammatical forms. Having now read nearly all that has been written on the subject, I am determined to avoid pedantry, dogmatism & vain theories. This is in my power. The rest must be left to Providence. O that it may only inspire me with somewhat of a tolerable Style to set off the dryness of this dryest of all dry subjects. This is the grand difficulty, but allons! the die is cast, and il faut marcher.[37]

In a letter to Jefferson, written a few months later, Du Ponceau explained his purposes with characteristic enthusiasm:

> No nation elsewhere can combine so many ideas together in one word by means of a most admirable combination of grammatical forms, by which they [the Indians] can write almost all the parts of speech into one, which is the verb. . . . What would not Anacreon, or Tibullus have given to have been able in their amatory poems to say in one single word "O thou who makest me happy!" . . . Those multiform compound parts of speech, in which there are fewer irregularities than in any other language, are peculiar to the Indian nations, are found with few variations in all their idioms thro' the Continent from North to South, & disappear in the adjacent countries of Asia & Europe. I am preparing a communication to the Society on this interesting subject; the study of languages has been too long confined to mere "word hunting" for the sake of finding affinities of sound. Perhaps a

comparison of the grammatical forms of the different nations may produce more successful results.[38]

The question of Indian origins Du Ponceau put completely aside. He had no opinion on this question, he told Heckewelder. His aim was "to take a general & a correct view of the character & physiognomy of the American languages as far as regards their grammatical forms and construction and the manner in which they *combine ideas together in the form of words.*"

> In this they appear to me to differ from the rest of the world . . . and offer a new & fruitful source of reflection to those who study the principles of language. . . . Mr. Zeisberger's grammar gave me the first idea of this research, and as I proceed in the translation of it, new prospects open upon my mind, & the conclusions which I am apt to draw from what I find in it . . . are always aided . . . by the valuable information which I obtain from you.[39]

Yet Du Ponceau could not help believing that his own approach to the study of Indian languages might eventually throw some light on the question of origins. "If something may be discovered by the mere similarity of words," he wrote to Heckewelder, "how much farther may not we proceed by studying and comparing the 'plans of men's ideas,' and the variety of modes by which they have contrived to give them body and shape through articulate sounds. This I consider to be the most truly philosophical view of human language generally considered, and before we decide upon the Tartar origin of the American Indians, we ought . . . to study the grammars of the Tartar languages, and ascertain whether their thoughts flow in the same course, and whether their languages are formed by similar associations of ideas, with those of their supposed descendants."[40]

Heckewelder's own opinion, like that of most of his fellow missionaries, was that the Indians were descendants of the ten lost tribes of Israel. But this opinion, he told Du Ponceau, was not for publication. "No one of us will undertake to *prove* the fact, and opinions without facts—without reasons for so believing, are worth nothing."[41] Heckewelder was in full accord with Du Ponceau's enthusiasm for Indian languages and his strong desire to exhibit their grammatical elegance and intricacy to the savants of Europe. "It often strikes me," he confided to Du Ponceau, "that if we had been acquainted with each other—and that for so long a time as Dr. Barton and I had been; we would have done wonders. I might while I was living with, or near the Indians have procured the names of every tribe of Indian, connected with the Lenape together with specimens of many of the languages of these—Such productions would have set the linguists in Philadelphia - Boston - Germany, Russia &c. going."[42]

In 1819, four years after its creation, the Historical and Literary Committee of the American Philosophical Society published the first volume of its *Transactions*. The main substance of the volume consisted of the Du Ponceau–Heckewelder correspondence, Heckewelder's "Account of the History, Manners, and Customs of the Indian Natives Who Once Inhabited Pennsylvania and the Neighbouring States," and Du Ponceau's "Report to the Committeee on the Languages of the American Indians." Heckewelder's "Account" was an important contribution to the history and ethnography of the tribes he had lived with, based as it was on close association with these tribes before they were exposed to extensive contacts with Europeans. In the opinion of Clark Wissler, formerly curator of anthropology at the American Museum of Natural History, Heckewelder and his associate Zeisberger were "the first great leaders in American ethnography."[43] The tribes Heckewelder described, both Delaware and Iroquois, could not have asked for a more sympathetic historian. "Alas!" wrote Heckewelder, "in a few years . . . they will have entirely disappeared from the face of the earth. . . . At least, let it not be said, that among the whole race of white Christian men, not one single individual could be found, who, rising above the cloud of prejudice [with] which the pride of civilisation has surrounded the original inhabitants of this land, would undertake the task of doing justice to their many excellent qualities."[44]

The results of the committee's linguistic researches were set forth by Du Ponceau in his "Report." He had tried, he said, to build on the earlier researches of Adelung and Vater and, like them, to go beyond etymological comparisons to an examination of grammatical forms. He had consulted every available book and manuscript and had subjected the testimony of various writers to the most searching scrutiny of which he was capable. He had no favorite theory to advance, but he felt justified in submitting three broad empirical generalizations to the learned world for their consideration:

> 1. That the American languages in general are rich in words and in grammatical forms, and that in their complicated construction, the greatest order, method and regularity prevail.
> 2. That these complicated forms, which I call polysynthetic, appear to exist in all those languages, from Greenland to Cape Horn.
> 3. That these forms appear to differ essentially from those of the ancient and modern languages of the old hemisphere.[45]

Du Ponceau proceeded to defend these generalizations with all the skill at his command. In support of the polysynthetic character of all the Indian languages he cited the works of Hans Egede and David Crantz on the language of the Greenlanders and the Eskimos and those of Heckewelder, Zeisberger, and other Moravian missionaries concerning the

Delaware and Iroquois tongues. For middle America he relied on Thomas
Gage's account of his voyage to New Spain and on the grammars of Mexican dialects by Tapia Zenteno, Father Antonio de Rincon, Father Diego
Basalenque, and the Abbé Neve y Molina. In South America the Carib-
bee grammar and dictionary by Father Breton and Molina's *History of
Chile* provided the information needed. As an example of polysynthetic
construction he cited the Araucanian word *iduancloclavin,* meaning "I do
not wish to eat with him," and matched it with the Delaware word
n'schwingiwipoma, "I do not like to eat with him," provided by Heckewelder. "Wherever the polysynthetic form of language prevails," Du Ponceau concluded, "it is natural to presume that it is accompanied with all
its inherent qualities. . . ."

The full meaning of the newly coined term *polysynthetic* was explained
in Du Ponceau's discussion of his second generalization. A polysynthetic
language, he declared, is "that in which the greatest number of ideas are
comprised in the least number of words," linguistic compression being
achieved primarily in two different ways: "by interweaving . . . the most
significant sounds or syllables of each simple word, so as to form a compound that will awaken in the mind at once all the ideas singly expressed
by the words from which they are taken" and by combining the various
parts of speech, particularly with the verb, "so that its various forms and
inflections will express not only the principal action, but the greatest possible number of the moral ideas and physical objects connected with it."[46]
At other times the possessive pronoun and various prepositions were combined with the transitive form of the verb so as to express in the same word
the ideas of the governing pronoun and that which was governed.

In his correspondence with Heckewelder, Du Ponceau contrasted the
polysynthetic class of languages with the *asyntactic* class found in China;
the *analytic* class exemplified by Icelandic, Danish, Swedish, German, and
English; the *synthetic* class represented by Latin, Greek, Slavonic, and
various oriental languages; and the *mixed* class, including French, Italian,
Spanish, and Portuguese and their dialects. In his "Report," however, he
declared that the time was not yet ripe for a comprehensive classification
of languages. The field of linguistics would require much more extensive
cultivation before its Linnaeus could appear.

Turning to his third generalization, Du Ponceau advanced arguments
to show that the Indian languages were fundamentally different from those
of the Old World.

> We find a *new* manner of compounding words from various roots so
> as to strike the mind at once with a whole mass of ideas; a *new* man-
> ner of expressing the cases of substantives by inflecting the verbs which
> govern them; a *new* number, (the particular plural,) applied to the

declension of nouns and conjugation of verbs; a *new* concordance in tense of the conjunction with the verb; we see not only pronouns, as in the Hebrew and some other languages, but adjectives, conjunctions, adverbs, combined with the principal part of speech, and producing an immense variety of verbal forms.[47]

Where in the Old World, Du Ponceau asked, could these linguistic traits be found? Vater had thought to find Indian-like combinations of the verb in the languages of the Basques, the Tschuktschi, and some of the Congo tribes, but Du Ponceau presented evidence to show that these tongues were essentially different in their construction from the Indian languages. As for the Basque language, Humboldt had shown in his essay in the fourth volume of the *Mithridates* that it was unlike any other in the world.

> It is preserved in a corner of Europe, by a few thousand mountaineers [wrote Du Ponceau], the sole remaining fragment of, perhaps, a hundred dialects, constructed on the same plan, which probably existed and were universally spoken at a remote period in that quarter of the globe. Like the bones of the Mammoth, and the shells of unknown fishes, the races of which have perished, it remains a frightful monument of the immense destruction produced by a succession of ages.[48]

Having presented his findings, Du Ponceau left it to his readers to judge for themselves whether it was true, as had long been thought, that the Indians had few ideas and few words to express them. "For my part," he declared, "I confess that I am lost in astonishment at the copiousness and admirable structure of their languages, for which I can only account by looking up to the GREAT FIRST CAUSE."[49]

The publication of Du Ponceau's "Report," the Du Ponceau–Heckewelder correspondence, and Heckewelder's history of the Indian tribes gave a strong impetus to Indian studies in both Europe and America. The entire volume was translated into German, and Heckewelder's "Account" into French. Du Ponceau's foreign correspondence increased rapidly. Among his European correspondents were Friedrich Adelung in St. Petersburg, Wilhelm von Humboldt in Berlin, Joseph von Hammer-Purgstall in Vienna, Johann A. Albers in Bremen, and Johann S. Vater in Königsberg. Vater sent Du Ponceau his *Index Linguarum* with a request for a number of Indian vocabularies. Du Ponceau, who was engaged in translating Vater's *Inquiry into the Peopling of America,* replied that the Long expedition to the Rocky Mountains had brought back vocabularies of the Winnebago, Shoshoni, Crow, Wahtoktata, Kansas, Omaha, Yankton Sioux, Pawnee, and Minnetaree Indians and that he was having copies made for his European friends. He requested a copy of the German translation of the *Transactions* of the Historical and Literary Committee, adding:

I am now employed about a complete Umarbeitung [reworking] of the late Dr. Barton's comparative vocabulary of Indian languages, to which I have already added 40 or 50 & corrected many errors in those he has given. I shall try to use an uniform mode of spelling the words, & shall give my authorities for every one. I shall avoid forming any theories, but merely give the facts, that the learned world may afterwards theorize upon them at their leisure & pleasure.[50]

Concerning the conditions under which American scholars worked, Du Ponceau wrote:

If Americans could devote as much time to literature as Europeans, you may be sure that there are talents in this country that would display themselves. But the Law, Physic, politics &c takes up the attention of most of our learned men. Every man here lives by his labour, with few exceptions, & there are not enough of rich men to encourage scientific investigations. . . . For my part, I have never received one cent for all I have written out of my profession; & whenever I have devoted much of my time to any literary pursuit, I have always found my book of receipts diminished by at least one half.[51]

Despite these handicaps, Du Ponceau pressed ahead with his researches, much encouraged by the "great excitement" (as he described it to Heckewelder) that the publication of the committee's *Transactions* had produced. Heckewelder and Jefferson continued to supply him with materials, and Heckewelder put him in touch with new informants among the missionaries. The Long expedition supplied vocabularies of several of the northern plains tribes, although those collected farther west were lost when some of Long's men deserted to the Indians, taking the baggage containing them. Further information about the Indian tribes and their languages was published in Jedidiah Morse's report to the secretary of war concerning his government-sponsored tour of the Indian tribes in the Great Lakes region in 1820. As a result of the Cass expedition in search of the sources of the Mississippi River in the same year, both Lewis Cass and his fellow-explorer Henry R. Schoolcraft became interested in Indian lore and languages. As governor of the Michigan Territory (covering the modern states of Michigan, Wisconsin, and a part of Minnesota), Cass sent out to Indian agents, traders, and military personnel in that region a wide-ranging Indian questionnaire that included word lists and other linguistic queries. Cass published only a small part of the material he collected in this way; but Schoolcraft, who married a half-breed Ojibway woman, soon became a major contributor of firsthand knowledge of Indian languages and legends.[52]

THE DU PONCEAU-PICKERING COLLABORATION

The publication of the Du Ponceau–Heckewelder correspondence and memoirs had awakened an interest in Indian languages in the mind of another lawyer-linguist, John Pickering of Salem, Mass. Pickering was the son of Timothy Pickering, secretary of state in John Adams's administration and a friend of Du Ponceau. The younger Pickering had been an avid student of languages from an early age. By the time he joined his father in Philadelphia in 1796 to pursue his legal studies, he had achieved a mastery of Greek, Latin, and French and begun the study of Spanish. He added Portuguese to his linguistic repertoire during two years service in the American embassy in Lisbon. In 1806 he was offered but declined the newly established Eliot Professorship of Greek at Harvard, choosing instead the practice of law in his native Salem. His first contact with Indian languages occurred in 1810, when a chief of the Oneida tribe visited Salem and Pickering obtained the alphabet and a small vocabulary of the Oneida language from him. In 1815 the American Academy of Arts and Sciences published Pickering's "Memoir on the Present State of the English Language in the United States, with a Vocabulary," and he began work on a Greek-English lexicon adapted from the Greek-Latin lexicon of Cornelius Schrevelius. Three years later his "Memoir on Greek Pronunciation" appeared and attracted much attention in both Europe and America.[53]

Stimulated by the publication of the Du Ponceau–Heckewelder memoirs and correspondence, Pickering now added Indian languages to his linguistic studies. He began with Roger Williams's vocabulary of the Narragansett language and a manuscript dictionary of the Norridgewock dialect written by the Jesuit missionary Sebastien Rasles and preserved at the Harvard College Library. He soon sensed the need for a uniform method of representing the sounds in these languages and in 1820 published his essay "On the Adoption of a Uniform Orthography for the Indian Languages of North America," described by a modern linguist as "nothing more nor less than a start towards an international phonetic alphabet."[54] The essay was published in the *Memoirs* of the American Academy of Arts and Sciences, "by way of exciting a little interest in the *Indian* languages," Pickering wrote to Du Ponceau. In the proposed orthography the vowel sounds were those of the German and Italian languages. The nasals were indicated by a comma or cedilla under each nasal vowel. The English *sh* was preserved and its correlative *zh* adopted for the sound of the French and Portuguese *j*. Compound consonant sounds were represented by their component signs, as *ks, ksh, ts, tz,* and the like. No new characters were introduced. In the opinion of a twentieth-century linguist, Franklin Edgerton of Yale University, Pickering's orthography was "crude and rudimen-

Four major contributors to the study of Indian linguistics in the United States: Peter S. Du Ponceau (by Thomas Sully) (upper left), John Heckewelder (attributed to Gustav von Seckendorff) (upper right), John Pickering (by Chester Harding) (lower left), and Albert Gallatin (by William H. Powell) (lower right). Courtesy of the American Philosophical Society (Du Ponceau and Heckewelder); the Essex Institute, Salem, Mass. (Pickering); and the New-York Historical Society (Gallatin)

tary when judged by modern standards" but highly creditable as a first effort at providing a much needed tool.

> His alphabet was adopted by missionary societies, and it exerted an important and useful influence. . . . It should be emphasized that he did not expect his symbols to suffice for all, or even fully for any Indian languages. He was merely making a praiseworthy attempt to introduce a minimal degree of order into the dreadful confusion which had prevailed up to then, and which still makes it so hard to know what sounds those early writers were trying to represent by the letters they used.[55]

Spurred on by Pickering, the Massachusetts Historical Society, which had earlier republished Roger Williams's vocabulary of the Narragansett dialect, undertook to republish other early New England writings on Indian languages with introductions and notes. In 1822 John Eliot's *Grammar of the Massachusetts Indian Language* appeared, in 1823 the *Observations on the Language of the Muhhekaneew Indians* of Jonathan Edwards, Jr., and in 1829 Josiah Cotton's *Vocabulary of the Massachusetts (or Natick) Indian Language.* In his "Introductory Observations" to the first of these, Pickering took occasion to reinforce what Du Ponceau had written earlier about the importance of the study of Indian languages for the science of linguistics. Citing Friedrich Adelung's *View of All the Known Languages and Their Dialects,* published in German in St. Petersburg in 1820, he noted that 1,214 dialects had been identified in the New World;[56] but these were not radically different languages, as Jefferson and others had supposed. In North America, said Pickering, the numerous Indian dialects could probably be reduced to four families: the Karalit spoken by the Greenlanders and Eskimos; the Delaware; the Iroquois; and the Floridian, embracing the languages spoken on the southern frontier of the United States. In New England the principal Indian nations were the Pequots, Narragansetts, Pawkunnawkuts or Wampanoags, Massachusetts, and Pawtuckets, all of whom spoke dialects of the Delaware or Lenape language.

In the "Notes and Observations" appended to this edition of Eliot's *Grammar,* Du Ponceau stressed the apparent fact that the American languages and dialects, although they numbered more than all those of Asia and Africa and almost as many as those of the Old World generally, nevertheless exhibited a common and peculiar grammatical structure, for which Du Ponceau had coined the adjective *polysynthetic.* "And yet," he added, " our American idioms, except where they can be traced to a common stock, differ so much from each other in point of *etymology,* that no affinity whatever has been yet discovered between them." As for linguistic connections between the New and Old World, the only link yet discovered was the language of the sedentary Tschuktschi inhabiting the vicinity of Norton Sound in northeastern Siberia. Their language was plainly a dialect of the Karalit tongue. It was to be hoped, said Du Ponceau, that Russian scholars would investigate the languages of other tribes in northeastern Asia to determine whether further links to the American languages could be found. Further study might also discover some trace in the Americas of the Malay (Polynesian) language, whose wide geographic spread from Madagascar to the Pacific islands had led James McCulloh, Jr., of Baltimore to suppose the former existence of a Pacific continent from which the New World had been peopled.[57]

Passing from the realm of speculation to that of grammatical analysis, Du Ponceau proceeded to demonstrate the essential affinity between the language of the Massachusetts Indians and the other Delaware dialects despite the letter changes among them. He then showed that the Delaware language contained no word equivalent to the English verb *to be,* drawing heavily on Heckewelder for support. These notes by Du Ponceau were followed by an alphabetical index of Indian words drawn from Eliot's *Grammar* and from his translation of the Bible. The seventy words in this index corresponding to those compared in Barton's *New Views* were printed in small capitals to enable the reader to compare them with Barton's vocabularies. As an addendum Pickering printed an extract of a letter from Du Ponceau showing that the supposed Nanticoke numerals published by Barton in the Appendix to his *New Views* were those of the Bambara tribe in Africa. Barton claimed to have received the list from the Moravian missionary Christopher Pyrlaeus through Heckewelder, but Heckewelder, while agreeing that the list was found among Pyrlaeus's papers, denied that the numerals were those of the Nanticoke Indians. A true list had been sent by him to Thomas Jefferson, who passed it on to Du Ponceau.

Jefferson was doubtless pleased that one of the vocabularies he had collected had proved useful in this way, but he apparently never accepted the view of Pickering and Du Ponceau that the Indian languages could be grouped into a few families. Writing to Pickering in February 1825, he declared:

> I am persuaded that among the tribes on our two continents a great number of languages, radically different, will be found. . . . I had once collected about 30 vocabularies, formed of the same English words. . . . These were unfortunately lost, but I remember that on a trial to arrange them into families and dialects, I found, in one instance, that about half a dozen might be so classed; in another three or four; but I am sure that a third at least . . . were perfectly insulated from each other. Yet this is the only index by which we can trace their filiation.[58]

In the ensuing years Pickering and Du Ponceau continued their linguistic researches in such time as they could spare from their professional duties. Both were in touch with leading European authorities, exchanging ideas, information, and publications freely. Pickering published Sebastien Rasles's dictionary of the Abnaki language with an introduction and notes in 1833 and contributed a substantial appendix on Indian languages to the sixth volume of the *Encyclopaedia Americana* (1831), in which he summarized the progress of research in Indian linguistics and presented a systematic analysis of Indian grammar as illustrated in the Delaware and Cherokee languages.[59] In 1830 Du Ponceau published his translation of Zeisberger's Delaware grammar in the *Transactions* of the American Philosophical Society. In the preface he paid his respects to the *Mithridates,* "the glory of our science," and, less enthusiastically, to Barton's *New Views.* If Barton had extended his word lists to all the words in Pallas's *Lexicons* instead of speculating about Indian origins, said Du Ponceau, he would have made a greater contribution to comparative linguistics. Nevertheless his was the first attempt to collect and compare specimens of the Indian languages: "As such it is useful to philologists and entitled to respect." The *Mithridates* was a much more impressive contribution, but it also had been eclipsed by the rapid progress of philological science.[60]

After reviewing the progress of philology to 1830, giving credit to the American Philosophical Society for having established the polysynthetic character of the American languages, Du Ponceau returned once again to the defense of so-called "barbarous" languages, arguing that they, like all languages, were the product of the "natural logic" with which the Creator had endowed every human mind. A careful study of Zeisberger's grammar, he declared, should convince any reasonable person that the structure, or "organization," of language was the product of nature and not of civilization and its arts. Civilization might cultivate and polish a language to a certain extent, but it could no more alter its organization "than the art of the gardener can change that of an *onion* or a *potato.*" To believe otherwise, Du Ponceau asserted, was to succumb to pride, "a passion inherent in our nature, and the greatest obstacle that exists to the investigation of truth."[61] Thus what the modern linguist Noam Chomsky attributed to genetic endowment Du Ponceau attributed to nature and nature's God.

The climax of Du Ponceau's work on Indian linguistics was his *Mémoire sur le système grammatical des langues de quelques nations indiennes de l'Amérique du Nord,* which was awarded the Volney Prize of the Royal Institute of France in 1835 and was published in Paris in 1838. The prize committee had set as the problem for investigation "to determine the grammatical character of the North American languages of the Lenni-

Lenape, Mohegan, and Chippeway tribes." Du Ponceau accomplished this with great skill, drawing heavily on Schoolcraft, however, for his account of the Chippewa language. To the family of languages he had earlier called "Delaware" he now gave the name Algonquin, "to distinguish it from those of the Iroquois, the Sioux, the Eskimo, and the Floridian." The Algonquin languages, he explained, were similar to each other both grammatically and etymologically. To prove the latter point, he submitted a comparative vocabulary of forty-five words in these languages or dialects. By contrast, a comparative vocabulary of 250 words drawn from an Algonquin and an Iroquois tribe revealed only one word in common. Du Ponceau stressed again the polysynthetic character of the Indian languages, although he now had to concede that one of them, that of the Othomi tribe in Mexico, had been shown to be monosyllabic. He did not jump to the conclusion that America had been invaded from China, however. What reliance, he asked, could be placed on comparisons between the languages of widely separated peoples when one knew that those of neighboring peoples like the Iroquois and the Algonquins, though similar in grammatical structure, had very few words in common?[62]

In the preface and introductory chapters of his memoir Du Ponceau ranged over the whole field of linguistics, advancing general conclusions of great interest, some of which had been foreshadowed in the article "Language" he had contributed to the same edition of the *Encyclopaedia Americana* in which Pickering's essay on Indian languages appeared. After reviewing the progress of comparative philology, Du Ponceau launched into an attack on speculative theories of the origin of language, the search for the primitive language of mankind, and efforts to construct a "philosophical" language and universal grammar. The original language of mankind had become extinct, probably in the confusion of tongues at the tower of Babel, he declared. In any case, the existing languages of the world were constructed on several quite different plans, each of which was at least 4,000 years old and probably older. As for attempts to construct a philosophical language, they were doomed to failure because languages, like flowers and fruits, were products of nature, not of art. If one were to try to construct such a language, which of the various linguistic structures nature had produced would be chosen as the model? The idea of a universal grammar was equally futile, said Du Ponceau. "We are in the century of comparative sciences" — comparative anatomy, not universal anatomy; comparative philology, not general grammar. The age of speculation had been superseded by the age of induction from solid facts.

But although Du Ponceau decried speculation in favor of empirically based research, he could not resist attempting to explain how the various types of language extant in the world had been formed after the extinction of the original language of mankind. As the various groups of people dispersed into the world from the tower of Babel, he conjectured, each group was forced to originate language anew, using their God-given powers of associating sounds and ideas to form language on some plan.

> Thus was produced the difference of languages according to the character of those who presided at their formation, modified to a certain extent by climate and other local circumstances; from this have sprung analytical, synthetic, monosyllabic, polysyllabic, and inverted languages, languages in which the words follow each other in a more or less natural order, inflected languages, languages with particles, prefixes, and suffixes, and all others which form the variety observable in different idioms. The first impulse given by the first creators [*faiseurs*] was followed by those who succeeded them, for man is more prone to imitate than to invent; thus all languages have a manifest tendency to preserve the grammatical character originally imprinted on them.[63]

Whatever the merit of these speculations, Du Ponceau could rest content in the knowledge that he had done more than anyone else to establish the study of Indian languages on a firm empirical basis; to call attention to their "polysynthetic" structure; to combat the widespread idea that the languages of uncivilized peoples were lacking in order, complexity, and richness of expression; and to create a respected place in the burgeoning science of comparative philology for American scholars. More specifically, twentieth-century students of Indian languages have credited him with recognizing that the language of the Osages belonged to the Sioux family and that of the Cherokees to the Iroquois family and that the languages of eastern Siberia were of the American Indian type and distinct from other Asiatic languages.[64]

THE CLIMAX OF THE JEFFERSONIAN METHOD: ALBERT GALLATIN'S *Synopsis*

By an odd and appropriate coincidence the first systematic comparative treatment of the Indian languages of North America, Albert Gallatin's *Synopsis of the Indian Tribes,* was the work of a friend and former colleague of Thomas Jefferson and was accomplished by the means Jefferson had recommended, namely, the comparison of Indian vocabularies. According to Gallatin, the stimulus for the production of this important work came from Alexander von Humboldt. The two men first met when Humboldt visited Washington, D.C., in 1804 on his way back to Europe from Latin America. He was "of small figure, well made, agreeable looks, simple unaffected manners, remarkably sprightly, vehement in conversation and sometimes eloquent," and loaded down with statistics, books, and maps of the greatest interest to both Jefferson and Gallatin.[65] Gallatin's acquaintance with Humboldt ripened into friendship when Gallatin served as American minister to France in 1816–1823 and entertained Humboldt frequently at his house. According to Gallatin's son James, "Mr. Gallatin never was so happy and never so thoroughly in his proper social sphere as when he lived in Paris and talked of Indian antiquities with Humboldt."[66]

Urged on by Humboldt, Gallatin plunged into Indian studies on his return to the United States. His first efforts, communicated to Humboldt in 1823, were passed on by Humboldt to the Italian geographer and ethnographer Adriano Balbi and published with suitable acknowledgments in his *Introduction à l'Atlas Ethnographique*.[67] Meanwhile, Gallatin was busy gathering new data. From a group of southern Indian chiefs who met in Washington in the winter of 1825–1826 he obtained vocabularies of the Muskhogee, Uchee, Natchez, Chickasaw, and Cherokee languages. Shortly thereafter Gallatin persuaded the War Department to circulate printed forms containing a list of 600 words for which Indian equivalents were requested. The circular also contained a number of verbal forms, selected sentences, and a series of grammatical queries. Du Ponceau generously supplied the vocabularies he had collected, and the American Philosophical Society and the historical societies of New York and Massachusetts provided further information. In 1836, thirteen years after his first essay for Humboldt, Gallatin's revised and expanded *Synopsis* was published in the *Transactions and Collections* of the American Antiquarian Society.

In this work the Indian tribes were classed in families "according to their respective languages." These linguistic families in turn were determined by comparison of vocabularies. A family of languages, said Gallatin, included "all those which contained a number of similar primitive words, sufficient to show that they must, at some remote epoch, have had a common origin."

> The eighty-one tribes . . . embraced by the Synopsis, have been divided into twenty-eight families. A single glance at the annexed Map will show, that, excluding the country west of the Rocky Mountains and south of the fifty-second degree of north latitude, almost the whole territory contained in the United States and in British and Russian America is or was occupied by only eight great families, each speaking a distinct language, subdivided . . . into a number of languages or dialects belonging to the same stock. These are the Esquimaux, the Athapascas (or Cheppeyans), the Black Feet, the Sioux, the Algonkin-Lenape, the Iroquois, the Cherokee, and the Mobilian or Chahta-Muskhog. . . . Of the nineteen others, ten are west of the Stony Mountains. . . . Six of the remaining nine families . . . are found among

the southern tribes, either annexed to the Creek confederacy, or in the swamps of West Louisiana. The three others are the Catawbas, the Pawnees, and the Fall or Rapid Indians. Some new families . . . will hereafter be found. . . . Many distinct languages or dialects of the Esquimaux, of the Athapascas, and of some of the other great families, will be added to the present enumeration. But I believe that the classification now submitted will, as far as it goes be found correct.[68]

Gallatin's confidence in his linguistic comparisons was not misplaced. He admitted the difficulty of comparing vocabularies collected by missionaries, travelers, Indian agents, soldiers, and others of different nationalities; but he insisted that, with practice gained by comparing several vocabularies of the same language, one could learn to use this tool effectively. Subsequent investigation has vindicated his confidence. When John W. Powell published his *Indian Linguistic Families of America North of Mexico* (1891) he paid tribute to Gallatin's achievement in the *Synopsis* and in his subsequent works.

As Linnaeus is to be regarded as the founder of biologic classification, so Gallatin may be considered the founder of systematic philology relating to the North American Indians. Before his time much linguistic work had been accomplished, and scholars owe a lasting debt of gratitude to Barton, Adelung, Pickering, and others. But Gallatin's work marks an era in American linguistic science from the fact that he so thoroughly introduced comparative methods, and because he circumscribed the boundaries of many families, so that a large part of his work remains and is still to be considered sound. . . . Gallatin's work has therefore been taken as the starting point, back of which we may not go in the historic consideration of the systematic philology of North America.[69]

Jefferson would no doubt have been pleased at his former secretary of the treasury's success in applying the method of comparative vocabularies to Indian languages, though he would have wished that his own vocabularies, especially those collected by Lewis, had been available to him. He would have been less satisfied, however, with some of Gallatin's conclusions with respect to the speculations advanced in Jefferson's *Notes*. Contrary to Jefferson's opinion, Gallatin thought that the number of linguistic families, distinct languages, and dialects among the Indians was no greater than among uncivilized tribes in other parts of the world, such as Africa, Siberia, and the Pacific islands. Such diversity as there was sprang from the nomadic life of these tribes, said Gallatin, and indicated no greater an antiquity for these tribes than was consistent with the biblical chronology. The proof of the common origin of the Indian tribes was to be found, not in etymological similarities, but in the common grammatical structure of their languages first noted by Du Ponceau. Like Pickering, Gallatin delineated this structure at some length. Its uniformity throughout the Americas, he observed, suggested that the American tribes had all come from one place in the Old World, probably northeastern Asia, although the grammatical structure of the languages spoken in that region was not sufficiently well known to render this certain.

> The much greater facility of communication, either across Behring's straits, or from Kamschatka or Japan by the Aleutian Islands, would alone, if sustained by a similarity of the physical type of men, render the opinion of an Asiatic origin not only probable, but almost certain. The rapidity with which the human species may be propagated under favorable circumstances removes any apparent inconsistency between that opinion and the early epoch . . . which must be assigned to the first appearance of man in America. . . .
>
> Assuming the central parts of Asia to have been the cradle of mankind, and since three couples would, in thirty periods of duplication, increase to more than six thousand millions of souls, we may fairly infer . . . that America began to be inhabited only five or six hundred years later than the other hemisphere.[70]

These observations, Gallatin added, were intended only to show "that there is nothing in the American languages and the early epoch which may thence be deduced of the American population . . . inconsistent with the opinion of an Asiatic origin and with the received chronology." Thus was the argument of Jefferson's *Notes* refuted.

There remained, however, the problem of accounting for the relatively civilized cultures of Mexico and Central and South America and for the constructions of the mound builders in the Mississippi valley. As to the latter, Gallatin thought that they could have been constructed by peoples not much more advanced than the present Indians, though evidently more populous and agricultural. Some of the mounds resembled those found in Mexico, but it was more likely that this mode of construction had spread northward from Mexico than southward from the Mississippi valley. As for the Mexican and Peruvian monuments, paintings, hieroglyphics, calendars, and the like, they should be compared carefully with those of eastern Asia, as Humboldt had done. But if the comparison failed to establish the Asian origin of these civilizations by way of the Pacific Ocean, there was no reason to suppose that the descendants of the primitive tribes that migrated to America across the Bering Strait were not capable of developing civilized institutions when they came into regions favorable to a settled agricultural existence.

> On the probable supposition, that the whole continent of America was inhabited one thousand years after the flood, or near four thousand

years ago, the faculties of man, gradually unfolded and improved, may, in the course of as long a period, have produced, without any extraneous aid, that more advanced state of society and of knowledge, which existed in some parts of America, when first discovered by the Europeans. Those centres of American civilization were all found precisely in those places, where we might have expected to find them, if that civilization was of domestic origin.[71]

Gallatin's time scheme, like that of most of his contemporaries, was badly foreshortened, but he saw clearly that the development of civilization and language must differ markedly in agricultural and in hunting and gathering societies. The latter, he observed, were precluded from developing civilization by the circumstances of their existence, which compelled them to separate into small independent communities engaged in perpetual warfare to defend their hunting grounds. The diversity of their languages stemmed from the same causes. The complexity of those languages, Gallatin argued, was no proof that they were products of profound philosophical meditation.

All those inflections, however varied, never contain, independent of the root of the verb, any other ideas, but those of two pronouns, respectively agent and object of the action. . . . The fact that, although the object in view was, in every known Indian language without exception, to concentrate in a single word those pronouns with the verb, yet the means used for that purpose are not the same in any two of them, shows that none of them was the result of philosophical researches and preconcerted design. . . . Those transitions, in their complexness and in the still visible amalgamation of the abbreviated pronouns with the verb, bear in fact the impress of primitive and unpolished languages.[72]

CONCLUSION

With the publications of Gallatin, Du Ponceau, and Pickering in the 1830s the formative period of American comparative linguistics, which began with Jefferson's *Notes* and Barton's *New Views,* may be said to have come to an end, although Gallatin continued to contribute to this field of research until his death in 1849.[73] Jefferson's project of collecting "all the languages spoken in North and South America, preserving their appellations of the most common objects in nature . . . , with the inflections of their nouns and verbs, their principles of regimen and concord" had been carried a long way by a band of investigators, none of whom was a professional linguist. Politicians, explorers, missionaries, Indian agents, lawyers, and a physician-naturalist had collaborated with each other and with a few scientific and learned societies in collecting, analyzing, and generalizing the linguistic data. The result was a substantial intellectual achievement that won for American science and scholarship the European recognition that these men coveted not only for themselves but more especially for their country.

15

The End of the Jeffersonian Era

IT is easier to specify the beginning of the Jeffersonian era in American science than to say when it ended. It began with the achievement of independence, the rejuvenation of the American Philosophical Society and the founding of the American Academy of Arts and Sciences, and Thomas Jefferson's *Notes on the State of Virginia.* In that wide-ranging compendium of information about Virginia's institutions, laws, and civil and natural history Jefferson set the tone and posed the problems for much of American science in the early national period. One by one he introduced topics that were to dominate scientific inquiry for several decades: American climate and geography; the problem of explaining gaps in the Appalachian chain and the presence of marine fossils high in the mountains; the mystery of the gigantic bones and teeth of the "mammoth"; the enumeration of American plants and animals and their comparison with European species; the races of man and their physical, mental, and moral peculiarities; the question of Indian origins in relation to Indian languages and the constructions of the mound builders. Physics, chemistry, and the medical sciences escaped his notice, but only because the queries to which he addressed himself offered little opportunity to discuss them.

Both natural history and natural philosophy found a home in the American Philosophical Society and its sister institution in Boston, the American Academy of Arts and Sciences. The medical sciences were welcome there too, although the medical men soon founded journals of their own, notably Samuel L. Mitchill's *Medical Repository,* that served the needs of both medicine and science. At the same time the medical schools began to supplement the traditional college curriculum in natural philosophy with courses in chemistry, mineralogy, botany, materia medica, and anatomy with its offshoots in comparative anatomy, paleontology, and physical anthropology. An assortment of natural history museums, tree nurseries, botanical gardens, and historical societies interested in natural as well as civil history completed the roster of institutions supporting scientific inquiry.

The era inaugurated by these activities drew to its close toward the end of the second decade of the nineteenth century. In 1815 Jefferson retired from the presidency of the American Philosophical Society. Soon afterward he sent what was left of his Indian vocabularies to the society and arranged for it to receive whatever documents and other materials relating to the Lewis and Clark expedition could be retrieved from the persons to whom they had been entrusted. Jefferson lived on to 1826, but most of the founding fathers of American science passed from the scene much earlier. Benjamin Rush and Alexander Wilson died in 1813, Benjamin S. Barton and Henry Muhlenberg in 1815, Archibald Bruce and Caspar Wistar in 1818, Andrew Ellicott in 1820, William Bartram in 1822. Charles W. Peale outlived Jefferson by a year, but his museum had long since passed its zenith. Mitchill remained active until his death in 1831, but by that time his *Medical Repository* had ceased to be the general scientific journal it once had been. Jedidiah Morse turned over responsibility for the seventh edition of his *Universal Geography* (1819) to his son Sidney. The future of American science lay with a new generation that had already begun to make itself felt: Nathaniel Bowditch in astronomy; Benjamin Silliman and Robert Hare in chemistry; William Maclure and Amos Eaton in geology; John Torrey, Constantine Rafinesque, Thomas Say, Thomas Nuttall, John Godman, and Richard Harlan in the various branches of natural history; Caleb Atwater and Ephraim G. Squier in archaeology; Peter S. Du Ponceau, John Pickering, and Albert Gallatin in Indian linguistics. Many of these individuals carried over Jeffersonian ideas and attitudes into the post-Jeffersonian period; but the political, economic, social, and intellectual environment in which they worked was altered. The transition was gradual, but by the 1830s the change of context was unmistakable.

Beginning about 1815 there was a noticeable increase in the pace of scientific activity. In rapid succession were founded the Academy of Natural Sciences of Philadelphia, the literary and philosophical societies of New York and South Carolina, the Linnaean Society of New England, the Lyceum of Natural History of New York, the American Geological Society, and the Western Museum Society. As the number of colleges multiplied; as state agricultural and geological surveys were established; as steamboats, canals, railroads, and mining enterprises created a demand for scientific and technical expertise; and as the federal government expanded its activities in western and overseas exploration and in ordnance and coast survey, jobs were created for professional scientists. The leading figures in American science were less and less frequently men whose main business was something other than science. An increasing number were Americans by birth. Before 1820 much of the best "American" science had been the contribution of visiting or immigrant scientists like Maclure, Wilson, Pursh, Nuttall, Rafinesque, and Lesueur. In the future, native-born Americans like Silliman, Henry, William and George Bond, Peirce, Hare, Eaton, Bache, Hitchcock, Hall, Gibbs, Morton, Torrey, and Gray would dominate the American scene. By the 1840s and 1850s they would begin to win international recognition for their scientific work: Henry in physics; the Bonds and Peirce in astronomy and celestial mechanics; Bache in terrestrial magnetism; Espy, Redfield, and Loomis in meteorology; Maury in oceanography; Dana and Hall in geology; Harlan, Hall, and Leidy in paleontology; Squier in archaeology.[1]

Meanwhile, the religious context of American science had undergone important changes. In the Jeffersonian period the debate between the Christians and deists was dominant. In the election of 1800 the Federalists tried to discredit Jefferson by linking his interest in science to the "infidel" ideas of Thomas Paine and Joseph Priestley, but for the most part the Christian-deist controversy did not focus on scientific discoveries and ideas. Peale found it prudent to stop calling his museum a "temple of nature," but since deistic natural theology was simply Christian natural theology divorced from its traditional marriage to revealed theology, it was difficult to distinguish deist rhetoric about nature from Christian rhetoric. The two sides were at loggerheads about the Bible, but science had not yet begun to present a major threat to received ideas about the inspiration of Scripture. Defenders of the Bible had long since accepted the Copernican and Newtonian revolutions in cosmology. Paine did his best to resurrect these issues in his *Age of Reason,* but in general the deists relied less on particular scientific discoveries to discredit the Bible than they did on the mechanical view of nature as a law-bound system of matter in motion (interpreted as precluding miracles) and on Locke's idea that all human knowledge is derived from sense experience (interpreted as precluding "revealed" knowledge). The Christians, for their part, were generally content with Bishop Watson's *Apology for the Bible* and William Paley's *View of the Evidences of Christianity* as sufficient refutations of deism.[2]

This is not to say that new discoveries and theories in science were of no concern to devout Christians, but only that the concern was incipient rather than dominant. Jedidiah Morse and Samuel Miller invoked the biblical flood to explain various geological phenomena and defended the traditional chronology as sufficient to encompass the total history of the earth (see Chap. 9). Morse's *Universal Geography* also discussed the difficulty of accommodating the ever-growing number of known animal species within Noah's Ark and explaining their subsequent dispersion over the surface of the earth. How had the animals inhabiting the New World arrived at their places of residence? According to Morse, various writers had suggested that these animals had been transported to their destinations by angels, had swum there, had been carried there by men to provide prey for the hunt, had been formed there by the Creator in the beginning, or had passed there by no longer existing land bridges. St. Augustine's resort to angels, Morse observed, would solve the difficulty neatly but "would not be satisfactory in the present age." Instead, Morse agreed with Acosta, Grotius, Buffon, and others that the passage to the New World had been made by way of some ancient land bridge across the South Atlantic. But why had certain species completely deserted the Old World for the New, while others never left their original haunts? This difficulty, Morse observed, lay against all explanations except St. Augustine's. In any case, the two hemispheres had not been explored sufficiently to warrant the conclusion that their fauna were substantially different.[3]

That this issue was not of major concern to apologists for the Bible is shown by the fact that Barton incurred no censure when he took the position that there had been separate creations of animals in the New World. He was willing to grant that Asia and North America had probably once been connected by land and that many tribes of animals and humans had passed back and forth. But if there had also been a land bridge between Africa and South America, as Clavigero supposed, why were there so few African species in South America and vice versa? Migration could explain much, but most animals were incapable of enduring great changes of climate. "In the present constitution of these animals, and in the present temperature of the globe," wrote Barton, "the musk-deer of Thibet could not have travelled to the forests of Mexico or Peru, and the pacos could not have reached the mountains of the Caucasus. It seems necessary, then, to have created different species of animals in different parts of the world."[4] Yet Barton, like Morse, seems to have been satisfied with "some thousands of years" for the history of life on earth. And Jefferson considered it an unanswerable and uninteresting question whether God had created the world in six days or in six million years.

The only scientific issue bearing on the credibility of traditional ideas about the inspiration of the Bible that seems to have stimulated much controversy was the monogenist-polygenist debate concerning the unity of the human species. Samuel S. Smith's *Essay on the Variety of Complexion and Figure in the Human Species* was written expressly to vindicate the scriptural doctrine of the unity of mankind in their descent from Adam and Eve and, more generally, the biblical account of early human history. But the skirmishing between Smith and medical polygenists like Charles White of Manchester, John A. Smith, and Charles Caldwell was a slight affair compared to the full-scale warfare over this issue in the period after 1820.[5]

With the revival of evangelical orthodoxy after the War of 1812, the religious setting of scientific activity was profoundly altered. Deism, whether of the moderate variety espoused by Franklin and Jefferson or of the crusading type favored by Thomas Paine and Ethan Allen, was thrown on the defensive. Geology, paleontology, archaeology, anthropology, Lamarckian biology, and German biblical criticism posed major challenges to the traditional conception of biblical inspiration, but the scientists themselves (some of them at least) shouldered the task of reconciling science and Scripture. When Silliman was considering launching a scientific journal in 1817, he asked Parker Cleaveland: "Would our infidels and light minded folk bear to have geology made to illustrate the truth of the Mosaic history of which I am fully persuaded it demonstrates the truth?"[6] Events were to show that Silliman's fears were unfounded. Not only in the pages of the *American Journal of Science and Arts* but also in college halls, on the public lecture platform, and in innumerable books and pamphlets the harmony of science and religion was proclaimed throughout the land. Dissenting voices were heard, especially in the monogenist-polygenist debate, but the easygoing religious liberalism of the Jeffersonian era no longer pervaded the scientific community.

In his stimulating book *Yankee Science in the Making* Dirk Struik has attempted to distinguish the science of the Jeffersonian period from that which followed by a Marxian analysis linking science and technology to class interest.[7] Struik argues that New England science in the early national period was predominantly "Newtonian" in character, emphasizing astronomy and natural philosophy in keeping with the mercantile interests of the ruling classes in that section of the country, and that the rise of manufacturing during and after the War of 1812 brought about a shift of scientific interest toward chemistry, geology, and natural history. The Jeffersonians, he maintains, were more democratic and hence more favorable to science than the Federalists. New England scientists and clergymen were "allies" of the dominant mercantile aristocracy and hence were interested in harmonizing science and religion in order to maintain the status quo.

This Marxian view of early developments in American science is not very convincing, either for New England (where Struik applies it) or for the United States as a whole. The emphasis on mathematics and natural philosophy in the colleges in New England and elsewhere was due primarily to the fact that these subjects were part of the traditional college curriculum brought to America from England. There is little evidence to show that the astronomical researches of John Winthrop IV at Harvard or David Rittenhouse in Philadelphia had much to do with navigation. Except in New England, astronomy was linked to surveying rather than to navigation, largely because there were many boundaries to be surveyed. Andrew Ellicott, who made much of his living as a surveyor, was well aware of the importance of astronomy for navigation, but he was never able to persuade politicians to recognize it. Struik himself concedes that New England sea captains continued to sail by dead reckoning well into the nineteenth century. It was 1839 before Harvard managed to establish an astronomical observatory, and the great comet of 1843, not the practical needs of Massachusetts commerce, convinced the Boston merchants that Harvard must have a telescope equal to the best European instruments. Yale acquired a respectable telescope in 1828 and Wesleyan and Williams colleges followed that example in the 1830s, but without reference to the needs of commerce and navigation.

As for chemistry, mineralogy, geology, botany, and natural history generally, these subjects made their way into the academic curriculum through the medical schools. As soon as Harvard had a medical school, it acquired courses in chemistry, botany, and mineralogy, and these subjects were eventually taught to undergraduates as well as medical students. In 1805 the merchants of Boston and the Massachusetts legislature contributed handsomely, not to establish an observatory and a chair in astronomy, but to found a professorship of natural history and a botanical garden. True, John Lowell was investigating the possibility of building an observatory at Harvard about this time, but nothing came of it, partly because of the great expense of the undertaking. The truth is that the supporters of Harvard were interested in making the college the best in every field of study, scientific or otherwise. They showed no special favoritism to "Newtonian" science over other sciences and endeavored to introduce new fields like chemistry, mineralogy, and geology as rapidly as seemed feasible. Bowditch was offered a professorship at Harvard, but so was Cleaveland. Unfortunately, neither man accepted. The offer to Cleaveland was probably inspired by envy of Silliman's success in promoting geology and mineralogy at Yale. Yet it is difficult to see any connection between the rapid rise of chemistry, mineralogy, and geology at Yale and the economic interests of the New Haven aristocracy. President Timothy Dwight felt sure that all these subjects would eventually have important practical applications, but he was equally concerned to have them demonstrate the harmony of science and religion.

Nor will Struik's generalizations about the proscience orientation of the Jeffersonian Republicans as contrasted to the Federalists bear close examination. It is true, as Linda Kerber has shown in her book *Federalists in Dissent,* that Jefferson's opponents tried to make political hay by ridiculing his interest in gadgets, fossil bones, and western geography.[8] But it would be a mistake to conclude from this that the Federalists were generally antiscience. Federalist Benjamin Moore, who was simultaneously rector of Trinity Church, bishop of the diocese of New York, and president of Columbia College, called Mitchill, who was a staunch Jeffersonian Republican, "a chaos of knowledge," and his son, Clement C. Moore, raised religious objections against Jefferson's speculations in the *Notes,* but neither Moore seems to have objected to Mitchill's activities as professor of chemistry and natural history at Columbia or as editor of the *Medical Repository*. Indeed, the younger Moore took pains to defend Columbia from charges of neglecting instruction in religion and devoting too much attention to the sciences as compared to the classics. In his address to the alumni of the college in 1825, long after his intemperate attack on Jefferson's *Notes,* he praised the trustees for insisting that the liberal arts and sciences be taught "without making the institution an instrument for the propagation of any peculiar tenets of religion" and declared, in reference to the question of the balance between science and the classics, that "there can be little doubt that the diffusion of general knowledge . . . is peculiarly well suited, if not to the real advantage, at least to the existing habits and wants of our country; where young men commonly step from the academic shade into the toils of active life, or the study of some profession upon which their future maintenance depends."⁹ Timothy Dwight was a Federalist also and fully as much concerned as the Moores about the inroads of "infidelity" among the younger generation, but his answer to the challenge was to appoint Silliman professor of chemistry and natural history and give him unstinting support in developing those sciences at Yale.

Nor is it true, as Kerber suggests, that the American Academy of Arts and Sciences, dominated by Federalists, tended to favor natural philosophy, whereas the American Philosophical Society, many of whose members were Jeffersonian Republicans, "encouraged descriptive work in botany, geology, and paleontology, placing comparatively less emphasis on the mathematical sciences." Both societies were eager to receive contributions in all the sciences. They published what they could get, but what they could get often depended on accidental circumstances. With the death of Rittenhouse, only Ellicott was left to carry on the astronomical tradition in Philadelphia, but the flourishing condition of the medical school at the University of Pennsylvania ensured numerous contributions in chemistry and natural history, as did the presence of foreign visitors like Priestley, Volney, and Maclure.

In Boston the American Academy of Arts and Sciences inherited an astronomical tradition based on the achievements of John Winthrop IV in the colonial period, but there was no lack of interest in the scientific study of the earth and its productions. On May 29, 1781, the Academy's Committee Respecting the Arrangement of Subjects recommended that separate "classes" be established to study the various soils of the country, describe its plants, collect and analyze mineral specimens, investigate the diseases of the region and collect vital statistics, undertake meteorological investigations, attend to mathematical disquisitions and astronomical observations, examine progress in the mechanic arts, study the antiquities and aborigines of America, inquire into the "rationale, genius and idiom" of the English language, and discover ways of promoting commerce.[10] This list of subjects for investigation scarcely bears out the idea that the Bostonians favored natural philosophy above other sciences. Manasseh Cutler admitted that natural history had hitherto been neglected in New England and suggested that this was because of an erroneous notion that it served no useful purpose; but if that was indeed the cause, the Bostonians soon changed their minds. Their contributions toward establishing a professorship of botany and natural history and a botanical garden for the express purpose of promoting scientific agriculture proved that.

Cutler might also have mentioned the tardy acquisition of a medical school at Harvard as a reason for the slow development of the non-mathematical sciences in the Boston area. If to this is added that Benjamin Waterhouse was not a field naturalist or mineralogist; William D. Peck, though a good naturalist, was of a retiring disposition; and John Gorham was too busy teaching and practicing medicine to do much research in the laboratory or the field, one has a sufficient explanation of the slow development of chemistry, natural history, mineralogy, and geology in the Boston area. That there was no prejudice against these sciences is shown by the formation of the Linnaean Society of New England in 1816 and the attention paid to them in the pages of the *Monthly Anthology,* the *New England Journal of Medicine,* and the *North American Review.* The emphasis on natural philosophy, especially astronomy, in the *Memoirs* of the American Academy of Arts and Sciences was due largely to the appearance of Bowditch as a successor to John Winthrop IV and Joseph Willard. Without Bowditch's contributions the academy would have been hard pressed to fill the successive volumes of its *Memoirs* and would have acquired no reputation worth mentioning in natural philosophy.

Over against Struik's Marxian interpretation of the early development of American science stands Perry Miller's psychological interpretation. In Miller's view, the "Newtonian" science of the eighteenth century, far from being a reflex of the interests of a mercantile ruling class, was a manifestation of the Enlightenment idea that science consists in the aesthetic contemplation of a perfected universe. By studying the Newtonian system of nature and exhibiting a proper awe of its majesty, says Miller, Americans gained entrance into the cosmopolitan culture of Western Europe. Unfortunately, Miller continues, American patriotism demanded that the new nation prove itself culturally by making significant contributions to the reigning sciences, whether Newtonian or Linnaean. When it became evident that the United States was not producing scientific works worthy of comparison with those of Europe, patriotic fervor gradually concocted a new, utilitarian conception of science, which by equating science with technology enabled Americans to claim a place in the scientific sun and rationalize their passionate devotion to the conquest of a continent. By 1830, says Miller, "the gospel of science was, in America, converted to stark utilitarianism."[11]

In constructing this argument, Miller completely overlooks the already stark utilitarianism of American science in the eighteenth and early nineteenth centuries. Benjamin Franklin's list of the subjects that were to occupy the attention of the American Philosophical Society included "all new-discovered plants, herbs, trees, roots, their virtues, uses, etc; methods of propagating them . . . ; new methods of curing or preventing diseases; all new-discovered fossils in different countries, as mines, minerals, and quarries; new and useful improvements in any branch of mathematics; new discoveries in chemistry, such as improvements in distillation, brewing, and assaying of ores; new mechanical inventions for saving labour . . .; all new arts, trades and manufactures that may be proposed or thought of; surveys, maps and charts of particular parts of the sea coasts or inland countries . . . ; new methods of improving the breed of useful animals; new improvements in planting, gardening and clearing land; and all philosophical experiments that let light into the nature of things, tend to increase the power of man over matter, and multiply the conveniences and pleasures of life."[12] When the society's first volume of *Transactions* appeared in 1771, the preface announced that the members would devote themselves chiefly to "such subjects as tend to the improvement of their country, and advancement of its interest and prosperity." Mathematics, physics, and astronomy were mentioned as examples of useful knowledge, but they were given less emphasis than improvements in agriculture, transportation, and manufactures.

In a similar vein the American Academy of Arts and Sciences recommended to the attention of its members improvements in agriculture; researches in natural history, especially those useful to medicine; astronomical observations, "particularly those . . . which will serve to perfect the geography of the country, and improve navigation"; mechanical arts, manufactures, and commerce, which would "enrich and aggrandize these confederated States"; and, in general, "useful experiments and improvements whereby the interest and happiness of the rising empire may be essentially advanced."[13]

"Useful knowlege" was what these men prized; but Franklin, Jefferson, John Adams, and their compatriots saw no inconsistency between a concern for utility and a due reverence for the works of the Creator. Christian theology taught that God had made the world for man's use as well as his admiration and had commanded him to multiply and have dominion over it. To gain command of nature by science and practical invention was as much a religious as a patriotic duty. But although everyone paid lip service to useful knowledge, it soon became apparent that the application of the sciences to practical affairs was a long, slow, expensive business and that practical ingenuity must fill in the gap between theory and practice.

By the second decade of the nineteenth century the leading scientific societies had largely ceased publishing essays on practical subjects, leaving these to societies for promoting agriculture and the mechanic arts. That Americans were increasingly proud of American technological triumphs cannot be denied, but this does not mean that they lost sight of the distinction between theoretical science, applied science, and trial-and-error technology or ceased to hope Americans would distinguish themselves in all three. Silliman struck a characteristic note in the preface to his *American Journal of Science and Arts*. "This Journal," he declared, "is intended to embrace the circle of the PHYSICAL SCIENCES, with their application to the Arts, and to every useful purpose. While SCIENCE will be cherished for its own sake, and with a due respect for its own *inherent* dignity; it will also be employed as the handmaid to the Arts."[14] The "inherent dignity" of science, Silliman believed, sprang from its concern to understand the wonderful works of God. Its utility was in turn a proof of God's goodness. In Silliman's view patriotism was perfectly consistent with piety and practicality. The leading object of his journal, he explained, was "to advance the interests of the rising empire, by exciting and concentrating original American effort, both in the sciences, and in the arts," thereby enabling American enterprise to make the most of the natural resources provided by a bountiful God. In this aim Jefferson, Mitchill, Peale, and a host of other American scientists would have concurred heartily.

When all the many issues and circumstances that tended to divide the devotees of science in the Jeffersonian era have been taken into account, the fact remains that they exhibited an astonishing unity of attitude and purpose. "To advance the interests of the rising empire, by exciting and concentrating original American effort, both in the sciences, and in the arts" was their common goal. Patriotism, utilitarianism, love of science and scientific reputation, and admiration of the Creator's wisdom, power, and goodness were the sentiments that in varying proportions inspired the labors of the first generation of American scientists. Some, like Mitchill and Barton, spread their efforts over many sciences, sacrificing mastery for range and variety but encouraging others by their enthusiasm and example as teachers, editors, organizers, and researchers. Some, like Maclure, combined substantial achievement in a single science with generous patronage of individuals and institutions. Some, like Cutler, Muhlenberg, and Schweinitz, found time for significant scientific achievement in the busy round of clerical duties. Some, like Silliman and Cleaveland, introduced new sciences into the undergraduate curriculum while pursuing their own researches as best they could. Some, like Peale and Griscom, created institutions for popular education in science. Some, like Bartram and Hosack, maintained botanical gardens where naturalists could find a delightful variety of plants, congenial colleagues, and a place to work. Some, like Gibbs and Gilmor, had abundant wealth and education but chose to play the part of patrons rather than movers and shakers of science. Others, like Wilson, Ellicott, Nuttall, and Rafinesque, struggled alone against overwhelming odds to gain international recognition for their contributions to science.

Finally, there was Thomas Jefferson, no great scientist himself but a tower of strength and encouragement to all who labored to advance the cause of science in the rising American nation. Author of the *Notes on the State of Virginia,* president of the American Philosophical Society, discoverer of the megalonyx, instigator and overseer of the Lewis and Clark expedition, friend of Humboldt and Volney, supporter of Peale, Wilson, Ellicott, Mansfield, and many other aspiring American scientists, he became a symbol of American respect for science and faith in its power to promote human progress. It was indeed the "Age of Jefferson," not an age of brilliant scientific achievement judged by European standards, but an interesting and formative one, worthy of respect by all who value honest effort inspired by love of country and of science.

NOTES

1: THE AMERICAN CONTEXT

1. Dirk J. Struik, *Yankee Science in the Making,* rev. ed. (New York, 1962), 426.

2. *Trans. APS,* 1 (1771), "Preface," xvii.

3. *Mem. Am. Acad. Arts Sci.,* 1 (1785), "Preface," viiiff.

4. Simeon De Witt, "Address Delivered Before the Society, . . . in the City of Albany, February 20, 1799 . . . ," *Trans. Soc. Prom. Agric. Arts Manuf.,* 1, pt. 4 (1799), 2. See also John C. Greene, "Science, Learning, and Utility: Patterns of Organization in the Early American Republic," in Alexandra Oleson and S. C. Brown, eds., *The Pursuit of Knowledge in the Early American Republic: American Scientific and Learned Societies from Colonial Times to the Civil War* (New York, 1976), 1-20.

5. John Gardiner, *An Oration Delivered July 4, 1785* (Boston, 1785), 10. More generally, see Chap. 12, "The Prophets of Glory and Their Temples of Science," in Brooke Hindle, *The Pursuit of Science in Revolutionary America 1735-1789* (Chapel Hill, N.C., 1956), 248-79.

6. Samuel Miller, *A Brief Retrospect of the Eighteenth Century . . . ,* 2 vols. (New York, 1970), 2:409-10. This edition is a facsimile of the original edition published in 1803. See also John C. Greene, "American Science Comes of Age, 1780-1820," *J. Am. Hist.,* 55 (1968), 22-41, on which the present account of relations between British and American science is based. For a partial view of relations with French science see John C. Greene, "Relations of American and French Mineralogists 1780-1820," *Actes XII Congrès International d'Histoire des Sciences* (Paris, 1969), 61-64. See also Sir Gavin de Beer, *The Sciences Were Never at War* (London, 1960); Michael Kraus, "Scientific Relations between Europe and America in the Eighteenth Century," *Sci. Mon.,* 55 (1942), 259-72.

7. Gilbert Chinard, "The American Philosophical Society and the World of Science, 1768-1800," *Proc. APS,* 87 (1943), 5.

8. As quoted in James J. Abraham, *Lettsom: His Life, Times, Friends and Descendants* (London, 1933), 364.

9. Robert Hare to Benjamin Silliman, Apr. 23, 1817 and Aug. 3, 1814, Hare Correspondence, Edgar Fahs Smith Collection, University of Pennsylvania.

10. Abraham, *Lettsom,* 357-83.

11. As quoted in Lady Smith, ed., *Memoir and Correspondence of the Late Sir James Edward Smith, M.D.,* 2 vols. (London, 1832), 2:185-87.

12. Jacob Green, "On the Botany of the United States," *Trans. Soc. Promot. Useful Arts,* 3 (1814), 90.

13. [Caleb Cushing], review of *American Medical Botany* by Jacob Bigelow, *North Am. Rev.,* 13 (1821), 100-101.

14. As quoted in Thomas J. Pettigrew, *Memoirs of the Life and Writings of the Late John Coakley Lettsom . . . with a Selection from His Correspondence,* 3 vols. (London, 1817), 2:428.

15. On deism in America see Henry F. May, *The Enlightenment in America* (New York, 1976), pt. 3; Herbert M. Morais, *Deism in Eighteenth Century America,* Columbia University Studies in the Social Sciences, no. 397 (New York, 1960). See also Gustav A. Koch, *Republican Religion: The American Revolution and the Cult of Reason* (New York, 1933).

16. Thomas Jefferson to Moses Robinson, Washington, D.C., Mar. 23, 1801, as quoted in A. A. Lipscomb, ed., *The Writings of Thomas Jefferson,* 20 vols. (Washington, D.C., 1903-1904), 10:237.

17. As quoted in John F. Fulton and Elizabeth H. Thomson, *Benjamin Silliman 1779–1864: Pathfinder in American Science* (New York, 1947), 13, 23. Dwight's ideas about science, as they were reflected in the work and writings of his protégé Benjamin Silliman, are set forth in John C. Greene, "Protestantism, Science, and American Enterprise: Benjamin Silliman's Moral Universe," in Leonard G. Wilson, ed., *Benjamin Silliman and His Circle: Studies on the Influence of Benjamin Silliman on Science in America* (New York, 1979), 11–28. See also Charles E. Cunningham, *Timothy Dwight, 1752–1817* (New York, 1942).

18. Timothy Dwight, *Theology Explained and Defended, in a Series of Sermons . . . with a Memoir of the Life of the Author,* 12th ed., 4 vols. (New York, 1846), 1:284.

19. Ibid., 1:300.

20. As quoted in Theodore Dwight, Jr., ed., *President Dwight's Decisions of Questions Discussed by the Senior Class in Yale College, in 1813 and 1814. From Stenographic Notes* (New York, 1833), 331–32.

21. Miller, *Brief Retrospect,* 1:xiii. Miller told the Literary and Philosophical Society of New Jersey, "Let none indulge the ignoble fear, that the progress of science – I mean real science – will be dangerous; that it will endanger any of the true interests of man; and least of all, that it will endanger the true interests of religion."

22. Review of *Brief Retrospect Mon. Anthol. Boston Rev.,* 1 (1804), 364–65.

23. Miller, *Brief Retrospect,* 2:440.

24. Benjamin Waterhouse to James Edward Smith, Cambridge, Mass., July 24, 1811, as quoted in Smith, ed., *Memoir and Correspondence,* 2:174. The section on public attitudes toward science in this chapter draws on my article "Science and the Public in the Age of Jefferson," *Isis,* 44 (1958), 13–25.

25. Benjamin Silliman, "Notes to the . . . American Edition of Henry's Chemistry," in William Henry, *An Epitome of Experimental Chemistry . . . ,* 2nd American ed. from the 5th English ed. (Boston, 1810), i.

26. See Daniel Mornet, *Les sciences de la nature en France, au XVIIIᵉ siècle . . .* (Paris, 1911); A. Laming, "The Origins of the Popularization of Science," *Impact Sci. Soc.,* 3 (1952), 233–58; Charles A. Browne, "The Life and Chemical Services of Frederick Accum," *J. Chem. Educ.,* 2 (1925), 1–58.

27. Manasseh Cutler to Gustav Paykull, Hamilton, Mass., Feb. 14, 1799, as quoted in William P. Cutler and Julia Cutler, eds., *Life, Journals, and Correspondence of Rev. Manasseh Cutler, LL.D.,* 2 vols. (Cincinnati, 1888), 2:298.

28. Silliman, "Notes to . . . Henry's Chemistry," ii.

29. Letters from Thomas Jefferson, Monticello, Aug. 1805, and from Bushrod Washington, Mt. Vernon, Sept. 15, 1805, to Thomas Ewell, quoted in Ewell, *Plain Discourses on the Laws or Properties of Matter . . .* (New York, 1806), 8–9.

30. Benjamin Waterhouse, *The Botanist. Being the Botanical Part of a Course of Lectures on Natural History, Delivered in the University of Cambridge* (Boston, 1811), "Advertisement," vi.

31. An American edition of Erasmus Darwin's *The Botanic Garden* appeared in New York in 1798, followed by an abridged version in 1805. His *The Temple of Nature* was published in both Baltimore and New York in 1804. Pt. 1 of Darwin's *Zoonomia: Or the Laws of Organic Life* was published in New York in 1796; pt. 2, 2 vols., appeared in Philadelphia in 1797. Other American editions followed in 1803 (Boston), 1809 (Boston), and 1818 (Philadelphia). [Jacob Green and Erskine Hazard], *An Epitome of Electricity and Galvanism. By Two Gentlemen of Philadelphia* (Philadelphia, 1809).

32. *The Emporium of Arts and Sciences,* n.s., 1 (1813), "Prospectus," 2. This magazine was conducted by John R. Coxe in 1812; it continued under Cooper's editorship during 1813 and 1814, after which it expired.

33. *Med. Repos.,* 9 (1808), 42; 7 (1804), 403.

34. See Edgar F. Smith, *John Griscom (1774–1852), Chemist* (Philadelphia, 1925).

35. Letter to the editors, *Mon. Anthol. Boston Rev.,* 6 (1809), 236–37.

36. *Med. Repos.,* 4 (1801), 291; also 3d hexade, 2 (1811), 89.

37. See Charles Coleman Sellers, *Mr. Peale's Museum. Charles Willson Peale and the First Popular Museum of Natural Science and Art* (New York, 1980), chap. 4.

38. Thomas Jefferson to G. C. de La Coste, Washington, D.C., May 24, 1807, as quoted in Lipscomb, ed., *Writings of Thomas Jefferson,* 11:206-7.

39. See Edwin T. Martin, *Thomas Jefferson: Scientist* (New York, 1961), 13ff. See also Charles A. Browne, "Thomas Jefferson and the Scientific Trends of His Time," *Chron. Bot.,* 8 (1944), 1-64; Harlow Shapley, "Notes on Thomas Jefferson as a Natural Philosopher," *Proc. APS,* 87 (1944), 274-76.

40. Thomas Jefferson, *Notes on the State of Virginia . . . ,* ed. William Peden (Chapel Hill, N.C., 1955), 43-58.

41. For a picture of Buffon at his country estate in the 1780s see the translation of Jean Marie H. de Séchelles, *Voyage à Montbard* (Paris, 1801) in John Lyon and Phillip Sloan, *From Natural History to the History of Nature: Readings from Buffon and His Critics* (Notre Dame, Ind., 1981), 357ff.

42. Jefferson, *Notes,* 31-33. See also Julian P. Boyd, ed., *The Papers of Thomas Jefferson,* 19 vols. (Princeton, 1950-), 8:565-66; (1954), 9:215-17 for the correspondence between Jefferson and David Rittenhouse about fossil shells.

43. Thomas Jefferson to John P. Emmett, Monticello, May 2, 1826, as quoted in Lipscomb, ed., *Writings of Thomas Jefferson,* 16:171.

44. Thomas Jefferson, "A Memoir on the Discovery of Certain Bones of a Quadruped of the Clawed Kind in the Western Parts of Virginia," *Trans. APS,* 4 (1799), 255-56.

45. Thomas Jefferson to James Currie, Paris, Dec. 20, 1788, Boyd, ed., *Papers of Thomas Jefferson,* 14:366.

46. Jefferson to John Manners, Monticello, Feb. 22, 1814, as quoted in Lipscomb, ed., *Writings of Thomas Jefferson,* 14:97-99.

47. Jefferson to Thomas Cooper, Oct. 7, 1814, as quoted in Lipscomb, ed., *Writings of Thomas Jefferson,* 14:201.

48. Jefferson to Dr. Thomas Ewell, Monticello, Aug. 1805, as quoted in Ewell, *Plain Discourses,* 8-9.

49. William Cullen Bryant, *The Embargo.* Facsimile reproductions of the editions of 1808 and 1809, with an Introduction and Notes by Thomas O. Mabbott (Gainesville, Fla., 1955), 40.

50. See Robert A. Halsey, *How the President Thomas Jefferson and Dr. Benjamin Waterhouse Established Vaccination as a Public Health Procedure* (New York, 1936). See also John B. Blake, *Benjamin Waterhouse and the Introduction of Vaccination: A Reappraisal* (Philadelphia, 1957).

51. Joseph E. Ewan, "How Many Botany Books Did Thomas Jefferson Own?" *Bull. Mo. Bot. Gard.* (June 1976), 1; Emily M. Sowerby, *Catalogue of the Library of Thomas Jefferson,* 5 vols. (Washington, D.C., 1952), vols. 1, 2.

52. See Richard Rathbun, *The Columbian Institute for the Promotion of Arts and Sciences . . . ,* Smithsonian Institution U.S. National Museum bull. 101 (Washington, D.C., 1917); G. Brown Goode, *The Origin of the National Scientific and Educational Institutions of the United States* (New York and London, 1890). The Columbian Institute was chartered by Congress in 1818 for a term of twenty years and empowered to "procure . . . a suitable building for the sittings of the said institution, and for the preservation and safe-keeping of a library and museum; and, also, a tract or parcel of land, for a botanic garden. . . ." It held its meetings successively in Blodget's Hotel, the Treasury Department, and City Hall until it was granted a "permanent" home in the western addition to the Capitol building in 1824. The chief organizers were Edward Cutbush, a surgeon in the U.S. Navy, and Thomas Law, a real estate dealer of English origins. The membership, which reached a total of 70 in 1826, consisted of government officials and employees, most of whom were not very active in the institute. A total of 85 communications were read during the institute's existence. Forty-four of these were contributed by William Lambert, a clerk in the Pension Office who was an amateur astronomer interested in determinations of latitude and longitude. In 1821 Lambert was selected by the secretary of state to make observations for determining the meridian of Washington. His proposal for establishing a national observatory in Washington was endorsed by the institute but failed to win congressional approval. Papers other than those contributed by Lambert concerned meteorological observations, improvements in shipbuilding, the plants of the District of Columbia, and monetary subjects. Work on the botanic

garden began in 1821. By the end of 1823 the tract of land granted by Congress had been drained and leveled, an elliptical pond with an island at its center constructed, and four graveled walks laid out. Trees and shrubs were planted, and the garden was maintained as well as scanty funds would permit until the institute expired in 1837, one year before the termination of its charter. By an odd coincidence the very tract of land the garden occupied is presently the site of the United States Botanic Garden, established thirteen years after the demise of the institute. The idea of establishing a botanic garden in Washington was also supported by the Washington Botanical Society, organized in 1817, many of whose members were also members of the Columbian Institute, but this society lasted only until 1826. Rathbun's *Columbian Institute,* 75–77, includes a brief account of the society.

2: THE PHILADELPHIA PATTERN

1. On scientific and cultural life in Philadelphia in the late eighteenth century see Brooke Hindle, *The Pursuit of Science in Revolutionary America, 1735–1789* (Chapel Hill, N.C., 1956); Carl Bridenbaugh, *Cities in Revolt: Urban Life in America, 1743–1776* (New York, 1955); Whitfield J. Bell, Jr., "The Scientific Environment of Philadelphia," *Proc. APS,* 92 (1948), 6–14.

2. As quoted in William P. Cutler and Julia Cutler, *Life, Journals, and Correspondence of Rev. Manasseh Cutler, LL.D.,* 2 vols. (Cincinnati, 1888), 1:262–63.

3. Ibid., 1:267–68.

4. Ibid., 1:269.

5. See Whitfield J. Bell, Jr., et al. "The Cabinet of the American Philosophical Society," in *A Cabinet of Curiosities: Five Episodes in the Evolution of American Museums* (Charlottesville, Va., 1967), 4.

6. *Proc. APS,* 22, pt. 3 (1885), 252–53. This volume contains the minutes of the society from 1743 to 1838, when it began publishing its proceedings.

7. Kenneth Roberts and Anna M. Roberts, trans. and eds., *Moreau de St. Méry's American Journey [1793–1798]* (New York, 1947), 350. See also Gilbert Chinard, "The American Philosophical Society and the World of Science (1768–1800)," *Proc. APS,* 87 (1943), 1–11.

8. Murphy D. Smith, *Oak from an Acorn: A History of the American Philosophical Society Library, 1770–1803* (Wilmington, Del., 1976), xi–xii, 6.

9. See Edward P. Cheney, *History of the University of Pennsylvania, 1740–1940* (Philadelphia, 1940); George W. Corner, *Two Centuries of Medicine: A History of the School of Medicine, University of Pennsylvania* (Philadelphia and Montreal, 1965); Joseph Carson, *History of the Medical Department of the University of Pennsylvania* (Philadelphia, 1869); William H. Williams, *America's First Hospital: The Pennsylvania Hospital, 1751–1844* (Wayne, Pa., 1976).

10. Barton and James Woodhouse are described in Charles Caldwell, *Autobiography of Charles Caldwell, M.D.,* with a Preface, Notes, and Appendix by Harriot W. Warner (New York, 1968), 129, 173. On Barton see also the letter from John Davis to Jeremy Belknap, Nov. 14, 1795, "Belknap Correspondence," *Collect. Mass. Hist. Soc.,* 6th ser., 4 (1891), 603. Joseph Ewan is writing a biography of Barton.

11. Cutler and Cutler, *Life, Journals, and Correspondence,* 2:144–45.

12. Zacchaeus Collins to Henry Muhlenberg, Aug. 27, 1813, as quoted in Francis Harper, ed., *The Travels of William Bartram,* Naturalist's ed. (New Haven, 1958), xxxii.

13. Cutler and Cutler, *Life, Journals, and Correspondence,* 2:144–45.

14. On Humphry Marshall see John W. Harshberger, *The Botanists of Philadelphia* (Philadelphia, 1899), 80ff. On Philadelphia gardens and botanists generally see also Jeannette E. Graustein, *Thomas Nuttall, Naturalist . . . Explorations in America, 1808–1841* (Cambridge, Mass., 1967), 19ff.

15. Henry Muhlenberg to Stephen Elliott, Lancaster, Pa., Nov. 8, 1809, as quoted in Joseph and Nesta Ewan, *John Lyon, Nurseryman and Plant Hunter, and His Journal, 1799–1814, Trans. APS,* 54, pt. 1 (1963), 9.

16. Johann D. Schoepf, *Travels in the Confederation* [*1783–1784*] . . . , trans. and ed. Alfred J. Morrison, 2 vols. (Philadelphia, 1911), 2:12.

17. Henry Muhlenberg to William Bartram, Lancaster, Pa., Jan. 29, 1810, Bartram Papers, Historical Society of Pennsylvania, vol. 4.

18. The fullest and best account of Peale's museum is Charles C. Sellers, *Mr. Peale's Museum: Charles Willson Peale and the First Popular Museum of Natural Science and Art* (New York, 1980). See also Toby A. Appel, "Science, Popular Culture, and Profit: Peale's Philadelphia Museum," *J. Soc. Bibliogr. Nat. Hist.,* 9 (1980), 619–34; Lillian B. Miller, ed., *The Selected Papers of Charles Willson Peale and His Family,* vol. 1, 1735–1791 (New Haven, 1981). See also Edgar P. Richardson, Brooke Hindle, and Lillian B. Miller, *Charles Willson Peale and His World* (New York, 1982).

19. Quoted from the typescript copy of Peale's autobiography at the library of the American Philosophical Society.

20. Cutler and Cutler, *Life, Journals, and Correspondence,* 1:259–62.

21. Peale's autobiography, as quoted in Charles C. Sellers, *Charles Willson Peale, Mem. APS,* vol. 23, pts. 1, 2 (Philadelphia, 1947).

22. Sellers, *Mr. Peale's Museum,* 101ff.

23. Charles W. Peale, "A Walk with a Friend through the Museum," Peale mss., American Philosophical Society Library.

24. See Edward J. Nolan, *A Short History of the Academy of Natural Sciences of Philadelphia* (Philadelphia, 1909); W. S. W. Ruschenberger, *Notice of the Origin, Progress, and Present Condition of the Academy of Natural Sciences of Philadelphia* (Philadelphia, 1852); Patsy A. Gerstner, "The Academy of Natural Sciences of Philadelphia, 1812–1850," in A. Oleson and S. C. Brown, eds., *The Pursuit of Knowledge in the Early American Republic* . . . (Baltimore and London, 1976), 174–93.

25. "Annual Report of the Treasurer . . . ," Dec. 31, 1815, microfilm minutes of the Academy of Natural Sciences of Philadelphia. See also Nolan, *Short History,* 9; Graustein, *Thomas Nuttall,* 152–53, 422–23.

26. *First Annual Report of the Proceedings of the Franklin Institute of the State of Pennsylvania, for the Promotion of the Mechanic Arts* . . . (Philadelphia, 1825), 4–63. See also Bruce Sinclair, *Philadelphia's Philosopher Mechanics: A History of the Franklin Institute, 1824–1865* (Baltimore, 1974).

3: SCIENTIFIC CENTERS IN NEW ENGLAND

1. Timothy Dwight, *Travels in New-England and New-York,* 4 vols. (London, 1823), 2:399–400. Dwight visited the various colleges at different times and never brought all the information together as of one date. In general, the figures given in the text refer to the situation about 1811–1812.

2. See Edward C. Herrick, "Historical Sketch of the Connecticut Academy of Arts and Sciences," *Am. Q. Regist.,* 13 (1840), 23–24; Brooke Hindle, *The Pursuit of Science in Revolutionary America, 1735–1789* (Chapel Hill, N.C., 1956), 272–73.

3. See John C. Greene and John G. Burke, *The Science of Minerals in the Age of Jefferson, Trans. APS,* 68, pt. 4, (1978), chap. 7.

4. Charles W. Upham, *Discourse at the Funeral of the Rev. John Prince* . . . (Salem, 1836), 13.

5. As quoted in William P. Cutler and Julia Cutler, *Life, Journals, and Correspondence of Rev. Manasseh Cutler, LL.D.,* 2 vols. (Cincinnati, 1888), 1:86. See also Harold L. Burstyn, "The Salem Philosophical Library . . . ," *Essex Inst. Hist. Collect.,* 96 (1960), 169–206.

6. See Ralph H. Brown, *Mirror for Americans: Likeness of the Eastern Seaboard, 1810* (New York, 1943), "Introduction."

7. Theophilus Parsons, *Memoir of Theophilus Parsons* . . . (Boston, 1859), 273.

8. Manasseh Cutler to Theophilus Parsons, Ipswich, Mass., Aug. 10, 1782, as quoted in Cutler and Cutler, *Life, Journals, and Correspondence,* 1:293–94.

9. John Adams to Abigail Adams, Philadelphia, Aug. 4, 1776, as quoted in L. H.

Butterfield, M. Friedlander, and Mary-Jo Kline, eds., *The Book of Abigail and John: Selected Letters of the Adams Family, 1762–1784* (Cambridge, Mass., 1975), 149. See also *Bull. Am. Acad. Arts Sci.* (January 1971), 3ff. I am indebted to the American Academy for allowing me to see the historical account of the academy prepared in manuscript by Nicholas Ziegler, a student at Harvard University. It is to be hoped the manuscript will eventually be published.

10. John Adams, *The Works of John Adams* . . . , ed. Charles F. Adams, 10 vols. (Boston, 1851–1856), 4:260–261n.

11. James Bowdoin, "A Philosophical Discourse, Publickly Addressed to the American Academy of Arts and Sciences . . . on the Eighth of November, M,DCC,LXXX: When the President Was Inducted into Office," *Mem. Am. Acad. Arts Sci.,* 1 (1785), 1–20. The weather notes for the early meetings of the academy are taken from the manuscript diary (1777–1783) of Caleb Gannett preserved at the Massachusetts Historical Society.

12. Records of the American Academy of Arts and Sciences, Boston Athenaeum. These "Records" consist of the minutes of academy meetings. The minutes of the council of the academy, 1781–1797, are bound at the end of vol. 1 of the "Records"; those for 1797–1823 have not survived. I am grateful to the American Academy and to the Boston Athenaeum for permission to examine the "Records" and other documents pertaining to the early history of the academy. For a detailed account of the formation and activities of the academy and of other learned and scientific societies in the Boston area, see Bruce W. Stone, "The Role of the Learned Societies in the Growth of Scientific Boston, 1780–1848" (Ph.D. diss., Boston University, 1974).

13. Cutler and Cutler, *Life, Journals, and Correspondence,* 1:85.

14. Ibid., 1:93–94.

15. Ibid., 1:449.

16. Ibid., 1:464–65.

17. Sidney Willard, *Memories of Youth and Manhood,* 2 vols. (Cambridge, Mass., 1855), 1:83ff. See also "Willard Letters," *Proc. Mass. Hist. Soc.,* 43 (1909–1910), 609–46.

18. Handwritten motion dated Jan. 29, 1806, by Loammi Baldwin in miscellaneous mss. at the Massachusetts Historical Society. The academy's communication to the state legislature urging surveys for an accurate map of the state is recorded in a handwritten communication preserved in the Manuscripts Division of the Boston Public Library. Letterbook no. 1 of the American Academy (p. 68), preserved at the Boston Athenaeum, contains a letter from Edward A. Holyoke to James Bowdoin, dated Salem, Nov. 29, 1788, urging the academy to solicit maps from the towns of Massachusetts as a basis for a general map of the state.

19. *The American Academy of Arts and Sciences, 1780–1940* (Cambridge, 1941), 10–11.

20. See Warren, *Life of John Collins Warren,* 1:79.

21. Benjamin Vaughan to Nathaniel Bowditch, Aug. 13, 1817, Bowditch Papers, Boston Public Library. In 1816 a committee representing the scientific and learned societies in the Boston area presented a petition to the Massachusetts legislature requesting the grant of $50,000 to build a building to house the libraries and other collections of these societies, which were to be incorporated into an organization known as the Massachusetts Institution. The petition was unsuccessful. See *North Am. Rev.* . . . , 2 (1816), 309–19.

22. Quoted in Cutler and Cutler, *Life, Journals, and Correspondence,* 2:354.

23. See *Centennial Year (1792–1892) of the Massachusetts Society for Promoting Agriculture* (Salem, Mass., 1892), 32–33. See also *Journal* of the society, 5 (1819).

24. S. T. Riley, *The Massachusetts Historical Society, 1791–1959* (Boston, 1959), 8–11.

25. Josiah Quincy, *The History of the Boston Athenaeum, with Biographical Notices of Its Deceased Founders* (Cambridge, 1859), 28.

26. Edward Warren, *Life of John Collins Warren* . . . , 2 vols. (Boston, 1860), 1:73. The Massachusetts Historical Society has a photostat of a handwritten legal document (1801) setting forth the obligations of members of "The Society for the Study of Natural Philosophy." See also Linda K. Kerber, "Science in the Early Republic: The Society for the Study of Natural Philosophy," *William and Mary Quart.,* 3rd ser., 29 (1972), 263–80.

27. James Jackson to John Pickering, Jan. 10, 1802, as quoted in James J. Putnam, *A Memoir of James Jackson* . . . (Boston and New York, 1904), 248–49.

28. The best account of the Linnaean Society of New England and the Boston Society of Natural History that succeeded it is Thomas T. Bouvé, "Historical Sketch of the Boston Society of Natural History: With a Notice of the Linnaean Society Which Preceded It," in *Anniversary Memoirs: 1830–1880* (Boston, 1880). A list of the papers read before the Linnaean Society may be found in Augustus A. Gould, "The Linnaean Society of New England," *Proc. Boston Soc. Nat. Hist.*, 9 (1863), 335–40.

29. "Museum of Natural History," *New Engl. J. Med. Surg.*, 5 (1816), 189–91. Passing mention should be made of Daniel Bowen's Columbian Museum, founded in the 1790s in imitation of Peale's Museum. Destroyed by fire in January 1803 and again in November 1806, it was reopened under new management in 1807, but there is no evidence that the museum exhibits were ever of scientific importance. See the notices of the Columbian Museum in *Mon. Anthol. Boston Rev.*, 1 (1804), 143, 240; see also Nathaniel Dearborn, *Dearborn's Reminiscences of Boston, and Guide through the City and Environs* (Boston, 1851); Charles Shaw, *A Topographical and Historical Description of Boston . . . with Some Account of Its Environs* (Boston, 1817), 286–87.

30. Bouvé, *Historical Sketch*, 14ff.

31. Willard, *Memories*, 1:208–11.

32. Ibid., 1:272–73.

33. As quoted in Andrew P. Peabody, *Harvard Reminiscences* (Boston, 1888), 70–71; see also [John G. Palfrey], "Notice of Professor Farrar," *Christ. Exam.*, 55 (July 1853), 8–9.

34. See Solon I. Bailey, *The History and Work of Harvard Observatory . . .* (New York and London, 1931), Chap. 1.

35. Edward Warren, *Life of John Warren, M.D., Surgeon General during the War of the Revolution; First Professor of Anatomy and Surgery in Harvard College . . .* (Boston, 1874), 222–43.

36. Warren, *Life of John Collins Warren*, 1:12.

37. Jacques P. B. de Warville, *New Travels in the United States of America*, 2nd ed., 2 vols. (London, 1794), 1:83, 85.

38. I. Bernard Cohen, *Some Early Tools of American Science . . .* (Cambridge, 1950), 83–85.

39. See Lloyd E. Hawes, "Benjamin Waterhouse, M.D., First Professor of the Theory and Practice of Physic at Harvard and Introducer of Cowpox Vaccination into America . . . Including a Concordance of Dr. Waterhouse's *Hortus Siccus* by J. Worth Estes, M.D.," *Boston Medical Library Studies* (Boston, 1974).

40. Benjamin Waterhouse, autobiographical manuscript at the Francis A. Countway Library of Medicine (rare books and manuscripts room), Harvard Medical School.

41. Willard, *Memories*, 1:169.

42. Waterhouse, autobiographical manuscript, Countway Library. See also "Cabinet of Ores and Other Minerals, in the University of Cambridge in New England," quoted in T. J. Pettigrew, *Memoirs of the Life and Writings of the Late John Coakley Lettsom*, 2 vols. (London, 1817), 2:115–23.

43. Cohen, *Some Early Tools*, 73.

44. See Jeannette Graustein, "Natural History at Harvard," *Cambridge Hist. Soc. Proc.*, 38 (1961), 69–86.

45. "Botanic Garden," *New Engl. J. Med. Surg.*, 5 (1816), 119–20.

46. William D. Peck, *A Catalogue of American and Foreign Plants, Cultivated in the Botanic Garden, Cambridge, Massachusetts* (Cambridge, 1818), "Advertisement."

47. G. B. Emerson, *Reminiscences of an Old Teacher* (Boston, 1871), 13, as quoted in Jeannette Graustein, *Thomas Nuttall Naturalist . . . Explorations in America, 1808–1841* (Cambridge, 1967), 174.

48. John C. Warren, "Reminiscences," as quoted in Warren, *Life of John Collins Warren*, 1:57.

49. James J. Putnam, *A Memoir of Dr. James Jackson with Sketches of His Father, Hon. Jonathan Jackson, and His Brother . . .* (Boston and New York, 1906), 241.

50. *Some Account of the Medical School in Boston, and of the Massachusetts General Hospital* (Boston, 1824), 5–6, 13.

4: SCIENCE ALONG THE HUDSON

1. De Witt Clinton, "An Introductory Discourse Delivered on the 4th of May, 1814," *Trans. Lit. Philos. Soc. N.Y.,* 1 (1815), 21ff.

2. *Transactions of the Society Instituted in the State of New-York, for the Promotion of Agriculture, Arts, and Manufactures,* pt. 1 (1792), ix–x. A full account of the society and its successor, the Society for the Promotion of Useful Arts, may be found in James M. Hobbins, "Shaping a Provincial Learned Society: The Early History of the Albany Institute," in Alexandra Oleson and S. C. Brown, eds., *The Pursuit of Knowledge in the Early American Republic: American Scientific and Learned Societies from Colonial Times to the Civil War* (Baltimore and London, 1976), 117–50. See also in the same volume Brooke Hindle, "The Underside of the Learned Society in New York, 1754–1854," 84–116.

3. Samuel L. Mitchill, "Address to the Agricultural Society and both Houses of the Legislature of the State of New York, at Their Annual Meeting in the Court House, in the City of Albany, February 7, 1798," *Trans. Soc. Promot. Agric. Arts Manuf.,* 1, pt. 3 (1798), xxviii. The full text of Mitchill, "A Sketch of the Mineralogical and Geological History of the State of New York . . ." may be found in the *Med. Repos.,* 1 (1797–1798), 293–314, 445–52; 3 (1799–1800), 325–35. The first section only of this "Sketch" was published in the *Transactions,* 1, pt. 4 (1799), 124–52.

4. Besides James Hobbins's account of the Albany Institute, cited above, see vol. 1 of Nathan Reingold, ed., *The Papers of Joseph Henry* (Washington, D.C., 1972) for a full view of the development of scientific institutions in Albany.

5. See William P. Cutler and Julia Cutler, *Life, Journals, and Correspondence of Rev. Manasseh Cutler, LL.D.,* 2 vols. (Cincinnati, 1888), 1:239–40. See also David C. Humphrey, *From King's College to Columbia 1746–1800* (New York, 1976), chaps. 15–16; Clement K. Moore, *The Early History of Columbia College . . . ,* facsimile of the 1825 edition (New York, 1940).

6. *A History of Columbia University, 1754–1904* (New York, 1904), 76.

7. James Renwick, *Life of De Witt Clinton* (New York, 1859), 32–33. John W. Francis remembered Kemp as "impulsive and domineering in his nature," adding, "Kemp was clever in his assigned duties, but had little ambition. . . . He was devoid of genius and lacked enterprise." See John W. Francis, *New York during the Last Half Century, A Historical Discourse,* 5 vols. (New York, 1857), 1:23–24.

8. *Trans. Soc. Promot. Agric. Arts Manuf.,* 1, pt. 2 (1794), xl–xlv. See also xxix.

9. "Sketch of the ELGIN BOTANIC GARDEN, in the vicinity of the City of New-York," *Am. Med. Philos. Regist.,* 2 (July 1811), 3–4. This sketch was presumably written by Hosack, who was one of the editors of the *Register.* A full account of Hosack's life and the development of the Elgin Botanic Garden may be found in Christine C. Robbins, *David Hosack, Citizen of New York* (Philadelphia, 1964).

10. See Joseph Ewan, "Frederick Pursh, 1774–1820, and His Botanical Associates," *Proc. APS,* 92 (1952), 599–628.

11. "Extrait d'une lettre de M. Delile . . . à M. Deleuze, New Yorck 22 septembre, 1806," *Annales du Muséum National d'Histoire Naturelle,* 8 (1806), 476–77. Delile later became professor of botany and director of the botanical garden at the University of Montpelier.

12. Thomas Blatchford, "Our Alma Mater Fifty Years Ago: An Oration Delivered Before the Alumni Association of the College of Physicians and Surgeons of Columbia College, N.Y.," *Columbia Alumni Q.* (May 14, 1861), 20–22.

13. John W. Francis, *Old New York, Reminiscences of the Past Sixty Years,* ed. Henry Tuckerman (New York, 1865), 31.

14. Francis, *New York during the last Half Century,* 1:54–55.

15. Paraphrased from the French by Robbins, *David Hosack,* 105, citing Jacques G. Milbert, *Itinéraire pittoresque du fleuve Hudson et des parties latérales de l'Amérique du Nord, d'après les dessins originaux . . . ,* 2 vols. (Paris, 1828–1829).

16. John W. Francis, *Reminiscences of Samuel Latham Mitchill, M.D., LL.D.* (New York, 1859), 15–16. See also "Historical Sketch of the Origin, Progress, and Present State of the College of Physicians and Surgeons of the University, New-York," *Am. Med. Philos.*

Regist., 4 (1814), 131ff.; Courtney R. Hall, *A Scientist in the Early Republic, Samuel Latham Mitchill, 1764–1831* (New York, 1934).

17. "Cultivation of Natural History in the University College of New-York. Communicated for the Register, by a Correspondent," *Am. Med. Philos. Regist.*, 2 (1811), 155, 159.

18. Claire Klein, "Columbia and the Elgin Botanic Garden Property," *Columbia Univ. Q.*, 31 (1939), 272–97.

19. See Walter M. Whitehill, ed., *Independent Historical Societies . . .* (Boston, 1962), 58; Robert H. Kelby, *The New-York Historical Society, 1804–1904* (New York, 1905); *Am. Mon. Mag. Crit. Rev.*, 1 (1817), 48, 124.

20. *Am. Mon. Mag. Crit. Rev.*, 1–4 (1817–1819); *Trans. Lit. Philos. Soc. N.Y.*, 1 (1815); 2 (1824). The minutes of the society are preserved at the New-York Historical Society Library. They show that most of the meetings were scantily attended and most of the papers were communicated by medical men.

21. *Med. Repos.*, n.s., 4 (1818), 98–99. For a full account of the Lyceum of Natural History see Herman L. Fairchild, *A History of the New York Academy of Sciences, Formerly the Lyceum of Natural History* (New York, 1887).

22. *Am. Mon. Mag. Crit. Rev.*, 1 (1817), 271; Hindle, "The Underside of the Learned Society," 99–102; "The New-York Institution," *Med. Repos.*, n.s., 4 (1818), 95–99.

23. David Hosack, "Extracts from the Introductory Discourse on Medical Science, Delivered . . . on the 2d of November, 1818," *Am. Mon. Mag. Crit. Rev.*, 4 (1818–1819), 114.

24. For a full list of the papers on natural history read before the Lyceum of Natural History see Max Meisel, *A Bibliography of American Natural History: The Pioneer Century, 1769–1865*, 3 vols. (New York and London, 1967), 2:244–73.

5: OUTPOSTS OF SCIENCE IN THE SOUTH AND WEST

1. See Brooke Hindle, *The Pursuit of Science in Revolutionary America, 1735–1789* (Chapel Hill, N.C., 1956), 50–56; William Smallwood and Mabel Smallwood, *Natural History and the American Mind* (New York, 1941), 102–11; Joseph I. Waring, *A History of Medicine in South Carolina, 1670–1825* (Columbia, S.C., 1964); G. E. Gifford, Jr., "The Charleston Physician-Naturalists," *Bull. Hist. Med.*, 49 (1975), 556–74.

2. As quoted in Laura M. Bragg, "The Birth of the Museum Idea in America," *Charleston Mus. Q.*, 1 (1923), 3–4.

3. As quoted in Robert L. Brunhouse, ed., *David Ramsay, 1749–1815: Selections from His Writings, Trans. APS*, 55, pt. 4 (Philadelphia, 1965), letter 310, 172–73. See also George C. Rogers, Jr., *Charleston in the Age of the Pinckneys* (Norman, Okla., 1969), 103ff.; James H. Easterby, *A History of the College of Charleston, Founded 1770* (Charleston, S.C., 1935); Daniel W. Hollis, *University of South Carolina*, vol. 1: *South Carolina College* (Columbia, S.C., 1951).

4. See W. C. Coker, "The Garden of André Michaux," *J. Elisha Mitchell Sci. Soc.*, 27 (1911), 65–72; Gilbert Chinard, "André and François-André Michaux and Their Predecessors, an Essay on Early Botanical Exchanges between America and France," *Proc. APS*, 101 (1957), 344–68.

5. See David H. Rembert, *Thomas Walter, Carolina Botanist* (Columbia, S. C., 1980); William R. Maxon, "Thomas Walter, Botanist," *Smithsonian Miscellaneous Collections*, 95, no. 8 (Washington, D.C., 1936); Thomas Walter, *Flora Caroliniana . . .* (London, 1788).

6. Margaret Babcock Meriwether, ed., *The Carolinian Florist of Governor John Drayton of South Carolina, 1766–1822, with Water-color Illustrations from the Author's Original Manuscript and an Autobiographical Introduction* (Columbia, S.C., 1943); John Drayton, *A View of South Carolina, As Respects Her Natural and Civil Concerns . . .* (Charleston, S.C., 1802); David Ramsay, *The History of South Carolina . . . to the Year 1808*, 2 vols. (Charleston, S.C., 1809), as reprinted by W. J. Duffie (Newberry, S.C., 1858) 2:187–95.

7. David Ramsay to Benjamin Rush, Charleston, S.C., Sept. 29, 1788, as quoted in

Brunhouse, *David Ramsay,* letter 158, 123. See also Chalmers S. Murray, *This Is Our Land: The Story of the Agricultural Society of South Carolina* (Charleston, S.C., 1949), 49–50.

8. *Address and Rules of the South-Carolina Society for Promoting and Improving Agriculture and Other Rural Concerns* (Charleston, S.C., 1821), 3. The society seems to have reverted temporarily to its earlier title. See also C. Irvine Walker, *History of the Agricultural Society of South Carolina . . .* (Charleston, S.C., 1919).

9. *Report of the Curators of the Agricultural Society of South Carolina, at Their Anniversary Meeting, Held in Columbia, in December, 1819 . . .* (Columbia, S.C., 1820).

10. Waring, *History of Medicine,* 136. See also *Excerpts from the Minutes of the Medical Society of South Carolina, 1789–1820* (Charleston, S.C., n.d.); Ramsay, *History of South Carolina,* 2:59.

11. See Joseph Ewan and Nesta Ewan, *John Lyon, Nurseryman and Plant Hunter, and His Journal, 1799–1814, Trans. APS,* 53, pt. 2 (1963), 45; Ramsay, *History of South Carolina,* 2:60; Waring, *History of Medicine,* 128, 153, 305.

12. Waring, *History of Medicine,* 153; John L. E. W. Shecut, *Medical and Philosophical Essays . . .* (Charleston, S.C., 1819), 44–45.

13. Shecut, *Medical and Philosophical Essays,* 47–52.

14. See James Moultrie, Jr., "An Eulogium on Stephen Elliott, M.D. and LL.D. . . . Delivered . . . on 8th November, 1830 . . . (Charleston, S.C., 1830); Joseph Ewan, "Introduction," in Stephen Elliott, *A Sketch of the Botany of South-Carolina and Georgia,* 2 vols. (Charleston, S.C., 1821–1824), vols. 6 and 7 of *Classica Botanica Americana* ser. (New York, 1971).

15. Stephen Elliott, *An Address to the Literary and Philosophical Society of South-Carolina, Delivered in Charleston, on Wednesday, the 10th of August, 1814* (Charleston, S.C., 1814).

16. Shecut, *Medical and Philosophical Essays,* 52. See also William G. Mazyck, *The Charleston Museum: Its Genesis and Development . . .* (Charleston, S.C., 1908); Eola Willis, "Sketch of the Life of Dr. J. L. E. W. Shecut and of the Origin of the Museum," *Bull. Coll. Charleston Mus.,* 2 (1906), 27–32; Paul M. Rea, "The Collections of the Charleston Library Society from 1798 to 1815," ibid., 2:47–54.

17. Shecut, *Medical and Philosophical Essays,* 52. Joseph Ewan, in his "The Growth of Learned and Scientific Societies in the Southeastern United States to 1860," in Alexandra Oleson and S. C. Brown, eds. *The Pursuit of Knowledge in the Early American Republic: American Scientific and Learned Societies from Colonial Times to the Civil War,* (Baltimore and London, 1976), 208–18, suggests that the Huguenot tradition in Charleston was more favorable to the development of science than the Catholic tradition in the larger city of New Orleans.

18. Daniel Drake, "Discourses on Early Physicians, Scenery and Society of Cincinnati," in Charles T. Greve, *Centennial History of Cincinnati and Representative Citizens,* 2 vols. (Chicago, 1904), 1:367. For a general account of early developments in the western towns see Richard C. Wade, *The Urban Frontier: Pioneer Life in Early Pittsburgh, Cincinnati, Lexington, Louisville, and St. Louis* (Chicago and London, 1964), published originally by the Harvard University Press (1959) as *The Urban Frontier: The Rise of Western Cities, 1790–1830.*

19. See Emmet F. Horine, *Daniel Drake (1785–1852), Pioneer Physician of the Midwest* (Philadelphia, 1961), 133; Greve, *Centennial History,* 1:492; R. C. Buley, *The Old Northwest Pioneer Period, 1815–1840,* 2 vols. (Indianapolis, 1950); Otto Juettner, *Daniel Drake and His Followers, 1785–1909 . . .* (Cincinnati, 1909); Edward D. Mansfield, *Memoirs of the Life and Services of Daniel Drake . . . with Notices of the Early Settlement of Cincinnati . . .* (Cincinnati, 1855).

20. As quoted in Maria R. Audubon, *Audubon and His Journals,* 2 vols. (New York, 1897), 1:37. See also Walter B. Hendrickson, "The Western Museum Society of Cincinnati," *Sci. Mon.,* 64 (1946), 66–72.

21. Daniel Drake, *An Anniversary Discourse, on the State and Prospects of the Western Museum Society: Delivered by Appointment, in the Chapel of the Cincinnati College, June 10th, 1820, at the Opening of the Museum* (Cincinnati, 1820), as reprinted in Henry D. Shapiro and Zane L. Miller, eds., *Physician to the West: Selected Writings of Daniel Drake on Science*

and Society (Lexington, Ky., 1970), 135. See also Donald C. Peattie, *Audubon's America* (Boston, 1940), 13–14.

22. Francis Mason, *The Story of a Working Man's Life* (New York, 1870), 100, as quoted in Horine, *Daniel Drake,* 147.

23. Western Museum handbill in the possession of the Cincinnati Historical Society Library, cited in Louis L. Tucker, " 'Ohio Show-Shop': The Western Museum of Cincinnati, 1820–1867," in Whitfield J. Bell, Jr., et al., *A Cabinet of Curiosities: Five Episodes in the Evolution of American Museums* (Charlottesville, Va., 1967), 73–105.

24. Daniel Drake, *Narrative of the Rise and Fall of the Medical College of Ohio* (Cincinnati, 1822), 7–8. See also Otto Juettner, "Rise of Medical Colleges in the Ohio Valley," *Ohio Archaeol. Hist. Publ.,* 22 (1913), 481–91.

25. Drake, *Narrative,* 37.

26. *Niles' Register,* Jan. 28, 1815, as quoted in Wade, *Urban Frontier,* 49. Wade gives a succinct account of the social and economic development of Lexington in this period. See also Niels H. Sonne, *Liberal Kentucky, 1780–1828* (New York, 1939); Ash Gofar and J. Hill Homon, *A Lamp in the Wilderness* Lexington, Ky., 1982).

27. Horace Holley to Mrs. Holley, Lexington, Ky., May 27, 1818, as quoted in Huntley Dupre, *Rafinesque in Lexington, 1819–1826* (Lexington, Ky., 1945), 10. See also Robert Peter, *Transylvania University: Its Origin, Rise, Decline, and Fall* (Louisville, Ky., 1896), 77.

28. François A. Michaux, *Travels to the Westward of the Alleghany Mountains . . . ,* 2nd ed. (London, 1805), 160–61.

29. Typescript copy of a letter from Daniel Drake to "Dear Sir" dated Lexington, Ky., Sept. 9, 1823, preserved at the Ohio Historical Center in Columbus, Ohio. See also Robert Peter, *The History of the Medical Department of Transylvania University,* Filson Club publ. no. 20 (Louisville, Ky., 1905); Juettner, "Rise of Medical Colleges," 481–91.

30. *Delineations of American Scenery and Character by John James Audubon, with An Introduction by Francis Hobart Herrick . . .* (New York, 1926), 96–97.

31. Constantine S. Rafinesque, "A Life of Travels by C. S. Rafinesque, Being a Verbatim and Literatim Reprint of the Original and Only Edition" (Philadelphia, 1816), in *Chron. Bot.,* 8 (1944), 318.

32. John D. Clifford to Robert Wickliffe Esq., Lexington, Ky., Feb. 11, 1819, Farris Rare Book Room, Transylvania University Library.

33. See Harry B. Weiss, *Rafinesque's Kentucky Friends,* published by the author (Highland Park, N.J., 1936). This description and other details about Rafinesque are quoted in "Rafinesque Memorial Papers, October 31, 1940," *Transylvania Coll. Bull.,* 15 (1942), 27.

34. Quoted from Rafinesque's own translation of his letter to M. Bory St. Vincent, Lexington, Ky., Jan. 7, 1821, published in C. S. Rafinesque, *Western Minerva or American Annals of Knowledge and Literature,* 72, facsimile copy in Farris Rare Book Room, Transylvania University Library.

35. The Farris Rare Book Room at the Transylvania University Library has some boxes of documents and letters relating to the Transylvania Botanic-Garden Company and the garden itself, which apparently never amounted to much.

36. Charles Boewe, in an article in *Names,* 10 (1962), 58–60, notes that on an Erie Canal barge Rafinesque encountered the scientist Amos Eaton and several students geologizing and botanizing and returned with them to Albany and Troy, meanwhile cementing a fast friendship with Eaton. The result: "At the northern edge of Troy, . . . near the campus of Rensselaer Polytechnic Institute, rises a hill of modest pretensions with the sonorous name *Mt. Rafinesque.*"

6: FROM NEWTON TO LAPLACE IN AMERICAN ASTRONOMY

1. See Samuel A. Mitchell, "Astronomy during the Early Years of the American Philosophical Society," *Proc. APS,* 86 (1943), 13–21; Trudy E. Bell, "The Beginnings of American Astronomy," *Sky Telesc.,* 52 (1976), 26–31. In the ms "Communications on Mathematics and Astronomy," vol. 1, at the American Philosophical Society Library is a

letter from the Astronomer Royal Nevil Maskelyne to William Smith, Rittenhouse's collaborator in the observations of the transit of Venus of 1769, Greenwich, Dec. 28, 1769, reporting that in England the sun was too low to permit observation of the first contact of Venus with the sun as Rittenhouse and Smith had done. "Your measures of the nearest distances of the Limbs of the Sun and Venus," wrote Maskelyne, "determine very well the nearest approach of Venus to the Sun's center, which was a very important observation, and could not be made here." See also Maskelyne, "Account of the Transit of Venus . . . As Observed at Norriton, in the . . . Province of Pennsylvania June 3, 1769 . . . Communicated by Dr. [William] Smith . . . ," *Philos. Trans. Roy. Soc. London,* 59, pt. 2 (1769), 289–326. An account of the Harvard expedition to St. John's, Newfoundland, to observe the transit of 1761 is given in John Winthrop IV, *Two Lectures on the Parallax and Distance of the Sun, As Deducible from the Transit of Venus* (Boston, 1769), 20–21. See also Harry Woolf, *The Transits of Venus: A Study of Eighteenth-Century Science* (Princeton, N.J., 1959); John C. Greene, "Some Aspects of American Astronomy, 1750–1815," *Isis,* 45 (1954), 339–58.

2. See W. C. Rufus, "Astronomical Observatories in the United States Prior to 1848," *Sci. Mon.,* 19 (1924), 120–39; Willis I. Milham, "Early American Observatories," *Pop. Astron.,* 45 (1937), 465–74; David F. Musto, "A Survey of the American Observatory Movement, 1800–1850," *Vistas Astron.,* 9 (1967), 37–92.

3. See David Eugene Smith and Jekuthiel Ginsburg, *A History of Mathematics in America before 1900,* Carus Mathematical Monographs, no. 5 (Chicago, 1934), chaps. 2–3; Florian Cajori, *The Early Mathematical Sciences in North and South America* (Boston, 1928).

4. [John G. Palfrey], "Professor Farrar," *Christ. Exam.,* 55 (July 1853), 126.

5. Sidney Willard, *Memories of Youth and Manhood,* 2 vols. (Cambridge, 1855), 1:272–73. Harvard students who wanted to go beyond the prescribed course of study could arrange to do so, and a considerable number took advantage of this opportunity. In 1797 John Pickering, later a distinguished lawyer and linguist, pursued his mathematical studies into the calculus and was commended by the Corporation and Overseers for his solutions of problems in this branch of mathematics.

6. Florian Cajori, *Teaching and History of Mathematics,* Bureau of Education Circular of Information no. 3 (Washington, D.C., 1890), 127–32. Nathaniel Bowditch seems not to have had a high opinion of Farrar's texts, but they served a useful purpose.

7. As quoted in Cajori, *Teaching and History,* 117–18, citing "American Educational Biography," *Barnard's J.,* 18 (1866), 141. See also Sidney Forman, *West Point: A History of the United States Military Academy* (New York, 1950); Peter M. Molloy, *Technical Education and the Young Republic: West Point as America's École Polytechnique, 1802–1833,* (Ph.D. diss., Brown University, 1975; Xerox, University Microfilms, 76–115, 673.

8. This account of Adrain follows Julian L. Coolidge, "Robert Adrain, and the Beginnings of American Mathematics," *Am. Math. Mon.,* 3 (1926), 67–69. See also "Robert Adrain LL.D.," *U.S. Mag. Democr. Rev.,* 14 (1844), 646–52, said to have been written by Adrain's son, Garnett B. Adrain. Coolidge gives Adrain's derivation of the law of errors as follows:

What general laws, he asks, will errors in measurement follow? What is the probability that we shall make an error ξ in measuring a length X? Suppose that we measure two quantities whose true values are X and Y. What is the probability that we shall make the respective errors ξ and η where $\xi + \eta = C$? Adrain took it as evident that it is most likely that these errors shall be proportional to the quantities measured. Now let $\phi(\xi, X)$ be the probability of making the error ξ in the first measurement. Assuming the two events are independent, which is contrary to this previous assumption, their compound probability is the product of their individual probabilities, so that the function we would maximize is $\phi(\xi, X)\phi(\eta, Y)$. Equating the logarithmic derivative to zero,

$$\frac{\phi'(\xi, x)}{\phi(\xi, x)} \, d\xi + \frac{\phi'(\eta, y)}{\phi(\eta, y)} \, d\eta = 0 \qquad d\xi + d\eta = 0$$

$$\frac{\phi'(\xi, x)}{\phi(\xi, x)} = \frac{\phi'(\eta, y)}{\phi(\eta, y)} \text{ when } \frac{\xi}{x} = \frac{\eta}{y}$$

The simplest solution is

$$\frac{\phi'(\xi,x)}{\phi(\xi,x)} = C\,\frac{\xi}{x} \qquad \phi(\xi,x) = re^{-d\xi^2/x}$$

Here we have the first known demonstration of the exponential law of error, published a year later by Gauss and usually associated with his name.

See also O. B. Sheynin, "C. F. Gauss and the Theory of Errors," *Arch. Hist. Exact Sci.,* 20 (1979), 21–72.

9. Coolidge, "Robert Adrain," 75–76. Not all scholars agree with Coolidge's evaluation of Adrain and his contemporaries. James Montgomery, currently making a study of Bowditch's scientific career, thinks that Bowditch's command of Continental mathematics and mathematical astronomy was superior to Adrain's and that Coolidge's assessment is colored by disdain for applied mathematics.

10. See Brooke Hindle, *David Rittenhouse* (Princeton, 1964), 302, 328–30; Simon Newcomb, "Aspects of American Astronomy," *Ann. Rep. Smithsonian Inst., 1897* (Washington, D.C., 1898), 89. Newcomb notes that the collimating telescope "has become almost a necessity wherever accurate observations are made."

11. Silvio A. Bedini, "Andrew Ellicott, Surveyor of the Wilderness," *Surveying and Mapping,* 36 (1976), 123–25. See also Catherine Mathews, *Andrew Ellicott, His Life and Letters* (New York, 1908).

12. Silvio A. Bedini, *The Life of Benjamin Banneker* (New York, 1972), 103–36; Bedini, *Thinkers and Tinkers* (New York, 1975), 320–30.

13. William Dunbar, "Account of the Commencement and Progress of the First Eighteen Miles of the line of Demarcation . . . between the Territories of Spain and the United States of America . . . ," as quoted in Mrs. Dunbar Rowland, comp., *Life, Letters, and Papers of William Dunbar of Elgin, Morayshire, Scotland, and Natchez, Mississippi, Pioneer Scientist of the Southern United States,* compiled and prepared from the original documents for the National Society of Colonial Dames in America (Jackson, Miss., 1930), 81–82.

14. Andrew Ellicott, "Astronomical, and Thermometrical Observations, Made on the Boundary between the United States and His Catholic Majesty," *Trans. APS,* 5 (1802), 236. Susan F. Cannon, *Science in Culture: The Early Victorian Period* (New York, 1978), 101, notes that "the new astronomical circle by the great London craftsman Troughton turned out to be more trouble than it was worth; whereas Rittenhouse's two zenith sectors, in Ellicott's hands, were stellar performers, as was a 7-inch sextant by Ramsden." Cannon thinks that Rittenhouse's large zenith sector was "the most accurate instrument of any kind made in North America before 1840." According to Cannon it was good to two seconds of arc in latitude measurements when used by a careful observer like Ellicott.

15. Andrew Ellicott to Thomas Jefferson, Philadelphia, May 16, 1801, and May 26, 1801, Papers of Thomas Jefferson, Library of Congress. I am indebted to Silvio Bedini of the Smithsonian Institution for letting me use his transcripts of Ellicott's correspondence.

16. Ellicott to Jefferson, Lancaster, Pa., Dec. 29, 1801. Silvio Bedini's copy of this letter bears the notation: "Copy of Press Copy of Letter."

17. Ellicott to Jefferson, Lancaster, Pa., May 11, 1802.

18. Ellicott to Jean-Baptiste J. Delambre, Lancaster, Pa., Mar. 10, 1802, Ellicott Papers, Library of Congress. Ellicott overlooked the existence and activities of other American practical astronomers such as Nathaniel Bowditch of Salem, John Garnett of New Brunswick, N.J., and Simeon De Witt of New York.

19. Ellicott to Benjamin Rush, Lancaster, Pa., Oct. 27, 1802, Ellicott Papers, Library of Congress.

20. José J. de Ferrer, "Observations of the Eclipse of the Sun, June 16, 1806 . . . ," *Trans. APS,* 6 (1809), 265; "Further Observations on the Eclipse of the 16th of June, 1806 . . . ," ibid., 293–99. I have been able to find very little information about Ferrer. J. C. Poggendorff, *Biographisch-Literarisches Handwoerterbuch,* vol. 1, describes him as "Hoeherer Officier in d. Span. Marine." For the 1806 solar eclipse Andrew Ellicott and Simeon De Witt used large zenith sectors made by David Rittenhouse; William Dunbar used a reflecting

telescope of 6-foot focus, a 2½-foot focus achromatic refracting telescope, a circle of reflection made by Troughton, an astronomical clock with a gridiron pendulum, and a portable chronometer (information supplied by Deborah Warner, Smithsonian Institution).

21. See Franz von Zach's *Monatliche Correspondenz zur Beförderung der Erd-und Himmels-kunde* . . . , 7 (January–June, 1803), 91–99 for his biographical notice of David Rittenhouse. The frontispiece to the volume is a reproduction of Houston's portrait of Rittenhouse (1792), reproduced from the *American Universal Magazine*. For Zach's notice of vol. 6, pt. 1 of the *Transactions* of the American Philosophical Society, see 11 (January –June, 1805), 251–67.

22. Ellicott's biographer, Catherine Mathews, states that Ellicott was elected a foreign member of the National Institute of France about 1808, but I have been able to find no evidence either in the Ellicott Papers at the Library of Congress or in the archives and *Procès-verbaux* of the institute to support this statement. I am grateful to the archivists at both these institutions for their assistance in determining this point.

23. See Edward D. Mansfield, "The United States Military Academy at West Point," *Am. J. Educ.*, 30 (1863), 2–48; Molloy, *Technical Education and the Young Republic,* 354–55, 368–71, 400–405. According to Molloy, Mansfield was known to the superintendent of the academy, Sylvanus Thayer, as "Old Grumbler." He "seems to have lectured very rarely during the sixteen years of his tenure at the Academy, and not at all until well into 1815. . . . Mansfield admitted to a total ignorance of French, and boasted in a letter to Calhoun that he would never teach from a text. Since at the time of his letter the prescribed text was Haüy's treatise on physics, in the original French, Mansfield's reluctance to use the text probably stemmed from reasons other than educational theories." From 1817–1820, Molloy adds, Ellicott's duties at the academy were probably handled to a considerable extent by his principal assistant, Charles Davies, since Ellicott was in poor health.

24. Samuel Williams, "A Memoir on the Latitude of the University at Cambridge: With Observations of the Variation and Dip of the Magnetic Needle," *Mem. Am. Acad. Arts Sci.,* 1 (1785), 67.

25. "A letter from Joseph Willard to the Rev. Dr. Maskelyne . . . Concerning the Longitude of Cambridge in New England," *Philos. Trans. Roy. Soc. London,* 71, pt. 2 (1781), 507.

26. Samuel Williams, "Astronomical Observations, Made in the State of Massachusetts," *Mem. Am. Acad. Arts Sci.,* 1 (1785), 93. See also Cajori, *Early Mathematical Sciences,* 57; I. Bernard Cohen, *Some Early Tools of American Science* . . . (Cambridge, 1950), 50ff.; Robert F. Rothschild, "What Went Wrong in 1780?" *Harvard Mag.,* 83 (1981), 20–27.

27. There is a sketch of Bowditch's career and achievements in Dirk J. Struik, *Yankee Science in the Making,* rev. ed. (New York, 1962), 108–14, 227–32. Struik (p. 456) lists the chief books and articles about Bowditch. There is no good study of Bowditch's work in mathematics and mathematical astronomy based on the Bowditch materials in the Boston Public Library, but James Montgomery is working on such a study for a doctoral dissertation at the University of Connecticut.

28. Nathaniel Bowditch, "On the Eclipse of the Sun of September 17, 1811, . . . " *Mem. Am. Acad. Arts Sci.,* 3, pt. 2 (1815), 258.

29. Ibid.

30. David Rittenhouse, "Account of a Meteor," *Trans. APS,* 2 (1786), 175.

31. Andrew Ellicott to Thomas Jefferson, Lancaster, Pa., Oct. 6, 1805, Jefferson Papers, Library of Congress; Jefferson to Ellicott, Washington, D.C., Oct. 25, 1805, ibid.

32. Thomas Clap, *Conjectures upon the Nature and Motion of Meteors, Which Are Above the Atmosphere* (Norwich, Conn., 1781), 11.

33. Jeremiah Day, "A View of the Theories Which Have Been Proposed, To Explain the Origin of Meteoric Stones," *Mem. Conn. Acad. Arts Sci.,* 1, pt. 1 (1810), 163–74.

34. Nathaniel Bowditch, "An Estimate of the Height, Direction, Velocity and Magnitude of the Meteor, That Exploded over Weston in Connecticut, December 14, 1807 . . . ," *Mem. Am. Acad. Arts Sci.,* 3, pt. 2 (1815), 236.

35. Nathaniel Bowditch, "On the Meteor Which Passed Over Wilmington in the State of Delaware, November 21, 1819," *Mem. Am. Acad. Arts Sci.,* 4, pt. 2 (1819), 295–305.

36. Nathaniel Bowditch, "Observations of the Comet of 1807," *Mem. Am. Acad. Arts Sci.,* 3, pt. 1 (1809), 1.

37. Ibid., 1–17.

38. *Two Lectures on Comets. by Professor Winthrop, Also, an Essay on Comets, by A. Oliver, Jun. Esq. with Sketches of the Lives of Professor Winthrop and Mr. Oliver, Likewise, a Supplement, Relative to the Present Comet of 1811* (Boston, 1811). The copy in the Harvard Library bears the notation "edited by John Davis." Oliver's *Essay* is described as having been "translated into the French language at Paris" and praised by Jean S. Bailly in his *Histoire de l'astronomie ancienne* (Paris, 1775). Oliver suggested that the tail of a comet results from the mutual repulsion between the comet's atmosphere and that of the sun. He supported his theory by describing some electrical experiments with an artificial comet, "a small, gilt cork ball, with a tail of leaf-gold, about two inches and an half in length." Bailly found Oliver's work *"infiniment ingénieux"* and his hypothesis *"très-vraisemblable."*

39. James Dean, "Investigation of the Apparent Motion of the Earth Viewed from the Moon, Arising from the Moon's Librations," *Mem. Am. Acad. Arts Sci.,* 3, pt. 2 (1815), 241–45; Nathaniel Bowditch, "On the Motion of a Pendulum Suspended from Two Points," ibid., 413–36. Vol. 4, pt. 1, of the academy's *Memoirs* carried "A New Investigation of Kepler's Problem," by Friedrich T. Schubert, a member of the St. Petersburg Academy of Sciences. It seems likely that this contribution from a distinguished Continental astronomer was arranged by John Q. Adams, who was then the American minister in St. Petersburg.

40. Nathaniel Bowditch, review of three German treatises on astronomy, *North Am. Rev.,* 14 (1822), 26–34. See also 10 (1820), 269–70.

41. "Lettre XIII. de M. Nathaniel Bowditch," *Correspondance Astronomique . . . de Baron de Zach,* 10 (1824), 223–49. Pages 321–30 of the same volume contain Zach's account of Bowditch's method of computing lunar distances. On pp. 495–96 Zach reproduces Bowditch's table of longitudes and latitudes computed from American astronomical observations and published in vol. 3 of the American Academy's *Memoirs.*

42. Pierre S. de Laplace, *"Mécanique céleste" by the Marquis de La Place . . . ,* trans. with commentary by Nathaniel Bowditch, 4 vols. (Boston, 1829–1839). The comments of European astronomers are quoted liberally in the memoir of Bowditch by his son Nathaniel I. Bowditch that opens the fourth volume of the translation. See also Joseph Lovering, "The *Mécanique céleste* of Laplace, and Its Translation with a Commentary by Bowditch," *Proc. Am. Acad. Arts Sci.,* 24 (1889), 185–201.

7: THE CHEMICAL REVOLUTION COMES TO AMERICA

1. David Rittenhouse, "An Optical Problem, Proposed by Mr. Hopkinson," *Trans. APS,* 2 (1786), 201.

2. Ibid., 204.

3. Ibid., 205.

4. See Brooke Hindle, *David Rittenhouse* (Princeton, 1946), 276–77; Edward Ford, *David Rittenhouse, Astronomer-Patriot, 1732–1796* (Philadelphia, 1946), 54, 114, 139–40, 203.

5. See Aaron Ihde, *Development of Modern Chemistry* (New York, 1964), chap. 3; Joseph Priestley, *A Scientific Autobiography . . . ; Selected Scientific Correspondence,* ed. with commentary by Robert E. Schofield (Cambridge, 1966); Henry Guerlac, *Lavoisier— The Crucial Year: The Background and Origin of His First Experiments on Combustion in 1772* (Ithaca, N.Y., 1961).

6. Thomas Jefferson to James Currie, Paris, Dec. 20, 1788 in Julian P. Boyd, ed., *The Papers of Thomas Jefferson,* (Princeton, 1950-), 14:366.

7. Samuel L. Mitchill, *The Present State of Medical Learning in the City of New York* (New York, 1797), 5. See also Denis L. Duveen and H. S. Klickstein, "The Introduction of Lavoisier's Chemical Nomenclature into America," *Isis,* 45 (1954), 278–92, 368–82.

8. Samuel L. Mitchill, *Nomenclature of the New Chemistry* (New York, 1794). In his preface "To the Students of Chemistry," 3, Mitchill observes: "I have been careful to

insert the German Nomenclature as well as the French in this arrangement, on account of the high opinion I have formed of the writings of many authors in that language. Ever since I began to reflect on this subject, it has been matter of regret to me, that so few of our countrymen have cultivated that very useful and important branch of literature. Much of the time spent in pursuing the customary plan of dead letter education, might, I think, be more profitably employed in acquiring a competent knowledge of a modern tongue, in which so many books of high utility to mankind are written, and which, from our ignorance of it, are not often read by us. . . . It is high time to unlock this magazine of science, and he who shall make himself master of the language may be considered as already in possession of the key."

9. Ibid., 10.

10. Joseph Priestley to John Vaughan, Northumberland, Pa., July 21, 1794, Dreer Collection (Chemists), vol. 3, Historical Society of Pennsylvania. See also Edgar F. Smith, *Priestly in America, 1794–1804* (Philadelphia, 1920).

11. Joseph Priestley, *Considerations on the Doctrine of Phlogiston, and the Decomposition of Water* (Philadelphia, 1796), preface.

12. John Maclean, *A Memoir of John Maclean, M.D.* . . . (Princeton, 1876).

13. John Maclean, "Two Lectures on Combustion," in Joseph Priestley and John Maclean, *Lectures on Combustion,* ed. William Foster (Princeton, 1929), 113–14. Original published in Philadelphia in 1797.

14. Ibid., 116.

15. James Woodhouse, "An Answer to Dr. Joseph Priestley's *Considerations on the Doctrine of Phlogiston* . . . Founded upon Demonstrative Experiments," *Trans. APS,* 4 (1799), 475. Priestley was not offended by Woodhouse's article. On June 25, 1799, he wrote to Benjamin S. Barton: "Give my respects to Dr. Woodhouse. I am pleased with the spirit and ingenuity of his defence of the *new system,* tho I have no doubt of it's being a mere fallacy, as I am pretty confident I shall make appear before I have done. When they have printed all that I have sent to the Medical Repository I shall reduce all that I have written on this subject into one work, and reply to all my opponents."

16. See Sidney Edelstein, "The Chemical Revolution in America, from the Pages of the *Medical Repository," Chymia,* 5 (1959), 155–79.

17. Samuel L. Mitchill, "An Attempt to Accommodate the Disputes among the Chemists Concerning Phlogiston, letter to Dr. Priestley dated Nov. 14, 1797," *Med. Repos.,* 1 (1798), 515.

18. Joseph Priestley, "Miscellaneous Observations Relating to the Doctrine of Air," *Med. Repos.,* 5 (1802), 266.

19. Mitchill continued to call hydrogen *phlogiston* for another decade. Woodhouse's successor in the chemical chair at the University of Pennsylvania also retained belief in a principle of inflammability. See Robert Siegfried, "An Attempt to Resolve the Differences between the Oxygen and Phlogiston Theories," *Isis,* 41 (December 1955), 151–60; Thomas D. Mitchell, "Analysis of Prof. Coxe's Essay on Combustion and Acidification," *Mem. Columbian Chem. Soc.,* 1 (1813), 179.

20. Thomas Cooper, *The Introductory Lecture of Thomas Cooper, Esq., Professor of Chemistry at Carlisle College, Pennsylvania . . . with Notes and References* (Carlisle, Pa., 1812), 94.

21. Thomas Cooper, "Prospectus of the Emporium of Arts and Sciences," *Emporium Arts Sci.,* n.s., 1 (1813), 6.

22. As quoted in Edgar F. Smith, *James Cutbush, an American Chemist, 1788–1823* (Philadelphia, 1919), 13–14. In 1808 Cutbush began publication of a short-lived periodical entitled *The Useful Cabinet.*

23. Thomas P. Smith, *Annual Oration Delivered Before the Chemical Society of Philadelphia, April 11, 1798* . . . (Philadelphia, 1798), as quoted in Edgar F. Smith, *Chemistry in America* (New York and London, 1914), 35–36.

24. See John C. Greene and John G. Burke, *The Science of Minerals in the Age of Jefferson, Trans. APS,* 68, pt. 4 (1978), 68, chap. 4.

25. Benjamin Silliman, "Notes" to William Henry, *An Epitome of Chemistry,* 1st Am. ed. (New York, 1808), note 1.

26. See I. Bernard Cohen, *Some Early Tools of American Science . . .* (Cambridge, 1950), chap. 4.

27. See Duveen and Klickstein, "Introduction of Lavoisier's Chemical Nomenclature," 277.

28. *Am. Philos. Med. Regist.,* 2 (1811), 204–8; *Trans. Lit. Philos. Soc. New York,* 1 (1815), 539–57; Charles A. Browne, "Lecture and Laboratory Notebooks of Three Early Irish-American Refugee Chemists," *J. Chem. Educ.,* 10 (1941), 155–56; Desmond Reilly, "An Irish-American Chemist, William James MacNeven, 1763-1841," *Chymia,* 2 (1949), 17–26.

29. William MacNeven, *Exposition of the Atomic Theory of Chymistry; And of the Doctrine of Definite Proportions* (New York, 1819), 1–2.

30. William MacNeven, *A Tabular View of the Modern Nomenclature, and System of Chemistry* (New York, 1821), 3.

31. See Edgar F. Smith, *Chemistry in Old Philadelphia* (Philadelphia, 1919), 23–24.

32. Benjamin Silliman, "Reminiscences," as quoted in George P. Fisher, *Life of Benjamin Silliman . . . ,* 2 vols. (New York, 1866), 1:100–101.

33. See Greene and Burke, *Science of Minerals,* 26–28. See also Edgar F. Smith, *The Life of Robert Hare, an American Chemist (1781-1858)* (Philadelphia and London, 1917).

34. See Fisher, *Life of Benjamin Silliman,* 1:195–96; Greene and Burke, *Science of Minerals,* 97–98.

35. Benjamin Silliman, "Experiments on the Fusion of Various Refractory Bodies, by the Compound Blow-Pipe of Mr. Hare," *Mem. Conn. Acad. Arts Sci.,* 1, pt. 3 (1813), 329–39. See also Silliman's "Notes" to the 3rd Am. ed., William Henry, *Elements of Experimental Chemistry,* 2 vols. (Boston, 1814), 1:380–88, note 18.

36. Benjamin Silliman, "Notice of the Fusion of Plumbago, or Graphite . . . in a Letter to Dr. Robert Hare . . . dated March 26, 1823," *Am. J. Sci. Arts,* 6 (1823), 341–49; Silliman, "Experiments upon Diamond, Anthracite and Plumbago with the Compound Blow Pipe, in a Letter Addressed to Prof. Robert Hare . . . ," ibid., 349–53; Silliman, "Fusion of Charcoal, by the Deflagrator, with Proofs of a Current between the Poles," ibid., 5 (1822), 108–12.

37. Charles U. Shepard, "Address of Presentation," *Memorial Addresses at the Unveiling of the Bronze Statue of Professor Benjamin Silliman, at Yale College, June 24, 1880* (New Haven, 1885), 9.

38. For excellent accounts of Silliman and some of his students, see Leonard G. Wilson, ed., *Benjamin Silliman and His Circle: Studies on the Influence of Benjamin Silliman on Science in America* (New York, 1979). See also John F. Fulton and Elizabeth H. Thomson, *Benjamin Silliman, 1779-1864; Pathfinder in American Science* (New York, 1947).

39. Robert Hare, "Account of New Eudiometers, &c. Invented by Robert Hare," *Am. J. Sci. Arts,* 2 (1820), 312–19.

40. Edgar F. Smith, *Chemistry in America: Chapters from the History of Science in the United States* (New York and London, 1914), 189. See also Robert Hare, "A New Theory of Galvanism, Supported by Some Experiments by Means of the Calorimotor," *Am. J. Sci. Arts,* 1 (1819), 413–23.

41. See Smith, *Life of Robert Hare,* 86–87.

42. Benjamin Silliman, "Notice of the Galvanic Deflagrator of Professor Robert Hare . . . , in a Letter to That Gentleman from the Editor, Yale College, October 23d, 1821," *Am. J. Sci. Arts,* 4 (1822), 201–3.

43. As quoted by Smith, *Life of Robert Hare,* 132.

44. Robert Hare to Benjamin Silliman, May 25, 1822, "Correspondence between Robert Hare . . . and the Editor, on the Subject of Dr. Hare's Calorimotor and Deflagrator, and the Phenomena Produced by Them," *Am. J. Sci. Arts,* 5 (1822), 106.

45. As quoted in Smith, *Life of Robert Hare,* 120.

46. As quoted by Wyndham Miles, "Robert Hare," in Edouard Farber, ed., *Great Chemists* (New York, 1961), 428.

47. As quoted in Miles, "Robert Hare," 431.

8: AMERICAN GEOGRAPHY

1. James Bowdoin, "A Philosophical Discourse, Publickly Addressed to the American Academy of Arts and Sciences, in Boston, on the Eighth of November, M,DCC,LXXX: When the President Was Inducted into Office," *Mem. Am. Acad. Arts Sci.,* 1(1785), 12–13.

2. Samuel Williams, *The Natural and Civil History of Vermont* (Walpole, N.H., 1794); Jeremy Belknap, *The History of New-Hampshire . . . ,* 3 vols. (Boston, 1792), 3:3.

3. David Ramsay, *The History of South Carolina from the First Settlement in 1670, to the Year 1808,* 2 vols. (Charleston, S.C., 1809); John Drayton, *A View of South Carolina, As Respects Her Natural and Civil Concerns . . .* (Charleston, S.C., 1802). See also Dr. John Brickell, *The Natural History of North Carolina* (Dublin, 1737). Reprinted by Johnson Publ. Co., Murfreesboro, N.C., 1968.

4. Gilbert Imlay, *A Topographical Description of the Western Territory of North America, Containing a Succinct Account of Its Soil, Climate, Natural History, Population, Agriculture, Manners & Customs,* 3rd ed. (London, 1797). Reprinted by Augustus M. Kelley Publ., New York, 1969. For an account of Thomas Hutchins and his work, see the "Introduction" to the facsimile reprint of Hutchins, *An Historical Narrative and Topographical Description of Louisiana and West-Florida* (Gainesville, Fla., 1968), v–xlviii.

5. Jedidiah Morse, *The American Geography; or, A View of the Present Situation of the United States of America* (Elizabethtown, N.J., 1789), 34–63.

6. Samuel L. Mitchill to Jedidiah Morse, New York, July 18, 1789, Gratz Collection, Historical Society of Pennsylvania.

7. Manasseh Cutler to Jedidiah Morse, Ipswich, Mass., Aug. 13, 1792, Gratz Collection, Historical Society of Pennsylvania.

8. For an evaluation of the various editions of Morse's *American Geography* and an account of their preparation and reception, see Ralph M. Brown, "The American Geographies of Jedidiah Morse," *Ann. Assoc. Am. Geogr.,* 31 (1941), 145–217.

9. Christoph D. Ebeling to Jedidiah Morse, October 1793, as quoted in William B. Sprague, *The Life of Jedidiah Morse, D.D.* (New York 1874), 205.

10. Quoted from *The Private Journal of Aaron Burr* (Rochester, N.Y., 1903), 292, in Ralph H. Brown, "Early Maps of the United States: The Ebeling-Sotzmann Maps of the Northern Seaboard States," *Geogr. Rev.,* 30 (July 1940), 471–79. According to Brown, Ebeling gradually acquired more than 20,000 maps of various parts of the world for his collection. Of the Ebeling-Sotzmann maps of the northern states Brown says, "The evidence points clearly to Ebeling as the dominant figure in the production of this series of maps, which from the first has been referred to as excellent. . . . Knowing that the maps were prepared under the direction of a discriminating scholar, the present-day investigator can consult them with more than ordinary confidence." For a brief account of Ebeling and a selection of his letters to American correspondents, see William. C. Lane, *Letters of Christoph Daniel Ebeling* (Worcester, Mass., 1926).

11. Jedidiah Morse to Christoph D. Ebeling, May 1794, as quoted in Sprague, *Life of Jedidiah Morse,* 212.

12. Thomas Jefferson to William Dunbar, Mar. 13, 1804, Jefferson Papers, Library of Congress.

13. Thomas Jefferson to André Michaux, Apr. 30, 1793, quoted in Donald Jackson, ed., *Letters of the Lewis and Clark Expedition with Related Documents, 1783–1854* (Urbana, Ill., 1962), 669–72.

14. Alexander Mackenzie, *Voyages from Montreal through the Continent of North America, to the Frozen and Pacific Oceans in the Years 1789 and 1793,* 2 vols. (London, 1801), 2:411.

15. Edward Thornton to Lord Hawkesbury, Philadelphia, Mar. 3, 1803, as quoted in Jackson, *Letters,* 25–27.

16. Thomas Jefferson to the Count de Lacépède, Washington, D.C., Feb. 24, 1803, as quoted in Jackson, *Letters,* 15–16. On May 13, 1803, Lacépède replied: "Whatever may be the success of the expedition . . . , it will be extremely useful for the progress of industry, the sciences, and especially natural history," ibid., 47.

17. Meriwether Lewis to Thomas Jefferson, Lancaster, Pa., Apr. 20, 1803, as quoted in Jackson, *Letters,* 37–38.

18. See Donald Jackson, *Thomas Jefferson & the Stony Mountains: Exploring the West from Monticello* (Urbana, Ill., Chicago, London, 1981), chaps. 11–12.

19. Meriwether Lewis to William Clark, Washington, D.C., June 19, 1803, as quoted in Jackson, *Letters,* 59–60. For Jefferson's instructions to Lewis, see 61–67.

20. See John L. Allen, *Passage through the Garden: Lewis and Clark and the Image of the American Northwest* (Urbana, Ill., Chicago, London, 1975), chaps. 3–4.

21. Amos Stoddard to Henry Dearborn, St. Louis, Mo., June 3, 1804, as quoted in Jackson, *Letters,* 196.

22. Meriwether Lewis to Thomas Jefferson, Fort Mandan, Louisiana Territory, Apr. 7, 1805, as quoted in Jackson, *Letters,* 231–34.

23. Thomas Jefferson to Étienne Lemaire, Monticello, Aug. 17, 1805, as quoted in Jackson, *Letters,* 255.

24. Jefferson to Charles W. Peale, Washington, D.C., Oct. 6 and 9, 1805, as quoted in Jackson, *Letters,* 260–61, 263. Peale replied on November 3 that the prairie dog and the magpie were flourishing but that he was having difficulty mounting some of the specimens because some of the bones were broken or missing and the skins had been damaged by insects. He added, "I am very much obliged to Capt. Lewis for his endeavors to increase our knowledge of the Animals of that new acquired Territory. I wish I could get one of the sheep that carry such large horns as those you have done me the favor of sending. It is more important to have this Museum supplied with the American Animals than those of other Countryes, yet for a comparative view it ought to possess those of every part of the Globe!" ibid., 268.

25. *Message from the President of the United States, Communicating Discoveries Made in Exploring the Missouri, Red River and Washita, by Captains Lewis and Clark, Doctor Sibley, and Mr. Dunbar; with a Statistical Account of the Countries Adjacent* (Washington, D.C., 1806). William Dunbar and George Hunter had been employed by the U.S. government to explore and survey the country traversed by the Washita River. Their journals contained useful and interesting information about the geography, mineralogy, flora, and fauna of that region. See William Dunbar, "Journal of a Voyage Commencing at St. Catherine's Landing . . . ," in *Documents Relating to the Purchase & Exploration of Louisiana . . . Printed from the Original Manuscripts in the Library of the American Philosophical Society . . .* (Boston and New York, 1904), 151ff. The same volume contains Thomas Jefferson, "The Limits and Bounds of Louisiana."

26. Thomas Jefferson to the Count de Lacépède, Washington, D.C., July 14, 1808, as quoted in Jackson, *Letters,* 443.

27. Thomas Jefferson to Alexander von Humboldt, Dec. 6, 1813, as quoted in Jackson, *Letters,* 596.

28. Frederick Pursh, *Flora Americae Septentrionalis; or, A Systematic Arrangement and Description of the Plants of North America . . . ,* 2 vols. (London, 1814), 1:x–xi. For a thorough and careful account of the fate of the plant specimens collected by Lewis and Clark and their subsequent discovery and description by botanists, see Paul R. Cutright, *Lewis and Clark: Pioneering Naturalists* (Urbana, Ill., 1969), 357–75. The Lewis and Clark specimens Pursh took with him to London were bought at auction in 1842 by Edward Tuckerman, an American botanist, and presented to the Academy of Natural Sciences of Philadelphia. In 1896 several packages of specimens from the expedition were discovered at the American Philosophical Society. These specimens, representing approximately 180 different species, were subsequently placed on loan at the Academy of Natural Sciences, where they and the Pursh specimens are preserved on 216 herbarium sheets. According to Cutright: "The majority of the plants came from west of the Continental Divide. Of approximately 200 dated sheets . . . , Lewis collected 135 on the Columbian watershed and 65 on the Missouri. . . . He obtained 60 on the journey from St. Louis to Fort Mandan, only 10 from Fort Mandan to the Pacific (due largely to the loss suffered at Great Falls and possibly at Shoshoni Cove) and 130 during the winter at Fort Clatsop and on the return to St. Louis. . . . About one-fourth of the entire Herbarium was collected in Idaho on the

return trip, the majority at Camp Chopunnish. The stay at this camp (From May 14 to June 10 [1806] happily coincided with the appearance of a host of local spring flowers, including . . . the beautiful ragged robin and mariposa lily," 369. See also Raymond D. Burroughs, *The Natural History of the Lewis and Clark Expedition* (Lansing, Mich., 1961); also his "The Lewis and Clark Expedition's Botanical Discoveries," *Nat. Hist.,* 75 (1966), 59; Susan D. McKelvey, *Botanical Exploration of the Trans-Mississippi West, 1790–1850* (Jamaica Plain, Mass., 1955), chap. 3; Velva E. Rudd, "Botanical Contributions of the Lewis and Clark Expedition," *J. Wash. Acad. Sci.,* 44 (1954), 351–56.

29. For further information about the zoological contributions of the Lewis and Clark expedition, see Cutright, *Lewis and Clark,* chap. 22 and appendix B; Jackson, *Letters,* 292–97; Burroughs, *Natural History;* Charles C. Sellers, *Mr. Peale's Museum: Charles Willson Peale and the First Popular Museum of Natural Science and Art* (New York, 1980), 171–91, 260–61. For an account of the fossils collected by Lewis and Clark, see Clifford M. Nelson, "Paleontology in the United States Federal Service, 1804–1904," *Earth Sci. Hist.,* 1 (1982), 48–57. Nelson writes: "Although collecting fossils . . . was necessarily a highly uneven activity, the Corps of Discovery secured a small number of vertebrate, invertebrate, and plant specimens, principally along the Missouri River. . . . Six of the fossils collected in 1804–1805 were among the 68 'mineralogical' specimens Jefferson donated to the Philosophical Society's cabinet in November 1805. The Society later transferred these fossils to Philadelphia's Academy of Natural Sciences. . . . The fossil invertebrates and teleost fish jaw . . . remained in the Philosophical Society's cabinet 'unknown and neglected' for nearly two decades, until their biological and geological significance was assessed by Philadelphians Thomas Say and Richard Harlan, and the latter's junior colleague Samuel George Morton, who by then had received additional fossils from the Big Bend and elsewhere along the Missouri.

30. Minnie C. Yarborough, ed., *The Reminiscences of William C. Preston* (Chapel Hill, N.C., 1933), 16. Clark's museum was described by Henry R. Schoolcraft in 1819 as including "skins of remarkable animals, minerals, fossil-bones, and other rare and interesting specimens" as well as Indian artifacts. It was visited by Herzog P. W. von Württemberg in 1823, by Lafayette and his party in 1825, and by the Duke of Saxe-Weimar-Eisenach in 1826, all of whom described the museum in complimentary terms. The collections were disposed of shortly before Clark's death in 1838. Its most lasting influence may have been on the artist George Catlin, who saw Indian portraits in the museum when he first visited it in the spring of 1830. Catlin became famous for his Indian portraits and exhibited them along with a large collection of Indian artifacts in his "Indian Gallery," touring the eastern cities of the United States and moving thence to London, Paris, and Brussels. For a history of Clark's museum, see John C. Ewers, "William Clark's Indian Museum in St. Louis, 1816–1838," in Whitfield J. Bell, Jr., et al., *A Cabinet of Curiosities* (Charlottesville, Va., 1967), 49–72.

31. Thomas Jefferson to Peter S. Du Ponceau, Monticello, Dec. 30, 1817, as quoted in *Documents Relating to the Purchase & Exploration of Louisiana,* 7. For a full account of the history of the Lewis and Clark journals, see Paul R. Cutright, *A History of the Lewis and Clark Journals* (Norman, Okla., 1976).

32. Verne F. Ray and Nancy O. Lurie, "The Contributions of Lewis and Clark to Ethnography," *J. Wash. Acad. Sci.,* 44 (1954), 358. See also Reuben G. Thwaites, ed., *Original Journals of the Lewis and Clark Expedition, 1804–1806,* 8 vols. (New York, 1904–1905).

33. Allen, *Passage through the Garden,* 125–26. Allen gives a full account of the historical development of geographical ideas about the trans-Mississippi West, the efforts of Jefferson and Lewis and Clark to obtain accurate information before the expedition set out, and the gradual changes in the explorers' conceptions during the course of their exploration.

34. Ibid., 382–98; Herman R. Friis, "Cartographic and Geographic Activites of the Lewis and Clark Expedition," *J. Wash. Acad. Sci.,* 44 (1954), 350–51; Carl I. Wheat, *Mapping the Trans-Mississippi West,* 5 vols. (San Francisco, 1958), 2:chaps. 13–15.

35. Jedidiah Morse, *American Universal Geography,* 6th ed., 2 vols. (1812), 1:122.

36. Ibid., 1:589. See also *The Travels of Capts. Lewis & Clarke, By Order of the Government of the United States, Performed in the Years 1804, 1805 & 1806, Being Upwards of Three Thousand Miles, from St. Louis, by Way of the Missouri, and Columbia Rivers, to the Pacific Ocean: Containing an Account of the Indian Tribes, Who Inhabit the Eastern Part of the Continent Unexplored, and Unknown Before with Copious Delineations of the*

Manners, Customs, Religion, &c. of the Indians. Compiled from Various Authentic Sources, and Documents . . . Embellished with a Map of the Country Inhabited by the Western Tribes of Indians, and Five Engravings of Indian Chiefs (Philadelphia, 1809). There was a London edition, printed for Longman, Hurst, Rees, and Orme, in the same year. For an account of this and other apocryphal accounts of the Lewis and Clark expedition, see Cutright, *History of the Lewis and Clark Journals,* chap. 3.

37. See Wheat, *Mapping the Trans-Mississippi West,* 2:chap. 15, for an account of the maps of Aaron Arrowsmith, John Melish, Étiene Brué, and others in the years immediately following the publication of the Biddle-Allen *History of the Expedition.* Wheat (p. 62) quotes Jefferson's letter to Melish, Dec. 31, 1816, praising Melish, *Map of the United States with the Contiguous British and Spanish Possessions* (Philadelphia, 1816). See also Charles O. Paullin, "American Explorations in the West, 1803–1852," *Atlas of the Historical Geography of the United States,* John K. Wright, ed. (Baltimore, 1932), 19–21. See also William Guthrie, *A Universal Geography; or, A View of the Present State of the Known World . . . ,* 3rd Am. ed., with extensive additions and alterations, by several American editors, 2 vols. (Philadelpha, 1820), 1:135–90. From the eloquent passages on American birds, it seems likely that George Ord, who supplied the scientific list of American animals for the second American edition of Guthrie's *New Geographical, Historical, and Commercial Grammar* (1815), was one of the American editors referred to in the title page of the *Universal Geography.* Ord had completed the last two volumes of Alexander Wilson's *American Ornithology* after Wilson's premature death in 1813.

38. John C. Calhoun to Major Stephen H. Long, Mar. 1819, "MS Communications to the American Philosophical Society," American Philosophical Society Library, 2:41.

9: FROM THE THEORY OF THE EARTH TO EARTH SCIENCE

1. Benjamin Franklin, "Conjectures Concerning the Formation of the Earth, &c. in a Letter from Dr. B. Franklin . . . to the Abbé Soulavie, Passy, September 22, 1782," *Trans. APS,* 3 (1793), 4. For a general view of theories of the earth in the eighteenth century, see John C. Greene, *The Death of Adam: Evolution and Its Impact on Western Thought* (Ames, Iowa, 1959), chap. 3; Katherine Collier, *Cosmogonies of Our Fathers. Some Theories of the Seventeenth and Eighteenth Centuries,* Columbia University Studies in History, Economics, and Public Law, no. 402 (New York, 1934).

2. As quoted in Thomas Pownall, *A Topographical Description of Such Parts of North America As Are Contained in the (Annexed) Map of the Middle British Colonies, &c. in North America* (London, 1776), 29–30. See also Lawrence H. Gipson, *Lewis Evans . . . together with Facsimiles of His Geographical, Historical, Political, Philosophical and Mechanical Essays . . .* (Philadelphia, 1939).

3. See Alexander Ospovat, "Introduction" to Abraham G. Werner, *Short Classification and Description of Various Rocks,* (New York, 1971), 1–35. For a general view of the development of mineralogy to the mid-nineteenth century, see John C. Greene and John G. Burke, *The Science of Minerals in the Age of Jefferson, Trans. APS,* 68, pt. 4 (1978), chaps. 1–2. See also Robert M. Hazen and Margaret H. Hazen, comps., *American Geological Literature, 1669 to 1850* (Stroudsburg, Pa., 1980); Robert M. Hazen, ed., *North American Geology: Early Writings,* Benchmark Papers in Geology Series, vol. 51, (Stroudsburg, Pa., 1979).

4. Benjamin Silliman to Mary Noyes, Edinburgh, Feb. 27, 1806, Silliman Correspondence, Beinecke Library, Yale University.

5. Thomas Jefferson, *Notes on the State of Virginia,* ed. William Peden (Chapel Hill, N.C., 1955), 19.

6. Jedidiah Morse, *The American Geography; or, A View of the Present Situation of the United States of America,* 2nd ed. (London, 1792), 52.

7. Samuel Miller, *A Brief Retrospect of the Eighteenth Century . . . ,* 2 vols. (New York, 1970), 1:183. Original edition, 1803.

8. Ibid., 1:155.

9. "Heads of a Course of Lectures Intended as an Introduction to Natural History. By B. Waterhouse, M.D. . . . ," *Mon. Anthol. Boston Rev.,* 3 (1806), 311–12.

10. Schoepf, Johann D., *Beyträge zur Mineralogischen Kenntniss des Östlichen Theils von Nord America und seiner Gebürge* (Erlangen, 1787); Edmund M. Spieker, "Translator's Introduction" Schoepf, *Geology of Eastern North America* (New York, 1972), 25. Spieker adds "That he [Schoepf] got as much of the regional geology as he did is nothing short of remarkable. He used well the equipment that has long since become standard for the geologist — hammer, hand lens, compass, acid bottle, streel, magnet, measuring tape, collecting bag, and notebook. That he appreciated strongly the importance of collecting specimens is shown by his bitter disappointment at the loss of a shipment he had confided in Philadelphia to a German merchant from Hamburg." See also Schoepf, *Travels through Some of the Middle and Southern . . . North American States . . . in the years 1783 and 1784,* 2 vols., (Erlangen, 1788). Original in German. English trans. A. J. Morrison (Philadelphia, 1911).

11. Samuel L. Mitchill, "A Sketch of the Mineralogical and Geological History of the State of New York . . . ," *Med. Repos.,* 1 (1797–1798), 293–314, 445–52; 3 (1799–1800), 325–35. Only the first section of this "Sketch" was published in *Trans. Soc. Promot. Agric. Arts Manuf.,* 1, pt. 4 (1799), 124–52. The quotation is from *Med. Repos.,* 1:284. See also Greene and Burke, *Science of Minerals,* chap. 4.

12. Samuel L. Mitchill to Jedidiah Morse, New York, Apr. 15, 1796, Gratz Collection, Historical Society of Pennsylvania. See also James Hutton, "Observations on Granite," *Trans. Roy. Soc. Edinburgh,* 3 (1794), 77–85.

13. See Gilbert Chinard, *Volney et l'Amérique d'après des Documents Inédits et Sa Correspondance avec Jefferson,* The Johns Hopkins Studies in Romance Literatures and Languages, vol. 1, (Baltimore, 1928).

14. George White, "Editor's Introduction" to C. F. C. de Volney, *A View of the Soil and Climate of the United States of America* (New York and London, 1968), viii. See also Spieker's evaluation in Schoepf, *Geology of Eastern North America,* 13.

15. Volney, *View,* 51.

16. John W. Wells, "Notes on the Earliest Geological Maps of the United States, 1756–1832," *J. Wash. Acad. Sci.,* 49 (1959), 199.

17. Volney, *View,* 99–100.

18. James Mease, *A Geological Account of the United States; Comprehending a Short Description of Their Animal, Vegetable and Mineral Descriptions, Antiquities and Curiosities . . .* (Philadelphia, 1807). In the preface Mease says, "In treating of the Climate, the Geology, and the Winds of the United States, the divisions and remarks of Mr. Volney have been assumed as the basis . . . but care has been taken . . . to correct some important errors . . . and to supply his defects." Mease (p. 50) notes that Volney, although he follows Mitchill's arrangement of the various formations, "has omitted . . . an account of the slate or shistus which is so abundant and characteristic in many places, particularly in New York."

19. *Med. Repos.,* 12 (1808–1809), 295–96.

20. Benjamin Silliman, "Reminiscences . . . ," 4:162, Silliman Papers, Yale University.

21. William Maclure, "Observations on the Geology of the United States, Explanatory of a Geological Map," *Trans. APS,* 6 (1809), 426.

22. The modern equivalents are taken from George W. White, "William Maclure's Maps of the Geology of the United States," *J. Soc. Bibliogr. Nat. Hist.,* 8 (1977), 266–69. Describing the map in the Paris edition (1811), White lists "LIGHT BLUE" (Floetz Rocks) and "DARK BLUE" (Old Red Sandstone), but Maclure's 1809 map shows only one shade of blue covering both.

23. Wells, "Notes on the Earliest Geological Maps," 201. Wells bases his suspicions on Gilbert Chinard's remark (*Volney et l'Amérique,* 101n.) that Maclure obtained much unacknowledged information from Volney during his visits to Paris and on the fact that the 1811 version of Maclure's map, published in the *Journal de Physique,* had as its base map Volney's map of 1803, "with re-engraved title but retaining all other details including Volney's route trace and contact lines. . . . Either Maclure sent Delamétherie [the editor of the *Journal*] a copy of Volney's map with his own geological coloring, or Delamétherie had Maclure's data transferred to pulls from the engraved plate of Volney's map." Wells adds, "Maclure rarely referred to the work of others."

24. Edmund M. Spieker, "Schoepf, Maclure, Werner, and the Earliest Work on American Geology," *Science,* 172 (1971), 1333–34. See also "Translator's Introduction" to Schoepf, *Geology of Eastern North America,* 14–16. Maclure, says Spieker (p. 15), "recognized individual formations where Schoepf lumped such ingredients into the categories of his major rock belts; but when we look at Schoepf's work as a whole we see a much more thorough and extensive treatment. . . . His [Schoepf's] descriptions of rocks and minerals are far fuller than those of Maclure, and he recognized some major geologic features, such as the Fall Line and the Schooley Peneplain, that Maclure does not mention. Further, Schoepf attempted, at least, to explain most of the things he saw, whereas Maclure modestly abstained from any such essay."

25. See Greene and Burke, *Science of Minerals,* chap. 3. See also Clifford Frondel, "An Overview of Crystallography in North America, in D. McLochlan and J. P. Glusker, eds., *Crystallography in North America* (American Crystallographic Assoc., 1983), 1–23.

26. William Maclure, "Essay on the Formation of Rocks, or an Inquiry into the Probable Origin of Their Present Form and Structure," *J. Acad. Nat. Sci. Phila.,* 1 (1818), 324.

27. John Doskey is editing for publication Maclure's travel journals in Europe, 1805–1815 and 1820–1825.

28. William Maclure, "Observations on the Geology of the United States," *Trans. APS,* n.s., 1 (1818), 22–23.

29. Maclure, "Essay," 340–41.

30. Maclure, "Observations," iv–v.

31. William Maclure, "Some Speculative Conjectures on the Probable Changes That May Have Taken Place in the Geology of the Continent of North-America East of the Stoney Mountains . . . Madrid, July 9, 1822," *Am. J. Sci. Arts,* 6 (1823), 99–100.

32. William Maclure, "Hints on Some of the Outlines of Geological Arrangement, with Particular Reference to the System of Werner, . . . Paris, 22d August, 1818," *Am. J. Sci. Arts,* 1 (1818), 211–12. For a convincing refutation of the idea that Maclure was a simple Wernerian, see George W. White, "William Maclure's Concept of Primitive Rocks ('Basement Complex')," in W. O. Kupsch and W. A. S. Sargeant, eds., *History of Concepts in Precambrian Geology,* Geological Association of Canada Special Paper 19 (1979), 251–59.

33. Maclure, "Observations," 87.

34. George Brush, *A Sketch of the Progress of American Mineralogy,* address delivered before the American Association for the Advancement of Science, Montreal, Aug. 28, 1882, New Haven, Conn., 10.

35. For details of these developments see Greene and Burke, *Science of Minerals,* chap. 4.

36. Cleaveland's work and contributions are discussed in ibid., chap. 6.

37. Robert Gilmor, Jr., to Parker Cleaveland, July 20, 1819, Cleaveland Correspondence, Bowdoin College Library.

38. See Greene and Burke, *Science of Minerals,* chap. 7. See also Stephen D. Beckham, "Colonel George Gibbs," in Leonard G. Wilson, ed., *Benjamin Silliman and His Circle: Studies on the Influence of Benjamin Silliman on Science in America Prepared in Honor of Elizabeth H. Thomson* (New York, 1979), 29–48.

39. As quoted in George Fisher, *Life of Benjamin Silliman . . . Chiefly from His Manuscript Reminiscences, Diaries, and Correspondence,* 2 vols. (New York, 1866), 1:259.

40. Wilson, *Silliman and His Circle,* 49–128.

41. Fisher, *Life of Benjamin Silliman,* 1:285–86.

42. Benjamin Silliman to Parker Cleaveland, Mar. 2, 1818, Cleaveland Correspondence, Bowdoin College Library.

43. See Edward S. Dana, *A Century of Science in America* (New Haven, 1918), chap. 1.

44. Charles Lyell to Benjamin Silliman, New York, Apr. 4, 1842, as quoted in Fisher, *Life of Benjamin Silliman,* 2:163–64.

45. Amos Eaton to John Torrey, Aug. 14, 1819, as quoted in Ethel McAllister, *Amos Eaton: Scientist and Educator, 1776–1842* (Philadelphia, 1941), 301.

46. Amos Eaton, *An Index to the Geology of the Northern States, with Transverse Sections, Extending from Susquehanna River to the Atlantic, Crossing Catskill Mountains.*

To Which Is Prefixed a Geological Grammar, 2nd ed. (Troy, N.Y., 1820), ix. First edition, 1818.

47. Ibid., 11–13. See also John Wells, "Early Investigations of the Devonian System in New York, 1656–1836," *Special Papers, Geological Society of America,* no. 74 (New York, 1974), 28.

48. Eaton, *Index* (1818), 44.

49. Eaton, *Index* (1820), 275–76.

50. Amos Eaton to Benjamin Silliman, Sept. 2, 1822, as quoted in McAllister, *Amos Eaton,* 303.

51. Eaton to Torrey, Feb. 4, 1824, as quoted in McAllister, *Amos Eaton,* 308.

52. Amos Eaton, *A Geological and Agricultural Survey of the District Adjoining the Erie Canal . . .* (Albany, 1824), 9.

53. Ibid., 155–56.

54. James Hall, *New York State Natural History Survey. Geology of New York. Comprising the Survey of the Fourth Geological District,* pt. 4 (Albany, 1843), 6.

55. Wells, "Early Investigations of the Devonian System," 43, 55.

10: NATURAL HISTORY IN A NEW WORLD: BOTANY

1. Thomas Jefferson to Joseph Willard, Mar. 24, 1789, Jefferson Papers, Library of Congress.

2. See Joseph Ewan, "Early History," in Ewan, ed., *A Short History of Botany in the United States* (New York and London, 1969), 27–48.

3. Henry Muhlenberg to Manasseh Cutler, Lancaster, Pa., Nov. 12, 1792, as quoted in William P. Cutler and Julia P. Cutler, *Life, Journals, and Correspondence of Rev. Manasseh Cutler, LL.D.,* 2 vols. (Cincinnati, 1898), 2:291. See also Paul A. W. Wallace, *The Muhlenbergs of Pennsylvania* (Philadelphia, 1950), 308–20.

4. Manasseh Cutler to Jedidiah Morse, Ipswich, Mass., Aug. 13, 1792, Gratz Collection, Historical Society of Pennsylvania.

5. Of Cutler's "Account of Some of the Vegetable Productions, Naturally Growing in This Part of America, Botanically Arranged" (1785) Ewan writes (*Short History of Botany,* 36): "No new names were proposed, hence the paper is not well known. Indeed, the names are *American* vernacular, the author adding that 'from want of botanical knowledge, the grossest mistakes have been made in the application of *English* names of *European* plants, to those of America.' Cutler offers no binomials even though the arrangment is Linnaean Sexual System."

6. For an appreciation of the contributions of the Bartrams to botany, see Ewan, *Short History,* 33–35 and Joseph Ewan, ed., *William Bartram. Botanical and Zoological Drawings, 1756–1788; Reproduced from the Fothergill Album in the British Museum (Natural History),* with an introduction and commentary by Joseph Ewan (Philadelphia, 1968). In the latter work Ewan says (p. 12), "Even with the appearance of the *Travels . . .* priorities were confused because sometimes Bartram's plant name and its description were widely separated in the text. . . . Numerous variable and garbled spellings of scientific names were due to faulty proofreading or none at all. . . . The fact that he [Bartram] did not cite authorities for most of the binomials, and did not often indicate that the given names for 123 new binomials in the book were newly proposed, does not invalidate all the instances where his binomial and description leave no doubt of the plant intended."

7. Henry Muhlenberg to William D. Peck, May 19, 1812, as quoted in Jeannette E. Graustein, "The Eminent Benjamin Smith Barton," *Pa. Mag. Hist. Biogr.,* 85 (1961), 431, citing Peck Papers, Harvard University Archives. See also Francis W. Pennell, "Botanical Collectors of the Philadelphia Local Area," *Bartonia,* no. 21 (1942), 38–57.

8. The successive editions of Barton's *Collections* were published in 1798, 1801, and 1810. See also Theodore W. Jeffries, "Barton's Unpublished Materia Medica," *Pharm. Hist.,* 17 (1975), 69–71.

9. Ibid., 141. See also 90–91, 139–40.

10. Ibid., 144. Barton was certainly acquainted with Jussieu's writings, for he cited them frequently throughout his text.

11. Benjamin S. Barton to William Barton, Apr. 18, 1806, Barton Papers, American Philosophical Society Library.

12. See Francis Pennell, "Benjamin Smith Barton as Naturalist," *Proc. APS,* 86 (1942), 108–22; also Pennell, "Travels and Scientific Collections of Thomas Nuttall," *Bartonia, no.* 18 (1936), 1–51.

13. See Jeannette E. Graustein, *Thomas Nuttall Naturalist: Explorations in America, 1808–1841* (Cambridge, 1967), 85–86; Pennell, "Benjamin Smith Barton as Naturalist," 108–22.

14. Henry Muhlenberg to William Bartram, Lancaster, Pa., Sept. 13, 1792, as quoted in William Darlington, *Memorials of John Bartram and Humphry Marshall. With Notices of Their Botanical Contemporaries* (Philadelphia, 1849), 466; Muhlenberg to Moses Marshall, Lancaster, Pa., Apr. 9, 1792, ibid., 576. Schreber named a genus of grasses *Muhlenbergia.*

15. Muhlenberg to James E. Smith, Lancaster, Pa., Dec. 1, 1792, as quoted in Lady Smith, ed., *Memoir and Correspondence of the Late Sir James Edward Smith, M.D.,* 2 vols. (London, 1832), 1:403–5.

16. Henry Muhlenberg, "Index Florae Lancastriensis," *Trans. APS,* o.s., 3 (1793), 157–84; Muhlenberg, "Supplementum Indicis Florae Lancastriensis," ibid., 4 (1799), 235–42.

17. André Michaux, *Histoire des chênes de l'Amérique* . . . (Paris, 1801); Michaux, *Flora Boreali-Americana, sisten caracteres plantarum quas in America Septentrionali collegit et detexit . . . ,* 2 vols. (Paris, 1803); François A. Michaux, *Histoire des arbres forestiers de l'Amérique Septentrionale . . . ,* 3 vols. (Paris, 1810–1813); the American edition was published in Paris in 1818–1819. See also Gilbert Chinard, "André and François-André Michaux and Their Predecessors: An Essay on Early Botanical Exchanges between America and France," *Proc. APS,* 101 (1957), 344–61; William J. Robbins, "French Botanists and the Flora of the Northeastern United States: J. G. Milbert and Elias Durand," ibid., 362–68.

18. Henry Muhlenberg to John Brickell, Lancaster, Pa., Feb. 7, 1803, Dreer Collection—Scientists, vol. 3, Historical Society of Pennsylvania. See also Brickell's articles in *Med. Repos.,* 1, no. 4 (1798), 573, and *Philadelphia Med. Phys. J.,* 2 (1804), 164, 176–77.

19. Muhlenberg to William Baldwin, Lancaster, Pa., Nov. 4, 1811, as quoted in William Darlington, *Reliquiae Baldwinianae,* ed. Joseph Ewan (New York, 1968), 53. The original edition was published in Philadelphia in 1843.

20. Henry Muhlenberg, *Catalogus Plantarum Americae Septentrionalis, Huc Usque Cognitarum Indigenarum et Circum; or, A Catalogue of the Hitherto Known Native and Naturalized Plants of North America, Arranged According to the Sexual System of Linnaeus* (Lancaster, Pa., 1813).

21. I am indebted to Ronald L. Stuckey of Ohio State University for calling my attention to the importance of Muhlenberg's short diagnostic phrases as validating some of his names of plants. Stuckey and his colleague Emanuel D. Rudolph gave me helpful criticisms of the present chapter, as did Joseph Ewan of Tulane University.

22. The second edition of Muhlenberg's *Catalogue* contained an appendix entitled "Reduction of All the Genera . . . in the Catalogus . . . of the late Dr. Muhlenberg to the Natural Families of Mr. De Jussieu's System." This appendix was the work of the Abbé Correa da Serra, who published it in pamphlet form in 1815. It was published again in somewhat altered form as an appendix to the New York edition of James E. Smith's *A Grammar of Botany* (1822).

23. A. S. Hitchcock, "The Grasses of the Muhlenberg Herbarium," *Bartonia, no.* 14 (1932), 27–52.

24. E. D. Merrill and Shiu-Ying Hu, "Work and Publications of Henry Muhlenberg, with Special Attention to Unrecorded or Incorrectly Recorded Binomials," *Bartonia, no.* 25 (1949), 1–66. See also C. Earle Smith, Jr., "Henry Muhlenberg—Botanical Pioneer," *Proc. APS,* 106 (1962), 443–60.

25. Personal communication from Joseph Ewan, Tulane University.

26. Concerning the contributors to Muhlenberg's herbarium and its importance for modern botanists, see James A. Mears, "Some Sources of the Herbarium of Henry

Muhlenberg (1753–1815)," *Proc. APS,* 122 (1978), 155–74; Francis W. Pennell, "Historic Botanical Collections of the American Philosophical Society and the Academy of Natural Sciences of Philadelphia," ibid., 94 (1950), 137–51. Mears writes (p. 173), "That Muhlenberg published several works on virtually all the North American material in his herbarium, that he received North American plant material from most of the North American collectors of the period 1790 to 1815, and that he received identifications and specimens from important European botanists of all fields of study make Muhlenberg's herbarium extremely rich in authentic material documenting North American and European botanical studies of the period 1780 to 1803. . . . Although the full significance of the specimens in the Muhlenberg Herbarium will not be known until the relevant Muhlenberg manuscripts are transcribed (and in some cases translated) and associated with the herbarium specimens, botanists will usually be able to locate desired material in the Muhlenberg Herbarium." Mears notes that Benjamin S. Barton's herbarium was also acquired by the American Philosophical Society and placed on loan at the Academy of Natural Sciences.

27. See Joseph Ewan's "Introduction" to the facsimile reprint of Pursh's *Flora Americae Septentrionalis; or, A Systematic Arrangement and Description of the Plants of North America, Containing, Besides What Have Been Described by Preceding Authors, Many New and Rare Species, Collected during Twelve Years Travel and Residence in That Country* (New York, 1973). This introduction is a mine of information about Pursh, his book, his sources, the new species described and named by him, etc. See also Ewan, "Frederick Pursh, 1774–1820, and His Botanical Associates," *Proc. APS,* 90 (1952), 599–628. See also Constantine S. Rafinesque's review of Pursh's *Flora* in *Am. Mon. Mag. Crit. Rev.,* 2 (1818), 170–76.

28. David P. Penhallow, "A Review of Canadian Botany from 1800 to 1895," pt. 2, *Trans. Roy. Soc. Canada,* ser. 2, 3, no. 4 (1897), 3–6.

29. See Pennell, "Travels and Scientific Collections of Thomas Nuttall," 25; Graustein, *Thomas Nuttall,* Chaps. 7–8.

30. Thomas Nuttall, *The Genera of North American Plants, and a Catalogue of the Species to the Year 1817 . . . ,* 2 vols. (Philadelphia, 1818). According to Graustein, *Thomas Nuttall,* 124–25, Nuttall gave Baldwin credit as the discoverer of about twenty-four new species. "Pragmatically," Graustein adds, "the publication of some of Baldwin's species in the *Genera* was useful since Baldwin's rapidly failing health made any major publication impossible for him." Ewan, *Short History of Botany,* 40, says, "Baldwin's treatment of a number of genera especially in the Cyperaceae showed penetrating observation, understanding, and diagnosis."

31. As quoted in Ewan, *Short History of Botany,* 40. See also Constantine S. Rafinesque, "Review of Nuttall's Genera," *Am. Mon. Mag. Crit. Rev.,* 4 (1818–1819), 184–96; William J. Hooker, "On the Botany of America," *Edinburgh J. Sci.,* 2 (1825), 122–23.

32. William P. C. Barton, *Compendium Florae Philadelphicae* (Philadelphia, 1818); also his *A Flora of North America . . . ,* 3 vols. (Philadelphia, 1820–1824).

33. As quoted in Hooker, "On the Botany of America," 120–21.

34. Ibid., 121.

35. Stephen Elliott, *A Sketch of the Botany of South Carolina and Georgia,* edited with an Introduction by Joseph Ewan, 2 vols. (New York, 1971). Original edition, Charleston, 1821–1824. Ewan gives the details of the piecemeal publication of this work, beginning in 1816. On Schweinitz see Francis W. Pennell, "The Botanist Schweinitz and His Herbarium," *Bartonia,* no. 16 (1935), 1–8; C. L. Shear and Neil E. Stevens, eds., "The Correspondence of Schweinitz and Torrey," *Mem. Torrey Bot. Club,* 16 (1920), 119–281; Alfred E. Schuyler, "Five Early Botanists [Muhlenberg, Pursh, Nuttall, von Schweinitz, and Rafinesque]," *Morris Arbor. Bull.,* 24, no. 2 (June 1973), 28–34.

36. Jacob Bigelow, *Florula Bostoniensis; A Collection of Plants of Boston and Its Environs, with Their Generic and Specific Characters, Synonyms, Descriptions, Places of Growth, and Time of Flowering, and Occasional Remarks* (Boston, 1814); James E. Smith, *An Introduction to Physiological and Systematical Botany,* 1st Am. ed. with notes by Jacob Bigelow (Boston, 1814). Correa da Serra was much impressed with Bigelow's talents as a botanist when he visited Boston in November 1813. He wrote to James E. Smith: "In the two years that I have lived and travelled in the United States, I have found only a man,

and a young one too, a physician of this town, who shows a true botanic genius and great zeal for science; he has already acquired a great scientific knowledge, only aided by his industry and the limited number of botanic books which the country affords. He will be I am confident an illustrious botanist, if he is put in correspondence with the chiefs of the science. . . ." But Bigelow, like Barton, had too many irons in the fire. He never distinguished himself as a botanist. See Smith, ed., *Memoir and Corrrespondence . . . Sir James Edward Smith,* 2:221-22.

37. Amos Eaton to John Torrey, Jan. 12, 1822, as quoted in Ethel McAllister, *Amos Eaton: Scientist and Educator, 1776-1842* (Philadelphia, 1941), 218.

38. [Caleb Cushing], review of Bigelow's *American Medical Botany* in *North Am. Rev.,* n.s., 4 (1821), 119.

39. John Torrey, ed., *A Catalogue of Plants Growing Spontaneously within Thirty Miles of the City of New York* (Albany, 1819); Torrey, "Descriptions of Some New or Rare Plants from the Rocky Mountains, Collected in July, 1820, by Dr. Edwin James," *Ann. Lyceum Nat. Hist. New York,* 1, pt. 1 (1824), 30-36; Torrey, "Description of Some New Grasses, Collected by E. James . . . during Major Long's Expedition to the Rocky Mountains in 1819-1820," ibid., 148-56.

40. E. D. Merrill, "Rafinesque's Publications from the Standpoint of World Botany," *Proc. APS,* 87 (1944), 110-19.

41. Francis W. Pennell, "The Life and Work of Rafinesque," *Rafinesque Mem. Pap., Transylvania Coll. Bull.,* 15 (1942), 10-70.

42. See Edwin M. Betts, "The Correspondence between Constantine Samuel Rafinesque and Thomas Jefferson," *Proc. APS,* 87 (1944), 368-80. The range of Rafinesque's interests may be seen from the list of his writings he submitted to Jefferson with his letter of Feb. 15, 1824. For a guide to the voluminous literature on Rafinesque, see T. J. Fitzpatrick, *Rafinesque: A Sketch of His Life with Bibliography,* revised and enlarged by Charles Boewe (Weston, Mass., 1982); E. D. Merrill, *Index Rafinesquianus* (Jamaica Plain, Mass., 1949); Ronald L. Stuckey and Marvin L. Roberts, "Additions to the Bibliotheca Rafinesquiana," *Taxon,* 23 (1974), 365-72. See also Stuckey, "C. S. Rafinesque's North American Vascular Plants at the Academy of Natural Sciences of Philadelphia," *Brittonia,* 23 (1971), 191-208; Stuckey, "The First Public Auction of an American Herbarium Including an Account of the Fate of the Baldwin, Collins, and Rafinesque Herbaria," *Taxon,* 20 (1971), 443-59. Charles Boewe of Transylvania University is editing the correspondence of Rafinesque and preparing a new biography of him.

11: NATURAL HISTORY IN A NEW WORLD: ZOOLOGY AND PALEONTOLOGY

1. See William Bartram, *Botanical and Zoological Drawings, 1756-1788,* ed. Joseph Ewan (Philadelphia, 1968), 19.

2. Elliott Coues, "Fasti ornithologiae redivivi No. 1 — Bartram's 'Travels,' " *Proc. ANSP,* 27 (1875), 338-58; Witmer Stone, "Some Philadelphia Ornithological Collections and Collectors, 1784-1850," *Auk,* 16 (1899), 166-77. See also Coues, *Key to North American Birds,* 3rd ed. (Boston, 1887); Francis Harper, ed., *The Travels of William Bartram,* naturalist's ed. (New Haven, 1958).

3. Harper, ed., *Travels,* 376-77.

4. Ibid., 377. John Edwards Holbrook, in his *North American Herpetology . . . ,* 4 vols. (Philadelphia, 1836), 4:15, was less enthusiastic about Bartram as a zoologist; he described him as "an honest, upright, though somewhat overcredulous naturalist."

5. Thomas Pennant to Benjamin S. Barton, Downing in Devonshire, Oct. 17, 1790, Correspondence and Papers of Benjamin S. Barton, 1778-1815, Historical Society of Pennsylvania.

6. Barton to Pennant, Philadelphia, June 29, 1793, as quoted in Benjamin S. Barton, *Notes on the Animals of North America,* ed. Keir B. Sterling (New York, 1974).

7. Pennant to Barton, Apr. 16, 1796, Historical Society of Pennsylvania.

8. Charles C. Sellers, *Mr. Peale's Museum: Charles Willson Peale and the First Popular Museum of Natural Science and Art* (New York, 1980), 106.

9. Ibid., 108.

10. See John C. Greene, *The Death of Adam: Evolution and Its Impact on Western Thought* (Ames, Iowa, 1959), 100–101.

11. Thomas Jefferson to David Rittenhouse, Monticello, July 3, 1796, "Ms Communications on Natural History," 1:12, American Philosophical Society Library.

12. Georges Cuvier, *Mémoire sur un squelette fossile trouvé sur les bords du Rio de la Plata* (Paris, 1796); *Mon. Mag. Brit. Regist.,* 2 (1796), 637–38. For a full account of the events and publications surrounding the discovery of the megalonyx and the megatherium, see Julian P. Boyd, "The Megalonyx, the Megatherium, and Thomas Jefferson's Lapse of Memory," *Proc. APS,* 102 (1958), 420–35. See also George G. Simpson, "The Beginnings of Vertebrate Paleontology in North America," ibid., 86 (1942), 130–85.

13. Thomas Jefferson, "A Memoir on the Discovery of Certain Bones of a Quadruped of the Clawed Kind in the Western Parts of Virginia," *Trans. APS,* 4 (1799), 255–56. For Wistar's account of the megalonyx, see ibid., 526–31.

14. Georges Cuvier, "Sur le Megatherium . . . ," *Ann. Mus. Natl. Hist. Nat. Paris,* 5 (1804), 387; Cuvier, "Sur le Megalonyx, animal de la famille des paresseux . . . dont les ossemens ont été découverts en Virginie, en 1796," ibid., 358–76.

15. *Trans. APS,* 4 (1799), xxxvii.

16. James G. Graham, "Further Account of the Fossil Bones in Orange and Ulster Counties: In a Letter . . . to Dr. Mitchill; dated Sept. 10, 1800," *Med. Repos.,* 4 (1801), 213–14. See also "Account of Large Bones Dug Up in Orange and Ulster Counties: . . . in a Letter from Sylvanus Miller, Esq. to Dr. Mitchill; dated Sept. 20, 1800," ibid., 211–13. These contributions may have been solicited by Mitchill, who wrote to Wistar in June, 1798, indicating that he had received the American Philosophical Society's circular letter urging a search for fossil bones and other antiquities. Said Mitchill, "Near the Walkill a few Years ago, some large Bones were found in a Marle-Pit. They were probably of the Mammoth; and were procured by a German [Dr. Frederick Michaelis] . . . about the close of the Revolutionary War and carried to Europe. I never saw them myself, nor do I know that they were ever examined by any American Zoologist or Anatomist—unless Mr. Annan . . . may be deemed to be such." (Mitchill to Wistar, New York, June 30, 1798, Wistar Letterbook, American Philosophical Society Library.)

17. Rembrandt Peale, *An Historical Disquisition on the Mammoth, or Great American Incognitum . . .* (London, 1803), 32–33. See also Charles C. Sellers, *Charles Willson Peale,* 2 vols., *Mem. APS,* 23 (1947), 2:127–37; Sellers, *Mr. Peale's Museum,* chap. 5.

18. Charles W. Peale, typescript autobiography, Peale Papers, 314, American Philosophical Society Library.

19. Peale, *An Historical Disquisition on the Mammoth,* 90–91.

20. See the various memoirs by Cuvier in *Ann. Mus. Natl. Hist. Nat.* Paris, 8 (1806), 1–58, 93–155, 249–69, 270–312, 401–20, 420–24.

21. Benjamin S. Barton, *A Discourse on Some of the Principal Desiderata in Natural History, and on the Best Means of Promoting the Study of This Science, in the United States* (Philadelphia, 1807), as reprinted in Keir B. Sterling, ed., *Contributions to the History of American Natural History* (New York, 1974), 53–54.

22. Simpson, "Beginnings of Vertebrate Paleontology," 154. According to Simpson, Wistar's diagnosis was confirmed when the Philadelphia paleontologist Joseph Leidy examined the same specimen in the latter part of the nineteenth century.

23. Thomas Jefferson to William Clark, Dec. 19, 1807, in A. A. Lipscomb, ed., *The Writings of Thomas Jefferson,* 20 vols. (Washington, D.C., 1903–1904), 11:404–5.

24. Jefferson to Caspar Wistar, Washington, D.C., Mar. 20, 1808, in Lipscomb, ed., *Writings of Thomas Jefferson,* 12:15–16.

25. Certificate of the Institut National de France, Classes des Sciences Physiques et Mathématiques, Wistar Letterbook, 1794–1817, American Philosophical Society Library.

26. Simpson, "Beginnings of Vertebrate Paleontology, 155.

27. Charles Robert Leslie, *Autobiographical Recollections,* ed. Tom Taylor (London,

1860), as quoted in Robert Cantwell, *Alexander Wilson, Naturalist and Pioneer* (Philadelphia and New York, 1961), 144.

28. Alexander Wilson to Alexander Lawson, Albany, N.Y., Nov. 3, 1808, as quoted in Letter 100 of Clark Hunter, *The Life and Letters of Alexander Wilson, Memoirs,* APS, 154(1983).

29. See Elsa G. Allen, "John Abbot, Pioneer Naturalist of Georgia," *Ga. Hist. Q.,* 41 (1957), 146–57.

30. William Dunbar to Alexander Wilson, May 20, 1810, as quoted in Cantwell, *Alexander Wilson,* 225.

31. Alexander Wilson to François A. Michaux, Philadelphia, June 6, 1812, Letter 132 in Hunter, *Life and Letters.*

32. Alexander Wilson, *American Ornithology; or, The Natural History of the Birds of the United States: Illustrated with Plates Engraved and Colored from Original Drawings Taken from Nature,* 9 vols. in 4 vols. (Philadelphia, 1808–1814), 5:vii.

33. Ibid., 1:4.

34. Ibid., 1:66.

35. Ibid., 1:71.

36. Charles L. Bonaparte, *American Ornithology; or, The Natural History of Birds Inhabiting the United States, Not Given by Wilson . . . ,* 4 vols. (Philadelphia, 1825–1833), 1:iii.

37. Emerson Stringham, *Alexander Wilson A Founder of Scientific Ornithology* (Kerrville, Tex., n.d.), 13–18, published by the author; Cantwell, *Alexander Wilson,* 253.

38. See the *Mass. Agric. Repos. J.,* 4 (1816–1817), 89–92, 205–11; 5 (1818–1819), 57–73, 307–13.

39. For Mitchill's "Report . . . on the Fishes of New-York," see *Am. Med. Philos. Regist.* 4 (1814), 402. See also *Med. Repos.,* n.s., 2 (1815), 280, where Mitchill acknowledges the aid of Samuel Akerly and Samuel G. Mott in his researches and also of Scudder's Museum, which possessed specimens of some rare species of fish. See also Samuel L. Mitchill, "Arrangement and Description of the Codfishes of New-York; Addressed to the Editors," *Am. Med. Philos. Regist.,* 4 (1814), 619.

40. Theodore Gill, "Introduction" to the reprint edition of Mitchill's "Report on the Fishes of New-York" (Washington, D.C., 1898), iv–v. Compare David Starr Jordan's estimate of two genera and 26 species named by Mitchill in his *Manual of Vertebrate Animals,* 13th ed. (Chicago, 1929).

41. *Kentucky Reporter,* Sept. 30, 1818, handwritten copy in Transylvania University Library. Many of these discoveries by Rafinesque were published in his short-lived periodical *Annals of Nature* (Lexington, Ky., 1820).

42. William G. Binney and George W. Tryon, Jr., eds., *The Complete Writings of Constantine Smaltz Rafinesque, on Recent and Fossil Conchology* (New York, London, Paris, Madrid, 1864), preface. See also Constantine S. Rafinesque, *Monograph of the Fluviatile Bivalve Shells of the River Ohio, Containing Twelve Genera & Sixty-eight Species,* trans. Charles A. Poulson (Philadelphia, 1832).

43. Constantine S. Rafinesque, *Ichthyologia Ohiensis . . . , Preceded by a Physical Description of the Ohio and Its Branches* (Lexington, Ky., 1820). The serialized version appeared in the *Western Review and Miscellaneous Magazine* from December 1819 to November 1820.

44. John Richardson, "Report on North American Zoology," *Sixth Annual Report of the British Association for the Advancement of Science* (London, 1827), 203, reprinted in Sterling, ed., *Contributions.*

45. David S. Jordan, "Review of Rafinesque's Memoirs on North American Fishes," *Contributions to North American Ichthyology. Based Primarily on the Collections of the United States National Museum* (Washington, D.C., 1877), 5.

46. Ibid., 6, quoting Louis Agassiz, "Notice of a Collection of Fishes . . . ," *Am. J. Sci.,* 67 (1854), 354. See also William Smallwood and Mabel Smallwood, *Natural History and the American Mind* (New York, 1941), 343: "In Jordan's *Manual of Vertebrate Animals* [13th ed., 1929], Linnaeus is credited with having established thirteen genera and fifty-six

species of American fishes; Rafinesque with thirty-four genera and twenty-one species; Storer, one genus and five species; and Holbrook with one genus and three species."

47. Charles L. Bonaparte, "Observations on the Nomenclature of Wilson's Ornithology," *J. Acad. Nat. Sci. Phila.,* 5, pt. 1 (1824), 57–106. Also published separately in Philadelphia in 1826.

48. Bonaparte, *American Ornithology.*

49. Harry B. Weiss and Grace M. Ziegler, *Thomas Say, Early American Naturalist* (Springfield, Ill., and Baltimore, 1931).

50. For a full listing of Say's many publications, see Max Meisel, *A Bibliography of American Natural History: The Pioneer Century, 1769–1865,* 3 vols. (New York and London, 1967), 3:650.

51. Thomas Say, *American Entomology; or, Descriptions of the Insects of North America, Illustrated by Colored Figures from Drawings Executed from Nature* (Philadelphia, 1817).

52. Say, *American Entomology,* 3 vols. (Philadelphia, 1824–1828).

53. Wilhelm Erichson, *Genera et Species Staphylinorum,* as quoted (in translation) in Weiss and Ziegler, *Thomas Say,* 201.

54. John L. Le Conte, ed., *The Complete Writings of Thomas Say on the Entomology of North America,* 2 vols. (New York, 1978), 1:vi. This is a reprint of the edition published by Baillière Bros., New York, 1859. See also John G. Morris, "Contributions toward a History of Entomology in the United States," *Am. J. Sci. Arts,* 2nd ser., 1 (1846), 20–24.

55. Say's article "Conchology" appeared in William Nicholson's *Encyclopaedia of Arts and Sciences* (Philadelphia, 1816–1817), 2:254ff. and in vol. 4 of the 2nd and 3rd eds., dated 1818 and 1819 respectively. See also Thomas Say, *American Conchology; or, Descriptions of the Shells of North America,* pt. 1, 1830; pt. 2, April 1831; pt. 3, September 1831; pt. 4, March 1832; pt. 5, August 1832; pt. 6, April 1834 (New Harmony, Ind., 1830–1834). A seventh part was published by T. A. Conrad in 1838.

56. See Weiss and Ziegler, *Thomas Say,* chaps. 13, 14. See also Henry W. Fowler, "The Crustacea of New Jersey," *Rep. N.J. State Mus.* (1911); W. G. Binney, ed., *Complete Writings of Thomas Say on the Conchology of the United States* (New York, 1858); G. D. Harris, comp., "A Reprint of the Paleontological Writings of Thomas Say," *Bull. Am. Paleontol.,* no. 5 (Dec. 7, 1896).

57. Charles A. Lesueur to Anselme Desmarest, Dec. 6, 1818, as quoted in E. T. Hamy, *The Travels of the Naturalist Charles A. Lesueur in North America, 1815–1837,* H. F. Raup, ed., trans. Milton Haber (Kent, Ohio, 1968), 31.

58. David S. Jordan, "Charles A. Lesueur," *Dictionary of American Biography,* 20 vols. (New York, 1928–1937), 11:190–91.

59. Samuel N. Rhoads, "Appendix" to the 1894 reprint of George Ord, *Zoology of North America* (Haddonfield, N.J., 1894), as reprinted in Keir B. Sterling, ed., *Early Nineteenth-Century Studies and Surveys* (New York, 1974). Rhoads says (p. 51): "Of all the Reptiles enumerated in Ord's table of 'Amphibia' and which . . . he restricts to the 'Zoology of the United States,' about ten per cent are Mexican or South American, three or four are exotic and two or more unidentifiable. Prof. E. D. Cope, after careful inspection of the table, informs me that all the newly named species to which sufficient reference is made in the text to merit examination, are either synonyms or unidentifiable." Rhoads's "Appendix" gives a thorough critical evaluation of Ord's performance.

60. Richard Harlan, *Fauna Americana: Being a Description of the Mammiferous Animals Inhabiting North America,* ed. Keir B. Sterling (New York, 1974), vii–viii. Originally published in Philadelphia, 1825.

61. Simpson, "Beginnings of Vertebrate Paleontology," 163.

62. John Godman, review of *Fauna Americana . . . ,* in *Franklin J. Am. Mech. Mag.,* 1 (1826), 120–136. This review and the resulting exchanges between Godman and Harlan are reprinted in Harlan, *Fauna Americana.*

63. John Godman, "Natural History," *Franklin J. Am. Mech. Mag.,* 1 (1826), 15, as reprinted in Harlan, *Fauna Americana.*

64. John Godman, *American Natural History. . . . Pt. 1. Mastology,* 2 vols. (Philadelphia, 1826). Also published in three volumes, 1826–1828.

65. Keir B. Sterling, "Introduction" to the reprint edition of Harlan, *Fauna Americana.*

12: THE SCIENCES OF MAN: PHYSICAL ANTHROPOLOGY

1. For a discussion of scientific issues concerning human races, see Alice M. Brues, *People and Races,* Macmillan Series in Physical Anthropology (New York and London, 1977), 1–2. In the light of modern genetics, writes Brues, a race is "a division of a species which differs from other divisions by the frequency with which certain hereditary traits appear among its members." Large racial groups are exemplified in the Caucasoid, Negroid, Mongoloid, and Australoid races, but each of these may be subdivided into smaller groups displaying different frequencies of hereditary traits. "This is a matter of choice, depending on how much detail is desired." Moving from races on the grand scale to subdivisions within each of these is like turning the turret of a microscope to change the magnification. As to origins, modern anthropologists regard all existing races of man as varieties of a single species, but some assign a greater antiquity to them than others do. Natural selection, sexual selection, hybridization, and the accidents of breeding in small populations are all regarded as contributory causes of racial differentiation, but the exact role played by these and other factors in the evolution of any specific race is still largely a matter of speculation. The existence of superior and inferior races is still asserted and denied, but the burden of proof has shifted to those who assert the reality of qualitative differences among the races. The white man's ethnocentrism no longer goes unchallenged. The question of whether there is a common human nature underlying apparent racial and cultural differences is still a live issue. Even the biblical question continues to excite controversy.

2. Thomas Jefferson, *Notes on the State of Virginia . . . ,* ed. William Peden (Chapel Hill, N.C., 1955), 143. See also John C. Greene, "The American Debate on the Negro's Place in Nature, 1780–1815," *J. Hist. Ideas,* 15 (1954), 384–96; Winthrop D. Jordan, *White over Black: American Attitudes toward the Negro, 1550–1812* (Chapel Hill, N.C., 1969).

3. Jefferson, *Notes,* 143.

4. Ibid., 276–77, n104 to Query 6.

5. Ibid., 282, n12 to Query 13.

6. For a general account of theories of the origin and characteristics of human races in the eighteenth and early nineteenth centuries see John C. Greene, *The Death of Adam: Evolution and Its Impact on Western Thought* (Ames, Iowa, 1959), chaps. 6, 8.

7. Michael Kraus, "Charles Nisbet and Samuel Stanhope Smith — Two 18th Century Educators," *Princeton Univ. Libr. Chron.,* 6 (1944), 22–23.

8. Samuel S. Smith, *An Essay on the Causes of the Variety of Complexion and Figure in the Human Species, to Which Are Added Strictures on Lord Kaims's Discourse, on the Original Diversity of Mankind* (Philadelphia, 1787), 30–31.

9. Ibid., 133–34.

10. See Frank Spencer, "Two Unpublished Essays on the Anthropology of North America by Benjamin Smith Barton," *Isis,* 56 (1977), 567–73, n26, 571: "Smith's 1787 essay was also published a year later in Edinburgh with Barton supplying an introduction" (Edinburgh and London, 1788).

11. Charles White, *An Account of the Regular Gradation in Man, and in Different Animals and Vegetables, and from the Former to the Latter* (London, 1799), 134–35.

12. John A. Smith, "A Lecture Introductory to the Second Course of Anatomical Instruction in the College of Physicians and Surgeons for the State of New York . . . ," *N.Y. Med. Philos. J. Rev.,* 1 (1809), 32–48.

13. Smith, *Essay,* 2nd ed., ed. Winthrop D. Jordan (Cambridge, 1965), 21. This reprint of the 1810 edition has a good introduction by the editor.

14. Ibid., 186.

15. *Med. Repos.,* 15 (1812), 159. See also 9 (1809), 64–70, in which the editors of the *Repository* review with approval Felix d'Azara's *Essay on the Natural History of the Quadrupeds of the Province of Paraguay,* trans. M. Moreau-Saint-Méry, 2 vols. (Paris, 1801) and observe: "He [Azara] traces the variegated forms of the hair, skin, and exterior parts of man and other animals, to a generative agency, or operation coeval with the production of the creature."

16. *Med. Repos.,* 13 (1810), 266.

17. Samuel L. Mitchill to the secretary of the American Antiquarian Society, *Archaeol. Am.,* 1 (1820), 319–20.

18. "Heads of That Part of the Introductory Discourse Delivered November 7, 1816, by Dr. Mitchill, in the College of Physicians at New York, Which Related to the Migration of Malays, Tartars, and Scandinavians to America," *Archaeol. Am.,* 343.

19. Samuel L. Mitchill, "The Original Inhabitants of America Shown to Be of the Same Family and Lineage with Those of Asia . . . ," *Archaeol. Am.,* 331. The communication begins: "The view which I took of the varieties of the human race, in my course of Natural History, delivered in the University of New York, differs in so many particulars from that entertained by the great geologists of the age, that I give you for information, and without delay, a summary of my yesterday's lecture to my class." The communication is in the form of a letter to De Witt Clinton.

20. Ibid., 327.

21. Ibid., 332.

22. Charles Caldwell, review of Smith's *Essay* in *Am. Rev. Hist. Polit.,* 2 (1811), 137.

23. Charles Caldwell, review of Smith's *Essay* in *Port Folio,* 4th ser., 4 (1814).

24. Charles Caldwell, *Thoughts on the Original Unity of the Human Race* (New York, 1830).

25. Hugh Williamson, *Observations on the Climate in Different Parts of America Compared with the Climate in Corresponding Parts of the Other Continent. To Which Are Added, Remarks on the Different Complexions of the Human Race; With Some Account of the Aborigines of America. Being an Introductory Discourse to the History of North-Carolina* (New York, 1811), 49n.

26. William C. Wells, "An Account of a Female of the White Race of Mankind, Part of Whose Skin Resembles That of a Negro; With Some Observations on the Causes of the Differences in Colour and Form between the White and Negro Races of Men," appended to Wells, *Two Essays: One upon Single Vision with Two Eyes; the Other on Dew . . .* (London, 1818), 435–36.

27. Ernst Mayr, "Descent of Man and Sexual Selection," in *Atti Del Colloquio Internazionale Sul Tema L'Origine Dell'uomo . . . (Roma, 28–30 Ottobre 1971)* (Rome, 1973), 35–48.

28. John C. Warren, "Biographical Notes," as quoted in Edward Warren, *Life of John Collins Warren . . . ,* 2 vols. (Boston, 1860), 1:11.

29. John C. Warren, *A Comparative View of the Sensorial and Nervous Systems in Men and Animals* (Boston, 1822), 105.

30. Ibid., 102–3, 129–30. Warren notes: "The machines used for producing the compression have been obtained from the northwest coast, by Colonel PERKINS of this place [Boston], and are in the hands of Dr. [Walter] Channing."

31. Ibid., 135–36.

32. Ibid., 137–38.

33. Aleš Hrdlička, *Physical Anthropology: Its Scope and Aims; Its History and Present Status in the United States* (Philadelphia, 1919), 31.

13: THE SCIENCES OF MAN: ARCHAEOLOGY

1. James Bowdoin, "A Philosophical Discourse, Publickly Addressed to the American Academy of Arts and Sciences . . . ," *Mem. Am. Acad. Arts Sci.,* 1 (1785), 4, 7. The only American "antiquities" Bowdoin was likely to be familiar with in 1780 were certain markings that had been observed on Dighton Rock in the Taunton River in Massachusetts, copies of which had been sent to various European antiquaries from the time of Cotton Mather on. See Michael Lort, "Account of an Ancient Inscription in North America," *Archaeol.,* 8 (1787), 290–391; also Antoine C. de Gébelin, *Monde primitif, analysé et comparé avec le monde moderne . . . ,* 9 vols. (Paris, 1774–1796), 8:13, 561–68. See also Edmund B. Delabarre, "Early Interest in Dighton Rock," *Trans. Colon. Soc. Mass.,* 18 (1915–1916), 235–99. For a recent historical account of the development of American archaeological research see James E. Fitting, ed., *The Development of North American Archaeology* (Garden City, N.Y., 1973), esp. chaps. 4–5.

2. John Filson, *The Discovery, Settlement, and Present State of Kentucke* (Wilmington Del., 1784), 33, 98.

3. J. Heart, "Description and Plan of Some Remains of Ancient Works on the Muskingum," *Columbian Mag.,* 1 (May 1787), 425–27; Noah Webster, "ANTIQUITY, LETTER III. From Mr. N. Webster to the Rev. Dr. Stiles . . . on the Remains of the Fortifications in the Western Country, Dated New York, January 20, 1788," *Am. Mag. . . .* (February 1788), 155; Ezra Stiles, "A Letter from Rev. Ezra Stiles, S.T.D., President of Yale College, to the Editor, Dated March 18, 1788," ibid. (April 1788), 294. Webster's hypothesis was also attacked in the *Columbian Magazine* for September 1788; Webster replied in the November issue. Both magazines carried occasional short articles and extracts on American antiquities during this period.

4. Thomas Jefferson to Ezra Stiles, Sept. 1, 1786, in Julian Boyd et al., eds., *The Papers of Thomas Jefferson,* 19 vols. (Princeton, 1950–), 9:476–78.

5. Jefferson to Charles Thomson, Sept. 20, 1787, Boyd et al. *Papers,* 12:159.

6. Thomas Jefferson, *Notes on the State of Virginia . . . ,* ed. William Peden (Chapel Hill, N.C., 1955), 97–100.

7. Ibid.

8. Benjamin S. Barton, *Observations on Some Parts of Natural History; to Which Is Prefixed an Account of Several Remarkable Vestiges of an Ancient Date, Which Have Been Discovered in Different Parts of North America,* pt. 1 (London, 1787), 66.

9. Winthrop Sargent, "A Letter from Colonel Winthrop Sargent to Dr. Benjamin Smith Barton, Accompanying Drawings and Some Account of Certain Articles, Which Were Taken Out of an Ancient Tumulus . . . in the Western-Country," *Trans. Am. Philos. Soc.,* 4 (1799), 177–78.

10. Benjamin S. Barton, "Observations and Conjectures Concerning Certain Articles Which Were Taken Out of an Ancient Tumulus, or Grave, at Cincinnati . . . in a letter from Benjamin Smith Barton, M.D. to the Reverend Joseph Priestley, LL.D., F.R.S. &c.," *Trans. Am. Philos. Soc.,* 4 (1799), 201.

11. William Bartram, "Observations on the Creek and Cherokee Indians . . . 1789. With Prefatory and Supplementary Notes by E. G. Squier," *Trans. Am. Ethnol. Soc.,* 3, pt. 1 (1853), 34–35. In his "Prefatory Note," Squier states that Bartram's manuscript, untitled, was placed in his hands by Samuel G. Morton of Philadelphia, who had obtained it from a man in Mobile, Ala. "The questions were evidently framed by a man of learning and research," Squier wrote, "and as Dr. B. S. BARTON . . . in his Memoir on the 'Origin of the American Nations' p. 46, refers to a MS. by Bartram, on these subjects, in his possession, it can scarcely be doubted that he was the author of the inquiries submitted to Bartram, and the original proprietor of the MS. in question."

12. Ibid., 51. In the letter transmitting this information to Barton, dated Philadelphia, Dec. 15, 1789, Bartram wrote: "I doubt not but you will readily excuse bad writing, composition and spelling. My weakness of sight, I hope, will plead for me, when I assure you I have been obliged to write the greater part of this with my eyes shut, and that with pain. I do not mention this to claim any sort of obligation from you, Sir, for all that I knew concerning these matters are due to you and to science."

13. John R. Swanton, "The Interpretation of Aboriginal Mounds by Means of Creek Indian Customs," *Annual Report of the Smithsonian Institution for 1927* (Washington, D.C., 1928), 506.

14. Henry M. Brackenridge, "On the Population and Tumuli of the Aborigines of North America. In a Letter from H. H. Brackenridge, Esq. to Thomas Jefferson. Read Oct. 1, 1813," *Trans. Am. Philos. Soc.,* n.s. 1 (1818), 151. The letter is dated Baton Rouge, July 25, 1813, and signed "H. M. Brackenridge."

15. Henry M. Brackenridge, *Views of Louisiana, Together with a Journal of a Voyage up the Missouri River in 1811* (Pittsburgh, 1814), 187–89.

16. Brackenridge, "On the Population and Tumuli," 156.

17. Alexander von Humboldt, *Political Essay on the Kingdom of New Spain . . . ,* trans. John Black, 2 vols. (New York, 1811), 1:116.

18. Alexander von Humboldt, *Researches, Concerning the Institutions and Monuments of the Ancient Inhabitants of America, with Descriptions and Views of Some of the Most*

Striking Scenes in the Cordilleras, trans. Helen M. Williams, 2 vols. (London, 1814), 1:196ff.

19. De Witt Clinton, "A Discourse Delivered before the New-York Historical Society, at Their Anniversary Meeting, 6th December, 1811," *Coll. N.-Y. Hist. Soc.,* 2 (1814), 96. See also Clinton, "A Memoir on the Antiquities of the Eastern Parts of the State of New-York, Addressed to the Honourable Samuel L. Mitchill . . . ," *Trans. Lit. Philos. Soc. N.Y.,* 1 (1818), 1–14.

20. Manasseh Cutler to Jeremy Belknap, Mar. 6, 1789, quoted in William P. Cutler and Julia P. Cutler, *Life, Journals, and Correspondence of Rev. Manasseh Cutler,* 2 vols. (Cincinnati, 1888), 2:249.

21. Manasseh Cutler, "Note to Dr. Cutler's Charge at the Ordination of Rev. Daniel Story, Aug. 15, 1798," in Cutler and Cutler, *Life, Journals, and Correspondence,* 2:15.

22. Ibid., 2:15–16.

23. Ibid., 2:17.

24. See Francis P. Weisenburger, "Caleb Atwater: Pioneer Politician and Historian," *Ohio Hist. Q.,* 68 (1959), 3–22. Atwater's reply to President Monroe's request for information on the antiquities of Ohio on Jan. 1, 1818, was published in *Am. Mon. Mag. Crit. Rev.,* 2 (1817–1818), 333–36. The same magazine (3:441ff) published a letter to the editor from Atwater, Feb. 13, 1819, setting forth Atwater's views on Indian antiquities and announcing his intention to publish a work entitled *Notes on the State of Ohio.* "I will here premise," Atwater added, "that his Excellency De Witt Clinton, a considerable time since, addressed a circular letter to me, from the Philosophical and Literary Society of your city, proposing certain queries; to which my Notes are an answer, with some additions of my own." Atwater's *A History of the State of Ohio, Natural and Civil,* projected in 1818, was finally published in 1838. See Philip Skardon, "Caleb Atwater as Historian," *Ohio Hist.,* 77 (1964), 27–33.

25. Caleb Atwater to Rejoice Newton, Aug. 21, 1818, Atwater Letters, Manuscripts Department, American Antiquarian Society.

26. Atwater to Isaiah Thomas, Feb. 24, 1819.

27. Ibid.

28. Ibid., Mar. 25, 1819.

29. Caleb Atwater, *The Writings of Caleb Atwater* (Columbus, Ohio, 1833), 6.

30. Daniel Drake, *Natural and Statistical View, Or Picture of Cincinnati and the Miami Country* (Cincinnati, 1815), 218.

31. John D. Clifford was born in England. He and his brother established a shipping concern in Philadelphia, where Rafinesque first met Clifford in 1802. Clifford subsequently moved to Lexington, Ky., where he engaged in trade and began forming his museum. In a letter to Isaiah Thomas, Feb. 24, 1819, Caleb Atwater said of Clifford: "He writes me, that he has 3000 specimens, assorted, labelled and now opened free of expense to visitors, in a large room fitted up for the purpose." A trustee of Transylvania, Clifford was responsible for bringing Rafinesque to the university. In a letter to the Philadelphia merchant-botanist Zacchaeus Collins in 1818, Rafinesque told how much he admired Clifford's museum, "where I observed, described and drew nearly 220 species of fossil Shells, Polyps, &c., all from the Western States, and nearly all new," adding, "Mr. Clifford . . . has declared himself my friend & mecenate, and has offered me to travel next year at his expence through Tennessee, Alabama, Missouri, &c., in order to collect for his museum." (Rafinesque to Collins, Nov. 6, 1818, Manuscripts Division, Academy of Natural Sciences of Philadelphia Library.) Clifford died in 1820.

32. Caleb Atwater to Isaiah Thomas, Sept. 4, 1819, Atwater Letters, Manuscripts Department, American Antiquarian Society.

33. Ibid., Nov. 13, 1819.

34. Ibid., Oct. 4, 1819.

35. Caleb Atwater, *Description of the Antiquities Discovered in the State of Ohio and Other Western States . . . Illustrated by Engravings of Ancient Fortifications, Mounds, &c. from Actual Survey. Archaeologia Americana* 1 (1820), 209–10.

36. Ibid., 129–30.

37. Ibid., 144.

38. Ibid., 226–27.
39. Ibid., 225–26.
40. Ibid., 235–36.
41. Ibid., 195.
42. Ibid., 240–41.
43. Ibid., 213.
44. Henry Clyde Shetrone, *The Mound-Builders: A Reconstruction of the Life of a Prehistoric American Race, through Exploration and Interpretation of Their Earth Mounds, Their Burials, and Their Cultural Remains* (New York and London, 1936), 484–88. Shetrone says (p. 22): "To Atwater may be attributed the earliest systematic examinations and descriptions of mounds and earthworks. . . . Atwater's contribution, considering the almost total lack of precedent, was a most creditable one."
45. Alexander von Humboldt to George Bancroft, Sept. 5, 1821, Manuscripts Department, American Antiquarian Society.
46. [Constantine S. Rafinesque], review of *Archaeologia Americana* . . . , vol. 1, in *West. Rev. Misc. Mag.* . . . , 3 (1820–1821), 104. A copy of this issue, with Atwater's marginal annotations of Rafinesque's review, is among the collections of the Cincinnati Historical Society. I am obliged to the society for sending me a Xerox copy of this document.
47. Constantine S. Rafinesque to Isaiah Thomas, Lexington, Ky., Sept. 18, 1820, Photostat copy, Farris Rare Books Room, Transylvania University Library.
48. Rafinesque to Samuel M. Burnside, Jan. 30, 1821, Photostat copy in Transylvania University Library. According to Rafinesque, half of Clifford's museum was shipped to Philadelphia, the other half to Cincinnati. Apparently Isaiah Thomas had also written Atwater about procuring the antiquities in Clifford's museum, for in a letter to Thomas on Oct. 23, 1820, Atwater wrote: "I shall not forget Mr. Clifford's cabinet, which Mr. [Henry] Clay informs me, is now, in a room adjoining his office, in a house of Mr. Clay's. Rafinesque, I suspect, does us more harm, than any other person. Mr. C. laughs about him a great deal and thinks he has more vanity and self conceit than any thing else." In his *Writings,* 6, Atwater observed: "Every article mentioned in this portion of my volume [devoted to Atwater's *Description*] as belonging to Mr. John D. Clifford, now belongs to Mr. Joseph Dorfeuille, of Cincinnati, and may be seen in his Western Museum. To him I am indebted, not only for opportunities of daily examining every article I have described, but for every book I have quoted in my work, and for the use, for several weeks, of his highly valuable library. It contains every work, of any note, relative to Mexico, the Pacific Islands, India, Java, and Egypt."
49. Rafinesque to Bory St. Vincent, Jan. 7, 1821, as published in the *Western Minerva* (1821), 74–75.
50. Caleb Atwater to Parker Cleaveland, Nov. 4, 1820, Manuscripts Department, Cincinnati Historical Society Library. The deterioration of Atwater's opinion of Rafinesque can be traced in his successive letters to Isaiah Thomas, preserved in the Manuscripts Department, American Antiquarian Society. In May 1820 he reported that Rafinesque was surveying forts in Kentucky and "doing all he can for me." In September, however, he wrote that Rafinesque had demanded that all the memoirs he had sent to Atwater be published as an appendix to Atwater's memoir without alteration or be returned to him immediately. "His surveys, which I thought he made, with compass and chain, are not so, but as I now learn, more from conjecture than any thing else. . . . He has got possession of drawings, heretofore sent by me, to Lexington to be copied, and now demands pay for these same original drawings, and, that I publish *his* remarks, with my drawings!!!" In October he accused Rafinesque of altering the review of *Archaeologia Americana* in the *Western Review* without the knowledge or consent of the editor (Atwater to Thomas, Sept. 2 and Oct. 18, 1820).
51. Constantine S. Rafinesque to Isaiah Thomas, July 5, 1824, Photostat copy in Farris Rare Books Room, Transylvania University Library.
52. Constantine S. Rafinesque, *Ancient History, or Annals of Kentucky; With a Survey of the Ancient Monuments of North America, and a Tabular View of the Principal Languages and Primitive Nations of the Whole Earth* (Frankfort, Ky., 1824), app., 37.
53. Several sketches of mounds and other earthworks in the vicinity of Lexington are to be found in the Farris Rare Books Room of the Transylvania University Library, with

an extensive collection of other Rafinesque manuscripts. Curator William Marshall was very helpful with respect to these. Charles Boewe is preparing an edition of Rafinesque's correspondence.

54. Caleb Atwater to Benjamin Silliman, Aug. 15, 1821, Simon Gratz Autograph Collection, Historical Society of Pennsylvania. I am grateful to Ronald Stuckey of Ohio State University for making a copy of this letter available to me. Apparently the engraver of Atwater's map of Ohio, Abner Reed, had been promised 700 copies of the map of Ohio in exchange for engraving it, but Atwater claimed to have had no knowledge of this. On July 11, 1821, Atwater wrote to Thomas: "The principal object which I had in view in writing what I did for the Society, was, to get 3000 copies of every engraving for my Notes on Ohio. . . . You agreed to furnish me with that no. \$150 in cash and 1000 copies of the work so reads your letter. I was to have all the profits of the work forever, the privilege of republishing my essay in my Notes &c. As the work progressed I gave away the 100 colls. [dollars?] the \$50 I never had I had 700 copies only and instead of 3000 of *every* drawing, I was told that the Map was mine, exclusively—that Mr. Reed . . . had the privilege of printing for himself 750 copies of the Map. . . . I wrote 5 letters to Deacon Read, who never answered one of them and *now,* he says the propositions in them, were not worth replying to. I, at an early day, took the proper steps to secure the copyright of the Map and now I learn that Reed has struck off 7000 copies—has sold and is selling these maps. . . . I beg leave to resign my office of counsellor in the society and feel no great anxiety, any longer, to be considered a member of it. If you accept this resignation, *as a last act,* I recommend Dr. [Samuel P.] Hildreth to fill the vacancy."

55. Karl Bernhard, Duke of Saxe-Weimar-Eisenach, *Travels through North America, during the Years 1825 and 1826,* 2 vols. (Philadelphia, 1828), 2:148–49.

56. Quoted in Weisenburger, "Caleb Atwater," 22.

57. Caleb Atwater to the president of the American Antiquarian Society, July 25, 1835, Manuscripts Division, American Antiquarian Society. Atwater wrote: "I feel a strong desire to complete my survey of the ancient works in the western states, soon, before I become too old and infirm. . . . I am entirely willing to accept the same compensation as heretofore—\$150 in cash, and, 750 copies of the work. . . . Dr. Thomas being dead, I know not, to whom, to address this . . . nor do I know, whether the society prospers, as it formerly did, during his lifetime. . . . I would be glad to visit Mexico and Peru, and examine those works, there, evidently, belonging to the same people, who erected our ancient works. . . . Had I lived under any government in Europe, I feel assured that the government itself would have patronised me. But, under a government of political demagogues, I expect nothing—ask nothing—and scorn to receive any thing, at the hands, of such creatures."

58. *Proc. Am. Antiq. Soc.* (1867), 26.

59. For an account of the museum of the American Antiquarian Society, see Clifford J. Shipton, "The Museum of the American Antiquarian Society," in Whitfield J. Bell, Jr., et al., *A Cabinet of Curiosities: Five Episodes in the Evolution of American Museums* (Charlottesville, Va., 1967), 35–48.

14: THE SCIENCES OF MAN: COMPARATIVE LINGUISTICS
AND THE PROBLEM OF INDIAN ORIGINS

1. Thomas Jefferson, *Notes on the State of Virginia . . . ,* ed. William Peden (Chapel Hill, N.C., 1955), 101.

2. Ibid., 102. Jefferson seems to have had misgivings about this argument for the superior antiquity of the American Indians, for in the manuscript notes containing the corrections he planned to include in a revised edition he proposed another solution to the problem of the radical diversity of Indian languages. Relying on the known fact that the Indians considered it dishonorable to use any language but their own, he suggested that whenever a part of a tribe quarreled with the main body and separated itself, the members of the seceded group might refuse to use the language of the tribe and proceed to invent their own! "They

have use but for few words and possess but few. It would require but a small effort of the mind to invent these and to acquire the habit of using them. Perhaps this hypothesis presents less difficulty than that of so many radically distinct languages preserved by such handfuls of men from an antiquity so remote that no data we possess will enable us to calculate it." Ibid., 282, n12.

3. *The Papers of Thomas Jefferson,* ed. Julian P. Boyd, 19 vols. (Princeton, 1950–), 10:240.

4. Ibid., 11:683.

5. In the Bartram Papers at the Historical Society of Pennsylvania is a volume entitled "William Bartram's answers to B. S. Barton's queries about Indians." Barton's queries to John Heckewelder may be found in a collection entitled "Vocabularies and other MSS Relating to Indian Languages, collected by Peter S. Du Ponceau" at the American Philosophical Society Library.

6. For an interesting bibliography and account of linguistic researches in Russia in the eighteenth century, see Janos Gulya, "Some Eighteenth Century Antecedents of Nineteenth Century Linguistics: The Discovery of Finno-Ugrian," in Dell Hymes, ed., *Studies in the History of Linguistics: Traditions and Paradigms* (Bloomington, Ind., and London, 1974), 258–76.

7. *Comparative Lexicons of All Languages and Dialects, Compiled by the Right Hand of the Most High Personage. First Section, Containing European and Asiatic Languages,* pt. 1, 2 vols. (St. Petersburg, 1787–1789). Text and title in Russian. Vol. 2 has a Latin title page as well: *Linguarum totius orbis vocabularia comparativa, Augustissimae, cura collecta.* Some copies contain a ten-page introduction in Latin, in which the work is described and the phonetics of the Russian alphabet are explained. Barton's copy seems to have contained this supplement. For a fuller account of this work and of the circumstances attending its publication, see Johann C. Adelung and Johann S. Vater, *Mithridates: Oder Allgemeine Sprachenkunde . . . ,* 4 vols. (Berlin, 1806–1817), 4:4–8; Wilhelm Thomsen, *Geschichte der Sprachwissenschaft bis zum Ausgang des 19. Jahrhunderts,* trans. Hans Pollak, (Halle, 1927), 39; John Pickering, *An Essay towards a Uniform Orthography of the Indian Languages of North America* (Cambridge, Mass., 1820), 3–4n. Gulya, "Some Eighteenth Century Antecedents," calls the *Comparative Lexicons* the product of "the dilettantism of a monarch."

8. Benjamin S. Barton, *New Views of the Origin of the Tribes and Nations of America,* 2nd ed. (Philadelphia, 1798), viii.

9. Ibid., ix. "The A has always the open sound, as in the words father, rather, and many others. The AA is to be sounded long. The E is always sounded as E in head, bed, &c. or like A in table, and Ay, in say. The soft sound which is given to this letter I have represented by the double EE, as in tree, bee, &c. The G is sounded hard, as in God, go, &c. The J is sounded as in just, and many others; or like G in giant. The I has the several sounds of this letter. The OO, which so frequently occurs in the Indian words, has a long sound, as in the word ooze. It appears to be nearly equivalent to the U of the Latins. The U always sounds like U in us, or in the vulgar word, fuss. My mode of pronunciation will, I believe, be obvious in all other instances." The rules for pronouncing words transcribed from Pallas's work are given on pages xix–xx.

10. Ibid., "Preliminary Discourse," lxxxviii.

11. Ibid., "Appendix," 19. For Barton's view of the linguistic affinities between South American and Asian tribes, see his "Preliminary Discourse," xcvii–xcxix.

12. Ibid., "Comparative Vocabularies," 36n. See also "Preliminary Discourse," lxxxix: "My inquiries seem to render it probable, that all the languages of the countries of America may . . . be traced to one or two great stocks. In Asia, I think, they may be confidently traced to one. For the language of the Mexicans, which is so different from that of other Americans, has some affinity to the languages of the Lesghis and the Persians: and I have already observed, that the languages of these two Asiatic nations are preserved among many American tribes."

13. Peter S. Du Ponceau, "Report of the Corresponding Secretary to the Committee . . . ," *Trans. Hist. Lit. Comm. Am. Philos. Soc.,* 1 (1819), xix–xx; Adelung and Vater, *Mithridates.*

14. Johann S. Vater, review of Barton, *New Views,* in *Göttingische Anzeigen von Gelehrten Sachen unter des Sussicht der Königlichen Gesellschaft der Wissenschaften,* 96th Stueck (June 17, 1799), 953–60.

15. Thomas Jefferson to Benjamin Hawkins, Philadelphia, Mar. 14, 1800, in *The Writings of Thomas Jefferson,* library edition, ed., Andrew A. Liscomb, 20 vols. (Washington, D.C., 1903–1904), 10:161.

16. Jefferson to William Dunbar, Washington, D.C., Feb. 1, 1801, ibid., 10:190–91.

17. Jefferson to John Sibley, May 27, 1805, Lipscomb, ed., *Writings,* 11:80–81.

18. It is not clear what became of Priestley's copy of the *Lexicons.* In the Barton papers at the Historical Society of Pennsylvania is a letter from Priestley to Barton dated Northumberland, Pa., Nov. 27, 1800, in which Priestley says: "The excellent use you have made of the *Russian Nomenclature* fully entitled you to it. At least it shall always remain in your custody. If I ever want it (which is not very probable) I know where it is. . . ." Presumably Barton kept the book, but in that case one wonders why neither Jefferson nor Peter S. Du Ponceau ever laid eyes on it, although both were eager to see it. That Jefferson and Barton exchanged Indian vocabularies is evident from a letter from Jefferson to Barton, Mar. 1, 1800, in the Barton papers at the Historical Society of Pennsylvania. Jefferson says of himself: "He is just now beginning to copy the Indian vocabularies lent him by Dr. Barton; but finds it necessary to know previously whether some of them may not already have been entered in the Vocabularies of Thomas Jefferson lent to Dr. B. [H]e will therefore thank him for them, & if Dr. B. has not made the uses of them which he wished, they shall be speedily sent back to him, with his own." In March 1802 President Jefferson directed the secretary of war to provide Barton with a document entitling him to visit the Cherokees and other Indian nations "for the purposes of recovering his health and of obtaining useful information respecting the manners, natural history & language of the Natives."

19. Jefferson to Benjamin S. Barton, Monticello, Sept. 21, 1809, Lipscomb, ed., *Writings,* 12:312–14.

20. Ibid.

21. Benjamin S. Barton, "Hints on the Etymology of Certain English Words, and on Their Affinity to Words in the Languages of the Different European, Asiatic, and American (Indian) Nations . . . ," *Trans. APS,* 6 (1809), 145–57. Present-day linguists find Barton's supposed affinities as set forth in this article farfetched and unconvincing.

22. Ibid., 155.

23. See Alexander von Humboldt, *Political Essay on the Kingdom of New Spain . . . ,* trans. John Black, 2 vols. (New York, 1811), 1:101–2 for references to Barton's *New Views;* see also Humboldt, *Researches Concerning the Institutions and Monuments of the Ancient Inhabitants of America . . . ,* trans. H. M. Williams, 2 vols. (London, 1814), 1:19–20. Humboldt concludes (pp. 249–50) that "the greater part of the nations of America belong to a race of men, who, isolated ever since the infancy of the world from the rest of mankind, exhibit in the nature and diversity of language, in their features and the conformation of their skull, incontestable proofs of an early and complete separation."

24. Johann S. Vater, *Untersuchungen über Amerika's Bevölkerung aus dem Alten Kontinente* (Leipzig, 1810), 45ff. Vater invariably refers to Barton as Smith-Barton. Jefferson's observations on Indian languages in his *Notes* are discussed by Vater on p. 92ff. Modern linguists find the grammatical characteristic Vater attributed to all Indian languages less universal among them than he supposed. They note that neither Barton nor Vater made use of language families as a matrix for comparison.

25. See H. R. Dippold and F. U. Kothe, eds., *Allgemeine Historische Archiv,* 1 pt. 2 (1811), 198ff.

26. Adelung and Vater, *Mithridates,* 3 pt. 2 (1813), 339.

27. See Mary R. Haas, "American Indian Linguistic Prehistory," *Curr. Trends Linguist.,* 10 (1973), 697: "We have made very little progress toward the answer to our original puzzlement as to where the Indian came from. We still look to the Orient as the clue to the ultimate answer, but proof is as elusive as ever." See also Lyle Campbell and Marianne Mithun, eds., *The Languages of Native America. Historical and Comparative Assessment* (Austin, Tex., 1979), especially the "Introduction."

28. Thomsen, *Geschichte der Sprachwissenschaft,* 40–42. See also John B. Carroll, *The Study of Language: A Survey of Linguistics and Related Disciplines in America* (Cambridge, 1953), 17.

29. Peter S. Du Ponceau to John Heckewelder, Philadelphia, June 10, 1816, "Correspondence between Mr. Heckewelder and Mr. Duponceau, on the Languages of the American Indians," *Trans. Hist. Lit. Comm. Am. Philos. Soc.,* 1 (1819), 373.

30. Peter S. Du Ponceau to W. W. Esq., Philadelphia, May 12, 1836, Du Ponceau Papers–Autobiographical Letters, American Philosophical Society Library.

31. Richard Peters to Robert R. Livingston, Oct. 19, 1781, as quoted in Robley Dunglison, *A Public Discourse in Commemoration of Peter S. Du Ponceau, LL.D., Late President of the American Philosophical Society, Delivered before the Society Pursuant to Appointment, on the 25th of October* (Philadelphia, 1844), 16–17.

32. "Constitution of the Historical and Literary Committee," *Trans. Hist. Lit. Comm. Am. Philos. Soc.,* 1 (1819), vi–vii.

33. Caspar Wistar to John Heckewelder, Philadelphia, Jan. 9, 1816, "Correspondence between Mr. Heckewelder and Mr. Duponceau," *Trans. Hist. Lit. Comm. Am. Philos. Soc.,* 1:359.

34. John Heckewelder, "Life of John Gottlieb Ernestus Heckewelder, a Moravian Missionary among the American Indians," trans. from the German, *Am. Missionary Regist.,* 5 (October 1824), 289–94. See also Edward Rendthaler, *Life of John Heckewelder* (Philadelphia, 1847).

35. John Heckewelder to Caspar Wistar, Bethlehem, Pa., Mar. 24, 1816, "Correspondence Between Mr. Heckewelder and Mr. Duponceau," *Trans. Hist. Lit. Comm. Am. Philos. Soc.,* 1:361–62.

36. Ibid., 370.

37. Peter Du Ponceau to John Vaughan, Aug. 8, 1816, American Philosophical Society Archives.

38. Du Ponceau to Thomas Jefferson, Philadelphia, Feb. 17, 1817, Letterbooks of the Historical and Literary Committee, American Philosophical Society Library, 1:57.

39. Du Ponceau to John Heckewelder, Philadelphia, Aug. 5, 1816, Letterbooks of the Historical and Literary Committee, 1:43.

40. Du Ponceau to Heckewelder, Aug. 30, 1816, "Correspondence between Mr. Heckewelder and Mr. Duponceau," 1:432.

41. Heckewelder to Du Ponceau, Aug. 6, 1818, 1:147.

42. Heckewelder to Du Ponceau, Bethlehem, Pa., July 18, 1819, 1:311. In an earlier letter, Feb. 7, 1819, Heckewelder expressed similar sentiments and offered the opinion that Barton "was *too* much inclined to draw a *similarity* in point of Words in the *Languages* spoken by the different Tribes & Nations on this Continent, & comparing them with the Oriental, in order to discover from *whence* the Aborigines of this Country had sprung—in so doing he even—if I mistake not found a great *affinity* between the two Languages, the Delaware & the Iroquois, where even Zeisberger could find none."

43. Clark Wissler, "The American Indian and the American Philosophical Society," *Proc. APS,* 86 (1942), 197.

44. John Heckewelder, "Account of the History, Manners, and Customs of the Indian Natives Who Once Inhabited Pennsylvania and the Neighbouring States," *Trans. Hist. Lit. Comm. Am. Philos. Soc.,* 1:25. Concerning this work, Du Ponceau wrote to John Vaughan: "I have read almost the whole of Mr Heckewelder's Acct of the Indians & I have no doubt it will be a popular work in this country & that it will have a prodigious success in Europe. . . . It is written with the most unaffected simplicity & yet with perfect clearness. . . . Without being so intended, it conveys the most Severe Satyre on Civilised Society, & all men love Satyre. See the Chapter on Ambition, p. 113. This will have a great effect in Europe. The more corrupt a nation is, the more it loves pictures of virtue. . . ." (Peter S. Du Ponceau to John Vaughan, Philadelphia, Apr. 30, 1818, Miscellaneous Manuscript Collection, American Philosophical Society Library.) Du Ponceau's brother translated Heckewelder's "Account" into French for the Paris edition, 1822.

45. Peter S. Du Ponceau, "Report of the Corresponding Secretary to the Committee,

of His Progress in the Investigation Committed to Him of the General Character and Forms of the Languages of the American Indians," *Trans. Hist. Lit. Comm. Am. Philos. Soc.,* 1:xxvi.

46. Ibid., xxvii, xxx–xxxi.

47. Ibid., xxxviii.

48. Ibid., xxxix–xl.

49. Ibid., xxix.

50. Peter S. Du Ponceau to Johann Vater, Philadelphia, Sept. 9, 1821, Letterbooks of the Historical and Literary Committee, American Philosophical Society Library, 2:51.

51. Du Ponceau to Vater, Philadelphia, Dec. 22, 1821, Letterbooks, 3:6.

52. Jedidiah Morse, *A Report to the Secretary of War of the United States on Indian Affairs, Comprising a Narrative of a Tour Performed in the Summer of 1820* . . . (New Haven, 1822), 128, 356–60; Mentor L. Williams, ed., *Schoolcraft's Indian Legends* (East Lansing, Mich., 1956), ix–xxii, 288–93; Henry R. Schoolcraft, *Narrative Journal of Travels through the Northwestern Regions of the United States Extending from Detroit through the Great Chain of American Lakes to the Sources of the Mississippi River in the Year 1820,* ed. Mentor L. Williams (East Lansing, Mich., 1953), 1–24; Edwin James, *Account of an Expedition [with Stephen H. Long] from Pittsburgh to the Rocky Mountains,* 3 vols., in Reuben G. Thwaites, ed., *Early Western Travels,* 32 vols. (Cleveland, 1904–1907), 17:289–308. Original edition, Philadelphia, 1823.

53. Mary O. Pickering, *Life of John Pickering* (Boston, 1887).

54. John Pickering, "On the Adoption of a Uniform Orthography for the Indian Languages of North America," *Mem. Am. Acad. Arts Sci.,* 4 (1818–1821), 319–60. This work was also published separately in 1820 in Cambridge as *An Essay on a Uniform Orthography.* . . . Franklin Edgerton's evaluation of this effort occurs in his "Notes on Early American Work in Linguistics," *Proc. APS,* 87 (1943), 27.

55. Edgerton, "Notes," 27–28. Pickering's orthography was adopted by the American Board of Commissioners for Foreign Missions and used in their many publications in Indian languages. See the *Missionary Herald* for July, 1836.

56. John Eliot, *A Grammar of the Massachusetts Indian Language. A New Edition: With Notes and Observations, by Peter S. Du Ponceau, LL.D. and an Introduction and Supplementary Observations, by John Pickering. As Published in the Massachusetts Historical Collections* (Boston, 1822). See also Jonathan Edwards, Jr., "Observations on the Language of the Muhhekaneew Indians . . . ," *Collect. Mass. Hist. Sco.,* 2nd ser., 10 (1823); Josiah Cotton, *Vocabulary of the Massachusetts (or Natick) Indian Language* (Cambridge, 1829). Friedrich Adelung's *Übersicht aller Bekannten Sprachen und Ihrer Dialekte* (St. Petersburg, 1820) was reviewed by John Pickering in *North Am. Rev.,* 14 (1822), 128–44.

57. Peter S. Du Ponceau, "Notes and Observations," in Eliot, *Grammar of the Massachusetts Indian Language,* iv. See also James H. McCulloh, Jr., *Researches on America, Being an Attempt to Settle Some Points Relative to the Aborigines of America, &c.* (Baltimore, 1817); a revised and expanded version of this work was published in Baltimore in 1829 under the title *Researches, Philosophical and Antiquarian, Concerning the Aboriginal History of America;* see especially chap. 2, "On the Languages of the American Indians."

58. Thomas Jefferson to John Pickering, Monticello, Feb. 20, 1825, as quoted in Thomas A. Kirby, "Jefferson's Letters to Pickering," *Philologica: The Malone Anniversary Studies,* ed. Thomas A. Kirby and Henry B. Woolf (Baltimore, 1949), 263.

59. Sebastien Rasles, "A Dictionary of the Abnaki Language in North America. Published from the Original Manuscript of the Author, with an Introductory Memoir and Notes [by John Pickering]," *Mem. Am. Acad. Arts Sci.,* n.s., 1 (1833), 375–574. [John Pickering], "Appendix: Indian Languages of America," *Encyclopaedia Americana. A Popular Dictionary of Arts, Sciences, Literature, History, Politics and Biography* . . . , ed. Francis Lieber, 13 vols. (Philadelphia, 1831), 6:581–600.

60. Peter S. Du Ponceau, "Translator's Preface," *A Grammar of the Language of the Lenni Lenape or Delaware Indians. Translated from the German Manuscript of the Late Rev. David Zeisberger* . . . , *Trans. Am. Philos. Soc.* n.s., 3 (1830), 66–67.

61. Peter S. Du Ponceau, "Concluding Note by the Translator," 249.

62. Peter S. Du Ponceau, *Mémoire sur le système grammatical des langues de quelques Indiennes de l'Amérique du Nord; ouvrage qui, à la séance publique annuelle de l'Institut*

Royal de France, le 2 Mai 1835, a remporté le prix fondé par M. le comte de Volney (Paris, 1838), 23–24.

63. Ibid., 84. See also Du Ponceau's unsigned article "Language" in the *Encyclopaedia Americana* (1831), 7:408–17.

64. Wissler, "The American Indian and the American Philosophical Society," 193.

65. Helmut de Terra, "Alexander von Humboldt's Correspondence with Jefferson, Madison, and Gallatin," *Proc. APS,* 103 (1959), 787.

66. James Gallatin, *The Diary of James Gallatin,* 2 vols. (New York, 1926), 1:565.

67. Adriano Balbi, *Introduction à l'atlas ethnographique* (Paris, 1826).

68. Albert Gallatin, "Synopsis of the Indian Tribes within the United States East of the Rocky Mountains, and in the British and Russian Possessions in North America," *Archaeol. Am.,* 2 (1836), 3–4.

69. John Wesley Powell, *Indian Linguistic Families of America North of Mexico,* 7th annual report, Bureau of Ethnology (Washington, D.C., 1891), 9–10.

70. Gallatin, *Synopsis,* 142–43.

71. Ibid., 145.

72. Ibid., 202–3.

73. See Albert Gallatin, "Notes on the Semi-Civilized Nations of Mexico, Yucatan and Central America," *Trans. Am. Ethnol. Soc.,* 1 (1845), 1–352. See also Gallatin's "Introduction" to Horatio Hale, "Hale's Indians of North-West America, and Vocabularies of North America," ibid., 2 (1848); this was Gallatin's final summation of data on North American Indian languages — in it thirty-two Indian linguistic families were distinguished. See A. Irving Hallowell, "The Beginnings of Anthropology in America," in Frederica de Laguna, ed., *Selected Papers from the American Anthropologist, 1888–1920* (Evanston, Ill., 1960) 29. Gallatin was a founding member of the American Ethnological Society and made substantial financial as well as literary contributions to the publication of its *Transactions.*

15: THE END OF THE JEFFERSONIAN ERA

1. See George H. Daniels, *American Science in the Age of Jackson* (New York and London, 1968), chaps. 1–2; Sally G. Kohlstedt, *The Formation of the American Scientific Community . . .* (Urbana, Ill., 1976).

2. At Harvard, Jeremy Belknap informed Benjamin Rush in a letter of July 29, 1789: "There is a special injunction laid on the professors to take frequent occasion to introduce reflections on the being, perfections, and providence of the Creator; and I believe this injunction is strictly attended to." See Jane Marcou, ed., *Life of Jeremy Belknap, D.D. . . . with Selections from His Correspondence and Other Writings* (New York, 1847), 154. On the relations of science and religion in this period see Henry F. May, *The Enlightenment in America* (New York, 1976), chaps. 3–4; Herbert Hovenkamp, *Science and Religion in America, 1800–1860* (Philadelphia, 1978), chaps. 1–2; Theodore Dwight Bozeman, *Protestants in an Age of Science . . .* (Chapel Hill, N.C., 1977); Herbert M. Morais, *Deism in Eighteenth Century America* (New York, 1934).

3. Jedidiah Morse, *American Universal Geography,* 2 vols. (Boston, 1793), 1:79ff.

4. Benjamin S. Barton, "Some Account of the Different Species and Varieties of Native American, or Indian Dogs," *Phila. Med. Phys. J.,* 1, pt. 2 (1805), 15ff.

5. See John C. Greene, "The American Debate on the Negro's Place in Nature, 1780–1815," *J. Hist. Ideas,* 15 (1954), 384–96; William Stanton, *The Leopard's Spots: Scientific Attitudes toward Race in America, 1815–59* (Chicago, 1960).

6. Benjamin Silliman to Parker Cleaveland, Oct. 6, 1817, Cleaveland Correspondence, Bowdoin College Library.

7. Dirk J. Struik, *Yankee Science in the Making,* rev. ed. (New York, 1962).

8. Linda K. Kerber, *Federalists in Dissent: Imagery and Ideology in Jeffersonian America* (Ithaca, N.Y., and London, 1970).

9. Clement C. Moore, *The Early History of Columbia College . . .* (New York, 1940), facsimile of the 1825 edition, 25, 27. Moore was only twenty-five years old when he pub-

lished his *Observations upon Certain Passages in Mr. Jefferson's Notes on Virginia, Which Appear to Have a Tendency to Subvert Religion, and Establish a False Philosophy* (1804).

10. Records of the American Academy of Arts and Sciences, 1:44–45, Boston Athenaeum.

11. Perry Miller, *The Life of the Mind in America from the Revolution to the Civil War* (New York, 1965), 290.

12. Benjamin Franklin, *Proposal for Promoting Useful Knowledge among the British Plantations in America,* as quoted in Albert H. Smyth, ed., *The Writings of Benjamin Franklin,* 10 vols. (New York, 1905–1907), 2:229–30.

13. *Mem. Am. Acad. Arts Sci.,* 1 (1785), xi; *Trans. APS,* 1 (1771), xvii. See also John C. Greene, "Science, Learning and Utility: Patterns of Organization in the Early American Republic," in Alexandra Oleson and S. C. Brown, eds., *The Pursuit of Knowledge in the Early American Republic . . .* (Baltimore and London, 1976), 1–20.

14. *Am. J. Sci. Arts,* 2 (1820), preface. See also the preface to vol. 1.

INDEX

472 INDEX

Indian mounds and mound
builders (*cont.*)
monograph, 373–74
— theories of origin: Atwater, 365, 367–
68; Barton, 346, 348; Brackenridge,
352; Clinton, 354; Shetrone, 368–69;
Stiles, 344–45; twentieth-century, 265;
Webster, 344
Indiana Territory, 199
Indiana University, 156
Indians, American (*see also* Indian
mounds and mound builders): anatomical studies, 330; early investigations, 193, 327, 348–49; intelligence
questioned by Warren, 338; Lewis
and Clark Expedition, 202–3, 205,
207, 209–11, 339
— languages, 376–408; American Philosophical Society study and publication, 389, 393; classification and
nomenclature, 394, 399, 400, 404–5;
early studies republished, 399; European investigations, 386–87, 395;
grammatical structure, 391–92, 394–
95, 399–400, 401, 458n.24; individual
tribes and nations, 390, 391, 393–95,
399–405; New World–Old World
connection, 379, 382, 386, 387, 399–
400; theoretical concepts, 380–81,
385, 401, 405–6, 407; vocabulary
collections, 376–77, 378, 400, 404,
407–8
— theories of origins, 320, 376; Barton,
380–81; Gallatin, 406;Heckewelder,
392; Jefferson, 322, 376–77; Mitchill,
331; monogenist-polygenist debate,
376; twentieth-century, 458n.27;
Warren, 341–42
Indians: Central and South American,
406; Mexican, 406, 457n.12
Industrialization, relation to popular
science movement, 20
Inflammability, principle, 161, 169
Institute of France, 225, 336
Institutions, scientific (*see also* Colleges
and universities; Historical societies;
Hospitals; Medical schools and colleges; Natural history museums;
Scientific societies): Boston area, 61–
90; Charleston, S.C., 107–15; Cincin-
nati, 115–20; lack, Lexington, Ky.,
120–26; New England, 60–90; New
Haven, 60–61; New York, 105; Philadelphia, 37–59; Washington, D.C.,
36
Isostasy, 219
Ives, Eli, 244

Jackson, James: Harvard medical school,
84–88; member of Boston scientific
institutions, 64, 72–73; solar eclipse,
73; and Waterhouse, 82
James, Edwin, 231, 273

Jameson, Robert, 220, 239
Jardin des Plantes (Paris), 85, 97, 287
Jefferson, Thomas: aims of science, 418;
amateur astronomer, 147, 149–50;
American Indians' origin and antiquity, 29, 34, 456–57n.2; American
Philosophical Society president, 33,
43–44, 282, 410; archaeology, 345,
350, 374; botanical opportunities in
America, 253; chemistry, 163, 170;
coinage design, 43; evolution of
species, 32; Federalists vs. his interest
in science, 34, 414; geographical
contributions, 217; geology and
mineralogy, 32, 220–21; government-
sponsored exploration, 195–96, 215;
Indian languages, 34–35, 321–22,
381–82, 383, 384, 407; Lewis and
Clark Expedition, 11, 196–217, 410;
Louisiana Territory expeditions, 198;
measurement mania, 30; National
Institute of France, 289–91; natural
history museums, 27; Negroes, 329;
New England clergy, 13–14; nomenclature and classification, 32–33;
paleontologist, 44, 285, 288–91, 318;
religion and science, 13–14; religious
beliefs, 12, 13, 412; Report on
Weights and Measures, 43; scientific
interests and influence, 27, 29, 31–
36, 419; as utilitarian, 22, 31–32, 33,
417
— *Notes on the State of Virginia:* American comparative linguistics, 407;
botanical information, 193; geographic contribution, 188; patriotism,
29; racial views, 321–22; religious
objections, 414–15; significance and
influence, 409; zoological content,
277
— relations with other scientists: Bacon,
27; Barton, 289, 458n.18; Bowditch,
153; Buffon, 10, 28, 29–30, 321;
Clark, 34; Cuvier, 34; Daubenton, 29;
Dunbar, 136; Du Ponceau, 390, 396;
Ellicott, 138, 139, 142–43, 419;
Ewell, 22; Gallatin, 403–4; Gantt, 34;
Hamilton, 50; Harvey, 33; Hassler,
131; Humboldt, 419; Jenner, 33;
Lacèpéde, 196, 438n.16; Lewis, 197,
202, 203, 204; Maclure, 227; M'Mahon, 51; Mansfield, 419; Morse, 190,
213; Peale, 419; Rafinesque, 447n.42;
Rittenhouse, 160; Samuel Smith, 330;
Thouin, 29; Volney, 419; Wilson, 292,
294, 419
Jeffersonian Republicans: compared to
Federalists, 413; dominate American
Philosophical Society, 415; frugal
economists, 36; in Pennsylvania, 143;
proscience orientation, 414–15;
religion, 19
Jenner, Edward, 33
Jessup Augustus, 58